Tayeb Blouhi, Amin Benaissa Cherif, Khaled Zennir

Stochastic Differential Equations

Also of Interest

Differential and Integral Calculus
Implicit Functions, Stieltjes Integrals and Curvilinear Integrals
Svetlin G. Georgiev, Khaled Zennir, 2025
ISBN 978-3-11-914462-9, e-ISBN (PDF) 978-3-11-221808-2

Multiple Integrals in Calculus
Improper Integrals, Line Integrals, Surface Integrals
Svetlin G. Georgiev, Khaled Zennir, 2025
ISBN 978-3-11-914345-5, e-ISBN (PDF) 978-3-11-221960-7

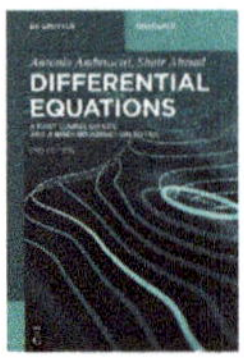

Differential Equations
A First Course on ODE and a Brief Introduction to PDE
Antonio Ambrosetti, Shair Ahmad, 2024
ISBN 978-3-11-118524-8, e-ISBN (PDF) 978-3-11-118567-5

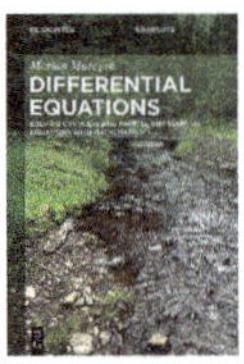

Differential Equations
Solving Ordinary and Partial Differential Equations with Mathematica®
Marian Mureşan, 2024
ISBN 978-3-11-141109-5, e-ISBN (PDF) 978-3-11-141139-2

Differential Equations, Fourier Series, and Hilbert Spaces
Lecture Notes at the University of Siena
Raffaele Chiappinelli, 2023
ISBN 978-3-11-129485-8, e-ISBN (PDF) 978-3-11-130252-2

Brownian Motion
A Guide to Random Processes and Stochastic Calculus
René Schilling, 2021
ISBN 978-3-11-074125-4, e-ISBN (PDF) 978-3-11-074127-8

Tayeb Blouhi, Amin Benaissa Cherif, Khaled Zennir

Stochastic Differential Equations

Inclusion of Impulses, Multiple Delay, Neutral Functional, Sweeping Process, Measure of Non-Compactness

DE GRUYTER

Mathematics Subject Classification 2020
Primary: 60H10, 34F05, 60H99; Secondary: 47H10, 34B37, 34K30

Authors
Tayeb Blouhi
University of Science and Technology of Oran
Mohamed-Boudiaf
Oran
Algeria

Amin Benaissa Cherif
University of Science and Technology of Oran
Mohamed-Boudiaf
Oran
Algeria

Khaled Zennir
Qassim University
Saudi Arabia

ISBN 978-3-11-225188-1
ISBN 978-3-11-225189-8 (PDF)
ISBN 978-3-11-225190-4 (E-PUB)
DOI https://doi.org/10.1515/9783112251898

Library of Congress Control Number: 2026936910

Bibliographic information published by the Deutsche Nationalbibliothek
The Deutsche Nationalbibliothek lists this publication in the Deutsche Nationalbibliografie; detailed bibliographic data are available on the Internet at https://dnb.dnb.de.

De Gruyter and Walter de Gruyter GmbH are part of De Gruyter Brill.
www.degruyterbrill.com

Questions about General Product Safety Regulation: productsafety@degruyterbrill.com

Cover illustration: AureliasDreams / iStock / Getty Images Plus

Contents

Preface — IX

1 Motivation — 1

2 Elements of stochastic analysis — 12
2.1 Some selected topics from probability theory — **12**
2.1.1 Some inequalities — **15**
2.1.2 Processes and filtration — **15**
2.1.3 Stopping times — **17**
2.1.4 Continuous time martingales — **17**
2.2 Brownian motion and the Wiener process — **20**
2.2.1 Functions of bounded variation — **26**
2.2.2 Itô stochastic integrals — **32**
2.2.3 Gaussian processes in Hilbert spaces — **39**
2.3 The stochastic integral in Hilbert spaces — **40**
2.3.1 Hilbert space valued Wiener processes — **40**
2.3.2 Definition of the stochastic integral — **42**
2.3.3 Stochastic integral for cylindrical Wiener processes — **48**
2.3.4 Stochastic integration with respect to continuous semimartingales — **50**
2.3.5 Itô's formula — **53**
2.3.6 The multidimensional Itô formula — **53**
2.4 Fractional Brownian motion — **61**
2.4.1 Correlation between two increments — **62**
2.4.2 Long-range dependence — **62**
2.4.3 Hölder continuity — **64**
2.4.4 Path differentiability — **65**
2.4.5 The fBm is not a semimartingale for $H \neq 1/2$ — **66**
2.4.6 Fractional integrals and derivatives — **69**
2.4.7 Integrals for fractional Brownian motion — **70**

3 Elements of functional analysis — 79
3.1 Generalized metric and Banach spaces — **80**
3.2 Compactness criteria — **89**
3.3 Measures of noncompactness — **91**
3.4 Fixed-point theory — **93**
3.5 Some properties of set-valued maps — **94**
3.6 Iterative results — **98**
3.7 Semigroups — **106**
3.7.1 Generators of uniformly continuous semigroups — **109**
3.7.2 C_0-Semigroups: General properties — **111**

3.7.3 The infinitesimal generator — **116**
3.7.4 Hille–Yosida theorem — **117**

4 Impulses stochastic differential equations — 125
4.1 Basic tools on stochastic calculus — **125**
4.2 Nonsmooth analysis — **126**
4.3 Unique solution via iterative methods — **128**
4.3.1 Nagumo type growth conditions — **132**
4.3.2 Example — **141**

5 Impulsive functional stochastic differential with multiple delay — 143
5.1 Multiple delays — **146**
5.2 Contraction principle — **147**
5.3 Existence with Nagumo-type growth conditions — **155**

6 Impulsive stochastic differential equations — 165
6.1 Mild solution — **166**
6.2 Weak solutions — **184**
6.3 Exponential stability — **188**
6.4 Example — **198**

7 Coupled impulsive neutral functional differential equations — 206
7.1 Existence and solution set — **211**
7.2 Example — **232**

8 Stochastic functional differential equations with measure of noncompactness — 242
8.1 Global existence result — **245**
8.2 Existence results — **249**

9 Differential inclusions stochastic — 266
9.1 Existence results — **267**
9.1.1 The convex case — **267**
9.1.2 The nonconvex case — **285**
9.1.3 Example — **291**

10 Second-order stochastic differential inclusions — 294
10.1 Stochastic impulsive differential inclusions — **295**
10.1.1 Mild solutions — **299**
10.1.2 Nonconvex case — **315**
10.1.3 A further result — **321**
10.1.4 Example — **328**

11 Sweeping process with fractional Brownian by time delay —— 332
11.1 Existence of unique solution —— **333**
11.2 Perturbation problem —— **343**

Bibliography —— 353

Index —— 355

Preface

The study of stochastic differential equations (SDEs) and stochastic inclusions has evolved into one of the most vibrant areas of modern mathematics, with wide-ranging applications in physics, finance, engineering, and the life sciences. The combination of randomness with differential dynamics not only enriches the theoretical framework but also offers a powerful tool to model complex systems influenced by uncertainty. In real-world applications, systems are often subject not only to continuous random fluctuations but also to sudden changes or impulses, which may arise from control actions, shocks, or abrupt environmental shifts.

This handbook, "Impulses stochastic differential equations and inclusions," aims to provide a comprehensive introduction to the fundamental concepts, analytical methods, and qualitative properties of certain stochastic systems governed by both equations and inclusions. While the theory of SDEs is well developed, stochastic inclusions—where uncertainty interacts with set-valued dynamics—remain an active field of research. The interplay of stochasticity, impulses, functional analysis, and differential inclusions presents unique challenges and opportunities that this text seeks to illuminate.

The exposition begins with essential elements of stochastic analysis, stochastic calculus, and elements of functional analysis. We then progress to detailed discussions on the existence, uniqueness, and stability of solutions, as well as qualitative behaviors under varying conditions. The text also incorporates selected applications, illustrating how stochastic models naturally arise in diverse scientific and engineering problems.

This work is intended for graduate students, researchers, and professionals who wish to deepen their understanding of stochastic systems. It may serve as both a handbook for beginner researchers and a reference for specialists pursuing further studies in the field. The presentation balances rigor with accessibility, combining mathematical depth with an emphasis on clarity.

We are indebted to the contributions of many mathematicians whose pioneering work laid the foundations of this subject. Our gratitude extends to colleagues and students whose questions and insights have helped shape the material presented here.

In the first two chapters, the book introduces selected topics from probability theory: Brownian motion and the Wiener process, the stochastic integral in Hilbert spaces, and fractional Brownian motion. It explains in detail the essential properties of functional analysis, such as generalized metrics and Banach spaces, compactness criteria, measures of noncompactness (MNC), fixed-point theory, some properties of set-valued maps, fixed-point results, and semigroup theory.

The question of the quantitative study of impulsive stochastic differential equations/systems is treated with particular attention in Chapter 4 and Chapter 5. With fixed moments and multiple delays, the existence of solutions with fixed moments and multiple delays is addressed through the application of Schaefer and Perov fixed- point theorems in generalized Banach spaces, driven by standard Brownian motion. Whereas in Chapter 6, sufficient conditions for the local and global existence and exponential stability

https://doi.org/10.1515/9783112251898-203

of mild solutions of semilinear systems of stochastic differential equations with infinite fractional Brownian motions and impulses are established with the Hurst index $H > 1/2$.

In Chapter 7, we discuss some results on the existence and uniqueness of mild solutions for systems of semilinear impulsive differential equations with infinite fractional Brownian motions and Wiener processes. The approach is based on a new version of the fixed-point theorem due to Krasnoselskii in generalized Banach spaces.

Chapter 8 deals with impulsive neutral stochastic functional differential equations driven by fBm with a noncompact semigroup.

In Chapter 9, we prove some existence results based on a nonlinear alternative of the Leray–Schauder-type theorem in generalized Banach spaces for the convex case; we establish a multivalued version of Perov's fixed-point theorem in a nonconvex setting.

In Chapter 10, we provide sufficient conditions for the existence of solutions for a class of second-order systems of stochastic impulsive differential inclusions.

In Chapter 11, it is devoted to the study of a convex stochastic sweeping process with fractional Brownian motion and time delay. The approach is based on discretizing stochastic functional differential inclusions.

This textbook stands out for the quality of its proofs, the clarity of its explanations, and the richness of its exercises, which include hints and solutions that are ideal for self-study. Each chapter has been designed to be both self-contained and in perfect continuity with the rest of the book, thus promoting a harmonious and coherent progression in the learning of mathematical analysis.

We hope that this book not only provides a useful guide to the mathematical theory of stochastic differential equations and inclusions but also inspires readers to further explore the rich interplay between randomness, dynamics, and analysis.

8/5/2025

Tayeb Blouhi (Oran)
Amin Benaissa Cherif (Algiers)
Khaled Zennir (Riyadh)

1 Motivation

The following examples provide a more concrete notion of the processes that can be described by impulsive stochastic differential equations.

Example 1. Consider a market consisting of one stock (risky asset) and one bond (riskless asset). The price process of the risky asset is assumed to be of the form

$$S_t = S_0 e^{H_t}, \quad t \in [0, T],$$

with

$$H_t = \int_0^t \left(\mu_s - \frac{\sigma_s^2}{2}\right) ds + \int_0^t \sigma_s dW_s, \tag{1.1}$$

where

$$W = \{W_t, t \in [0, T]\},$$

is a Brownian motion defined in a complete probability space $(\Omega, \mathcal{F}, \mathbb{P})$. We will denote by $\mathcal{F}_t$, $t \in [0, T]$ the filtration generated by the Brownian motion and completed by the $\mathbb{P}$-null sets. The mean rate of return μ_t and the volatility process σ_t are supposed to be measurable and adapted processes satisfying the following integrability conditions.

By Itô, we obtain that S_t satisfies a stochastic differential equation with impulses

$$\begin{cases} dS_t = S_t \mu_t dt + S_t \sigma_t dW_t, & t \neq t_k, \\ S(t_k^+) - S(t_k^-) = I_k(x(t_k)), & t = t_k, \\ S(0) = S_0, \end{cases} \tag{1.2}$$

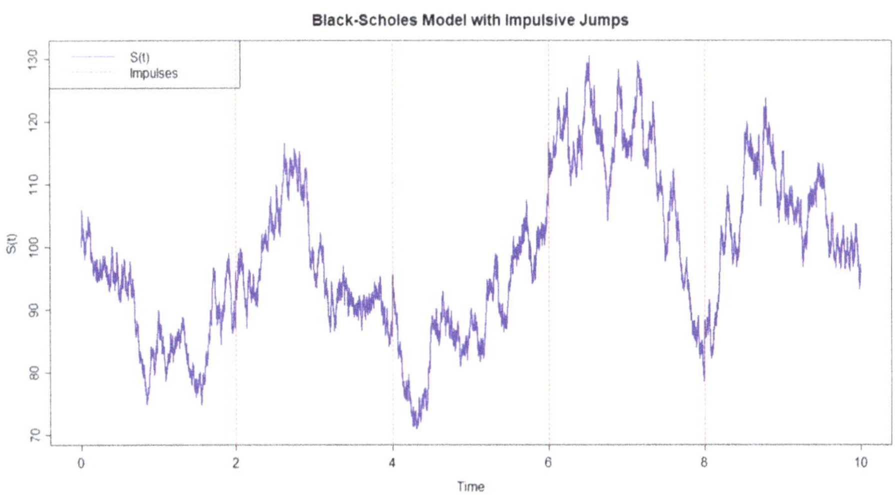

Figure 1.1: Black–Scholes model.

https://doi.org/10.1515/9783112251898-001

where S_t the price action at time t and $I_k(x(t_k))$ changes the price action at time t_k. The price of the bond $B_t, t \in [0,T]$ evolves according to the differential equation

$$dB_t = r_t B_t dt, \quad B_0 = 1,$$

where the interest rate process is a nonnegative, measurable, and adapted process satisfying the integrability condition (see Figure 1.1):

$$\int_0^T r_t dt < \infty,$$

almost surely. That is,

$$B_t = \exp\left(\int_0^T r_t dt\right).$$

Imagine an investor who starts with some initial endowment $x \geq 0$ and invests in the assets described above. Let α_t be the number of nonrisky assets and β_t the number of stocks owned by the investor at time t. The couple

$$\phi_t = (\alpha_t, \beta_t), \quad t \in [0,T],$$

is called a portfolio or trading strategy, and we assume that α_t and β_t are measurable and adapted processes such that

$$\int_0^T |\beta_t \mu_t| dt < \infty, \quad \int_0^T \beta_t^2 \sigma_t dt, \quad \int_0^T |\alpha_t| r_t < \infty, \tag{1.3}$$

almost surely. Then

$$x = \alpha_0 + \beta_0 S_0,$$

and the investor's wealth at time t (also called the value of the portfolio) are

$$V_t(\phi) = \alpha_t B_t + \beta_t S_t.$$

The gain $G_t(\phi)$ made by the investor via the portfolio ϕ up to time t is given by

$$G_t(\phi) = \int_0^T \alpha_s dB_s + \int_0^T \beta_s dS_s.$$

Notice that both integrals are well-defined thanks to condition (1.3).

We say that the portfolio ϕ is self-financing if there is no fresh investment and no consumption. This means that the value is equal to the initial investment plus the gain:

$$V_t(\phi) = x + \int_0^t \alpha_s dB_s + \int_0^t \beta_s dS_s.$$

From now on, we will consider only self-financing portfolios.

Completeness and hedging: A derivative is a contract on a risky asset that produces a payoff H at maturity time T. The payoff is, in general, a $\mathcal{F}_T$ measurable nonnegative random variable H.

European call option with maturity T and exercise price $K > 0$: The buyer of this contract has the option to buy, at time T, one share of the stock at the specified price K. If $S_T \le K$, the contract is worthless to him; he does not exercise his option. If $S_T > K$, the seller is forced to sell one share of the stock at the price K, and thus the buyer can make a profit $S_T - K$ by selling the share at its market price. As a consequence, this contract effectively obligates the seller to a payment of $H = (S_T - K)^+$ at time T.

Black–Scholes formula. Suppose that the parameters $\sigma_t = \sigma$, $\mu_t = \mu$, and $r_t = r$ are constant. In that case, we obtain that the dynamics of the stock price are described by a geometric Brownian motion:

$$S_t = S_0 \exp\left(\left(\mu - \frac{\sigma^2}{2}\right)t + \sigma W_t\right).$$

Moreover,

$$\theta_t = \theta = \frac{\mu - r}{\sigma},$$

and

$$Z_t = \exp\left(-\theta W_t - \frac{\theta^2}{2}t\right).$$

So, $\mathbb{E}(Z_T) = 1$, and $\widetilde{W}_t = W_t + \theta$ is a Brownian motion under Q, with $\frac{dQ}{dP} = Z_t$ on the time interval $[0, T]$.

This model is complete in the sense that any payoff $H \ge 0$ satisfying $\mathbb{E}_Q(H^2) < \infty$ is replicable. In this case, we simply apply the integral representation theorem to the random variable $e^{-rT}H, L^2(\Omega, \mathcal{F}_T, Q)$ with respect to the Wiener process $\widetilde{W}$. In this way, we obtain

$$e^{-rT}H = \mathbb{E}_Q\left(e^{-rT}H\right) + \int_0^T u_s d\widetilde{W}_s,$$

and the self-financing replicating portfolio is given by

$$\beta_t = \frac{\mu_t}{\sigma \widetilde{S}_t},$$

and

$$\alpha_t = M_t - \beta_t \widetilde{S}_t,$$

where

$$M_t = \mathbb{E}_Q\left(e^{-rT}H|\mathcal{F}_t\right) = \mathbb{E}_Q\left(e^{-rT}H\right) + \int_0^T u_s d\widetilde{W}_s.$$

Consider the particular case of a European option, that is, $H = \Phi(S_T)$, where Φ is a measurable function with linear growth. The value of this derivative at time t will be

$$\begin{aligned} V_t(\phi) &= \mathbb{E}_Q\left(e^{-r(T-t)}\Phi(S_T) \mid \mathcal{F}_t\right) \\ &= e^{-r(T-t)}\mathbb{E}_Q\left(\Phi\left(S_t e^{-r(T-t)} e^{\sigma(\widetilde{W}_T - \widetilde{W}_t) - \frac{\sigma^2}{2(T-t)}}\right) \mid \mathcal{F}_t\right). \end{aligned}$$

Hence,

$$V_t = F(t, S_t),$$

where

$$F(t,x) = e^{-r(T-t)}\mathbb{E}_Q\Big(\Phi\Big(x_t e^{-r(T-t)} e^{\sigma(\widetilde{W}_T-\widetilde{W}_t)-\frac{\sigma^2}{2(T-t)}}\Big)\Big). \tag{1.4}$$

Under general hypotheses on Φ (for instance, if Φ has linear growth, is continuous, and piecewise differentiable) that include the cases

$$\Phi(x) = (x-K)^+,$$

and

$$\Phi(x) = (K-x)^+.$$

Formula (1.4) can be written as

$$\begin{aligned} F(t,x) &= e^{-r(T-t)}\mathbb{E}_Q\Big(\Phi\Big(x_t e^{-r(T-t)} e^{\sigma(\widetilde{W}_T-\widetilde{W}_t)-\frac{\sigma^2}{2(T-t)}}\Big)\Big) \\ &= e^{-r\tau}\frac{1}{\sqrt{2\pi}}\int_{-\infty}^{+\infty} e^{-y^2/2}\Phi\Big(xe^{r\tau-\frac{\sigma^2}{2}\tau+\sigma\sqrt{\tau}y}\Big)dy, \end{aligned}$$

where $\tau = T-t$ is the time to maturity.

In the particular case of a European call option with an exercise price K and maturity T (see Figure 1.2),

$$\Phi(x) = (x-K)^+,$$

and we get

$$\begin{aligned} F(t,x) &= \frac{1}{\sqrt{2\pi}}\int_{-\infty}^{+\infty} e^{-y^2/2}\Phi\Big(xe^{-\frac{\sigma^2}{2}\tau+\sigma\sqrt{\tau}y} - Ke^{-r\theta}\Big)^+ dy \\ &= x\Phi(d_+) - Ke^{-r(T-t)}\Phi(d_-), \end{aligned}$$

where

$$d_+ = \frac{\log\frac{x}{k} + (r+\sigma^2/2)\tau}{\sigma\sqrt{\tau}},$$

and

$$d_- = \frac{\log\frac{x}{k} + (r-\sigma^2/2)\tau}{\sigma\sqrt{\tau}}.$$

Example 2. Fluids obey the general laws of continuum mechanics: conservation of mass, energy, and linear momentum. They can be expressed as mathematical equations once a representation of the state of a fluid is chosen. In the context of mathematics, there are two classical representations. One is the so-called Lagrangian representation, where the state of a fluid "particle" at a given time is described with reference to its initial position. The other representation is the so-called Eulerian representation, where

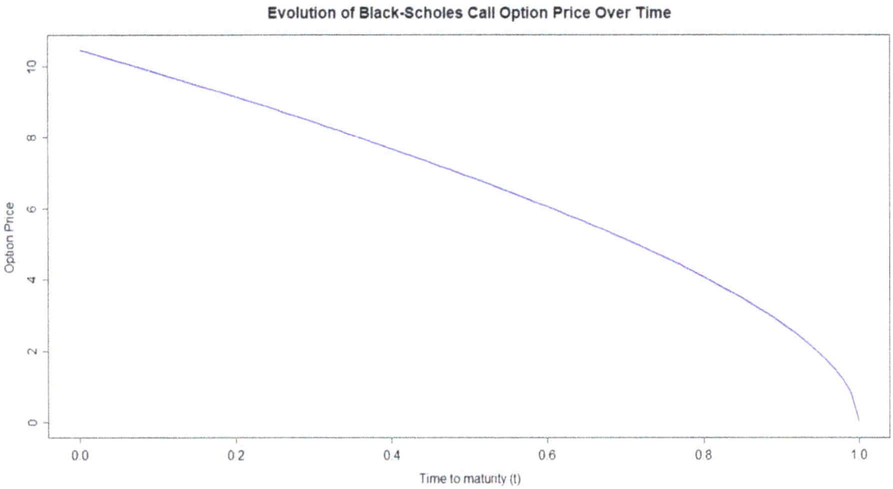

Figure 1.2: European call option with exercise price K.

at each time t and position x in space, the state, in particular, the velocity $u(x,t)$ of the fluid "particle" at that position and time, is given.

In the Eulerian representation of the flow, we also represent the density $\rho(x,t)$ as a function of the position x and time t. The conservation of mass is expressed by the continuity equation

$$\frac{\partial \rho}{\partial t} + \operatorname{div}(\rho u) = 0. \tag{1.5}$$

The conservation of momentum is expressed in terms of the acceleration γ and the Cauchy stress tensor σ:

$$\rho \gamma_i = \sum_{j=1}^{3} \frac{\partial \sigma_{ij}}{\partial x_j} + f_i, \quad i = 1,2,3. \tag{1.6}$$

Here,

$$\gamma = (\gamma_1, \gamma_2, \gamma_3),$$

and $\sigma = \sigma_{i,j=1,2,3}$, componentwise in the 3-dimensional case. Moreover,

$$f = (f_i, f_2, f_3),$$

represents volume forces applied to the fluid. The acceleration vector $\gamma = \gamma(x,t)$ of the fluid at position x and time t can be expressed, using purely kinematic arguments, by the so-called material derivative:

$$\gamma = \frac{Du}{Dt} = \frac{\partial u}{\partial t} + (u, \nabla)u,$$

or componentwise,

$$\gamma_i = \frac{\partial u}{\partial t} + \sum_{j=1}^{3} u_j \frac{\partial u_i}{\partial x_j}, \quad i = 1,2,3.$$

Inserting this expression into the left-hand side (LHS) of equation (1.6) yields the term $\rho(u.\nabla)u$, which is the only nonlinear term in the Navier–Stokes equations; this term is also called the inertial term. The Navier–Stokes equations are among the very few equations of mathematical physics for which the nonlinearity arises not from the physical attributes of the system but rather from the mathematical (kinematical) aspects of the problem.

Further transformations of the conservation of momentum equation necessitate additional physical arguments and assumptions. Rheology theory relates the stress tensor to the velocity field for different materials through the so-called stress-strain law and other constitutive equations. Assuming that the fluid is Newtonian, which is the case of interest to us, amounts to assuming that the stress-strain relationship is linear. More precisely, for Newtonian fluids, the stress tensor is expressed in terms of the velocity field by the formula

$$\sigma_{ij} = \mu\left(\frac{\partial u_i}{\partial x_j} + \frac{\partial u_j}{\partial x_i}\right) + (\lambda \operatorname{div} u - p)\delta_{ij}, \tag{1.7}$$

where $p = p(x,t)$ is the pressure. Here, δ_{ij} is the Kronecker symbol, and μ, λ are constants. The constant μ is called the shear viscosity coefficient, and $3\lambda + 2\mu$ is the dilation viscosity coefficient. For thermodynamic reasons, $\mu > 0$ and $3\lambda + 2\mu \geq 0$. Inserting the stress-strain law (1.7) into the momentum equation (1.6), we obtain

$$\rho\left(\frac{\partial u}{\partial t} + (u.\nabla)u\right) = u \bigtriangleup u + (\lambda + \mu)\nabla \operatorname{div} u - \nabla p + f. \tag{1.8}$$

If we also assume that the fluid is incompressible and homogeneous, then the density is constant in space and time: $\rho(x,t) \equiv \rho_0$. In this case, the continuity equation is reduced to the divergence-free condition:

$$\operatorname{div} u = 0. \tag{1.9}$$

Because the density is constant, we may divide the momentum equation (1.8) by ρ and consider the so-called kinematic viscosity $\nu = \mu/\rho_0$. We may then replace the pressure p and the volume force f with the kinetic pressure p/ρ_0 and the mass density of body forces f/ρ_0, respectively. In doing so, and taking into consideration the divergence-free condition (1.9), we obtain the Navier–Stokes equations for a viscous, incompressible, homogeneous flow:

$$\frac{\partial u}{\partial t} - \nu\Delta u + (u.\nabla)u + \nabla p = f, \tag{1.10}$$

$$\nabla.u = 0, \tag{1.11}$$

where for notational simplicity, we represent the divergence of u as ∇u. For all practical purposes, the density has actually been normalized to unity; even so, we may sometimes replace (1.10) with (1.11), remembering that $\nabla u = 0$ and ρ are constant.

For a random variable ξ, we therefore consider the stochastic Navier–Stokes equations with stochasticity introduced in the body force by a Brownian motion,

$$f = \sigma\frac{dW(t)}{dt},$$

with $\sigma = \sigma(x)$, or

$$\frac{\partial u}{\partial t} - \nu\Delta u + (u.\nabla)u + \nabla p = \xi.$$

The noise $\xi = \xi(x,t)$ will be white (delta-correlated) in time, with some degree of correlation in space. In the usual language of applied sciences,

$$\langle \xi_a(x,t), \xi_a(y,t) \rangle = (t \wedge s) Q_{a\beta}(x,y), \quad a, \beta = 1, \dots d,$$

where $Q(x,y)$ is the matrix-valued space-covariance of the noise. Assume

$$Q(x,y) = Q(x-y).$$

A rigorous mathematical model of ξ is given by a probability space $(\Omega, \mathcal{F}, \mathbb{P})$, a sequence of independent Brownian motions $W_k(t)$ on $(\Omega, \mathcal{F}, \mathbb{P})$, and $k = 1, \dots,$ is a sequence of divergence-free vector fields such that

$$Q_{a\beta}(x,y) = \sum_{k=0}^{\infty} \sigma_a^k \sigma_\beta^k.$$

Assumptions of summability and regularity of σ^k are required, depending on the results. With this data, the time-distribution

$$\xi(x,t) = \sum_{k=0}^{\infty} \sigma^k(x) \frac{dW(t)}{dt},$$

is white noise in time, with covariance $Q(x,y)$ in space. Transport type noise

$$du + (u.\nabla u + \nabla p - \nu \Delta u)dt = \sum_{k=0}^{\infty} \sigma^k(x) dW(t),$$

(interpreted weakly: integrated in time and against smooth test functions; see Figure 1.3).

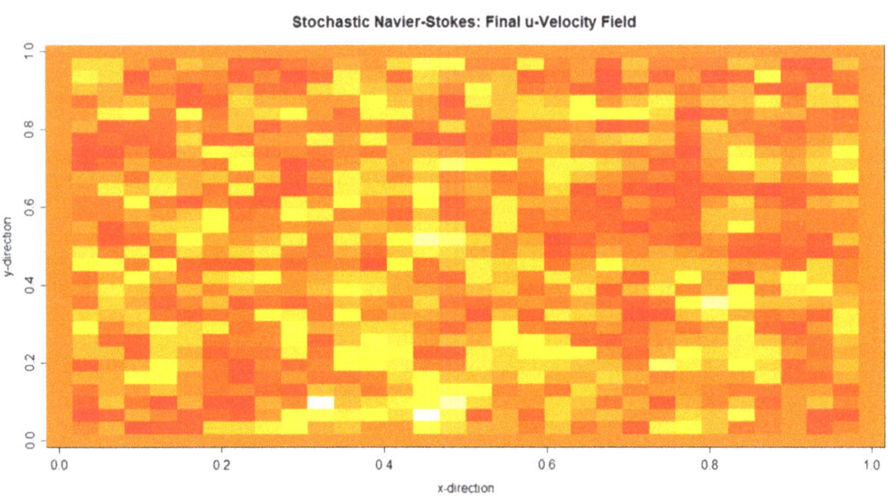

Figure 1.3: Stochastic Navier–Stokes equations.

Example 1.1 (Analysis of a predator prey model with stochastic perturbation). One of the first mathematical models that incorporate the interaction between two species (Holling–Tanner predator–prey). The relationship between predator and prey has been, and will long be, one of the most important topics in mathematical biology. The famous predator–prey system, as first introduced

$$\begin{cases} \frac{dx}{dt} = x(t)(a_1 - b_1 x(t)) - p(x)y(t), \\ \frac{dy}{dt} = y(t)(a_2 - b_2 \frac{y(t)}{x(t)}), \end{cases} \tag{1.12}$$

where $x(t)$, $y(t)$ are the densities of prey and predators at time t, respectively. $p(x)$ is the functional response of the predator to the prey. Here, the predator still grows logistically with an intrinsic growth rate a_2 and a carrying capacity $x(t)/b_2$, and the prey variable appears in the denominator when

$$p(x) = \alpha x/(b + x),$$

is the Holling II functional response. The population dynamics of the type (1.12) are the Holling–Tanner predator–prey system:

$$\begin{cases} \frac{dx}{dt} = x(t)(a_1 - b_1 x(t)) - \frac{\alpha x(t)y(t)}{b+x(t)}, \\ \frac{dy}{dt} = y(t)(a_2 - b_2 \frac{y(t)}{x(t)}), \end{cases} \tag{1.13}$$

where $\alpha, b > 0$ denotes the capturing rate and the half-capturing saturation constant, respectively. In addition, in the natural world, the birth rate, death rate, carrying capacity of species, and other parameters will not remain constant but will exhibit more or less periodicity as a result of seasonal change, life cycles, and man-made factors. There have been many studies in the literature that investigate the population dynamics of the type (1.13) models. However, in the study of the dynamic relationship between species, the effects of some impulsive factors, which exist widely in the real world, have been ignored. For example, the birth of many species is characterized by an annual birth pulse or harvest. Moreover, human beings have been harvesting or stocking species at certain times, and then the species is affected by another impulsive type. Additionally, the impulsive reduction of the population density of a given species is possible following its partial destruction through catching or poisoning with chemicals used during certain transitory slots in fishing or agriculture. Such factors have a significant impact on population growth. If we incorporate these impulsive factors into the model of population interaction, the model must be governed by an impulsive differential system.

For example, if at the moment $t = t_k$ the population density of the predator changes, then we can assume that

$$x(t_k^+) - x(t_k^-) = \alpha_k x(t_k), \quad y(t_k^+) - y(t_k^-) = \beta_k y(t_k), \quad k = 1, 2, \ldots,$$

where $(x(t_k^-), y(t_k^-)) = (x(t_k), x(t_k))$ and $(x(t_k^+), y(t_k^+))$ are the population densities of prey and predators before and after the perturbation. The corresponding periodic system with impulses is as follows:

$$\begin{cases} \frac{dx}{dt} = x(t)(a_1(t) - b_1(t)x(t)) - \frac{\alpha(t)y(t)}{b(t)+x(t)}, \\ \frac{dy}{dt} = y(t)(a_2(t) - b_2(t)\frac{y(t)}{x(t)}), \\ x(t_k^+) - x(t_k^-) = \alpha_k x(t_k), \quad k = 1, 2, \ldots, \\ y(t_k^+) - y(t_k^-) = \beta_k y(t_k), \quad k = 1, 2, \ldots. \end{cases}$$

Simulated by the work above, we assume the intrinsic growth rates $a_1(t)$ and $a_2(t)$ of the prey and the predator are disturbed with

$$a_1(t) \to a_1(t) + \sigma_1(t)dB_1(t),$$
$$a_2(t) \to a_2(t) + \sigma_2(t)dB_2(t),$$

where $B_i(t), i = 1,2$ are independent Brownian motions. σ_1^2, σ_2^2 denotes the intensity of the white noise. That is, we consider the following stochastic nonautonomous predator–prey system with impulsive effects:

$$\begin{cases} \frac{dx}{dt} = x(t)(a_1(t) - b_1(t)x(t)) - \frac{a(t)y(t)}{b(t)+x(t)} + x(t)\sigma_1(t)dB_1(t), \\ \frac{dy}{dt} = y(t)(a_2(t) - b_2(t)\frac{y(t)}{x(t)}) + y(t)\sigma_2(t)dB_2(t), \\ x(t_k^+) - x(t_k^-) = \alpha_k x(t_k), \quad k = 1,2,\dots, \\ y(t_k^+) - y(t_k^-) = \beta_k y(t_k), \quad k = 1,2,\dots. \end{cases} \tag{1.14}$$

where

$$a_i(t), b_i(t), \sigma_1(t) \ (i = 1,2), a(t), b(t),$$

are positive and continuous T periodic functions and

$$0 < t_1 < t_2 < \cdots < t_k < \cdots,$$

where

$$\lim_{k\to+\infty} t_k = +\infty.$$

For biological purposes, we are only concerned with the positive solution of the equation. Therefore, it is a natural constraint that

$$1 + \alpha_k > 0, \quad 1 + \beta_k > 0, \quad k = 1,2\dots.$$

For example, if $\alpha_k, \beta_k > 0$, the impulsive effects denote the planting of the species, while $\alpha_k, \beta_k < 0$ denotes harvesting.

In mathematical ecology, the system (1.14) denotes a model of the dynamics of a stochastic predator–prey system, which is subject to impulsive effects at certain moments in time. By means of such models, it is possible to take into account the possible environmental changes or other external effects due to which the population density of the predator is momentarily changed; see Figure 1.4.

Example 1.2. Vaccination strategies are designed and implemented to control or eradicate infectious diseases, including continuous vaccination and pulse vaccination. The pulse vaccination strategy (PVS) consists of periodic repetitions of impulsive vaccinations in a population.

Suppose an epidemic model with constant vaccination is given. We study a population that is composed of three classes of individuals: susceptible (S), infective (I), and recovered (R). The SIRS model is described by the following ordinary differential equation:

$$\begin{cases} \frac{dS(t)}{dt} = b - dS(t) - g(S(t), I(t)) + \gamma R(t), \\ \frac{dI(t)}{dt} = g(S(t), I(t)) - (d + \mu + \delta)I(t), \\ \frac{dR(t)}{dt} = \mu I(t) - (d + \gamma)R(t), \end{cases} \tag{1.15}$$

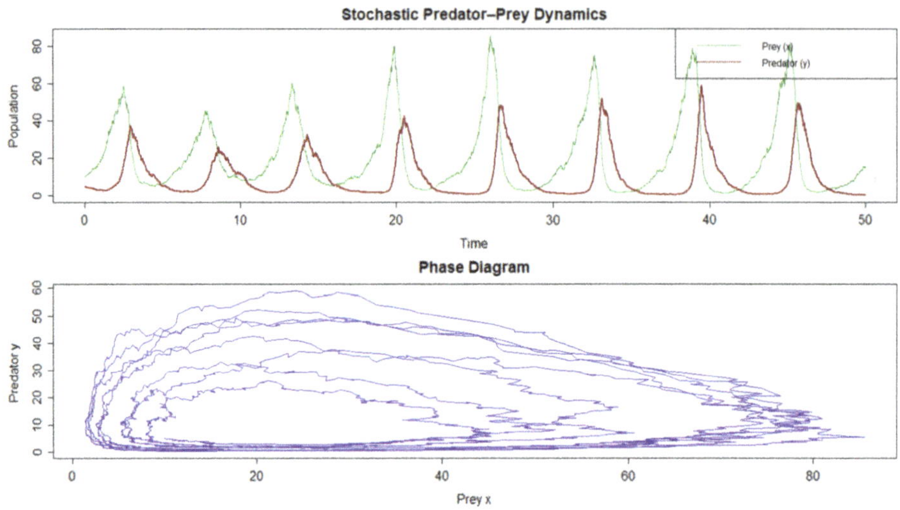

Figure 1.4: Analysis of a predator–prey model with stochastic perturbation.

where b is the recruitment rate of the population, d is the natural death rate of the population, μ is the natural recovery rate of the infective individuals, γ is the rate at which recovered individuals lose immunity and return to the susceptible class, δ is the disease inducing death rate, and $g(S(t), I(t))$ is the transmission of the infection, also called the incidence rate. The transmission function $g(S(t), I(t))$ plays a key role in determining disease dynamics. There are many different approaches to deriving a stochastic version of the deterministic SIRS model with the ratio-dependent transmission function $g(S(t), I(t))$. Thus, the stochastic perturbation in our model is a white noise type that is directly proportional to $S(t)$, $I(t)$, $R(t)$, and is influenced by $\frac{dS(t)}{dt}$, $\frac{dI(t)}{dt}$, and $\frac{dR(t)}{dt}$, respectively. Following this approach, we obtain the following SDE epidemic model (1.16) that is analogous to its deterministic version (1.15) by introducing stochastic perturbation terms into the growth equations of susceptible, infective, and recovered individuals to incorporate the effect of randomly fluctuating environments:

$$\begin{cases} \frac{dS(t)}{dt} = b - dS(t) - g(S(t), I(t)) + \gamma R(t) + \sigma_1 S(t) dB_1(t), \\ \frac{dI(t)}{dt} = g(S(t), I(t)) - (d + \mu + \delta) I(t) + \sigma_2 I(t) dB_2(t), \\ \frac{dR(t)}{dt} = \mu I(t) - (d + \gamma) R(t) + \sigma_3 R(t) dB_3(t), \end{cases} \tag{1.16}$$

where σ_1 $(i = 1, 2, 3)$ are real constants known as the intensity of environmental fluctuations, $B_i(t)$ $(i = 1, 2, 3)$ independent standard Brownian motions, and the total population

$$N = S + I + R.$$

Assume the pulse scheme proposes to vaccinate a fraction p $(0 < p < 1)$, of the entire susceptible population in a single pulse, applied every year. When pulse vaccination is incorporated into the SIR model (1.16) (see Figure 1.5), the system can be rewritten as

$$\begin{cases} \frac{dS(t)}{dt} = b - dS(t) - g(S(t), I(t)) + \gamma R(t) + \sigma_1 S(t) dB_1(t), \\ \frac{dI(t)}{dt} = g(S(t), I(t)) - (d + \mu + \delta) I(t) + \sigma_2 I(t) dB_2(t), \\ \frac{dR(t)}{dt} = \mu I(t) - (d + \gamma) R(t) + \sigma_3 R(t) dB_3(t), \\ S(k^+) = (1 - \rho) S(k), \\ I(k^+) = I(k), \quad t = k, \\ R(k^+) = R(k) + \rho S(k), \quad k = 0, 1, 2, \ldots. \end{cases}$$

Here,

$$f(k^+) = \lim_{t \to k^+} f(t),$$
$$f(k) = \lim_{t \to k^-} f(t).$$

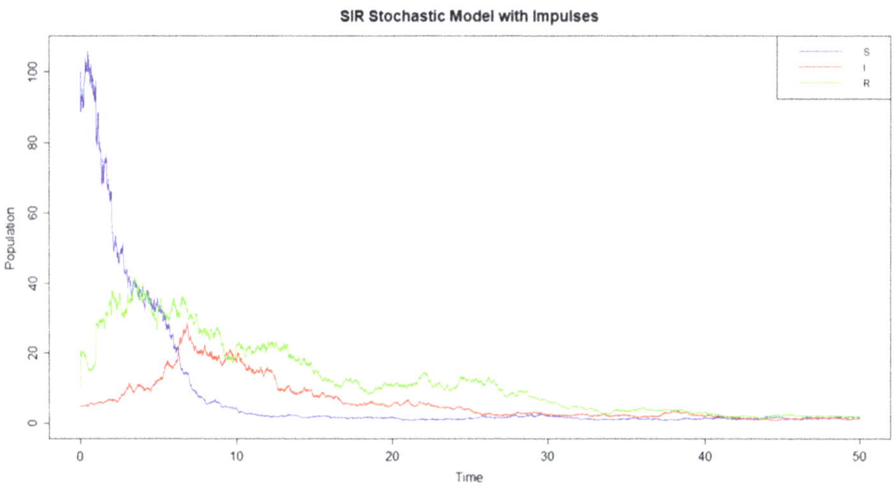

Figure 1.5: Stochastic SIR epidemic model.

2 Elements of stochastic analysis

2.1 Some selected topics from probability theory

The purpose here is to recall some standard notions and facts and to introduce the tools to be used. We introduce the background material used herein, such as stochastic calculus and some properties of generalized Brownian motion, as well as an application we are interested in: the definition of the stochastic integral corresponding to *fBm* with values in a Hilbert space. We show some recent applications of the Malliavin calculus to develop a stochastic calculus with respect to fractional Brownian motion.

Definition 2.1. Let Ω be a set and $\mathcal{F}$ a collection of its subsets. We say that $\mathcal{F}$ is a σ-field (or σ-algebra) if:
1. $\Omega \in \mathcal{F}$.
2. For every $A_1, \dots, a_n, \dots$ such that $A_n \in \mathcal{F}$, we have

$$\bigcup_n A_n \in \mathcal{F}.$$

3. If $A \in \mathcal{F}$, then

$$A^c = \Omega \setminus A \in \mathcal{F}.$$

In the case when $\mathcal{F}$ is a σ-field, the couple $(\Omega, \mathcal{F})$ is called a measurable space, and elements of $\mathcal{F}$ are called events.

Example 2.1. Let Ω be a set and

$$\mathcal{F} = \{\Omega, \emptyset\},$$

is a σ-field, which is called the trivial σ-field.

Definition 2.2. $(E, \mathcal{O})$ is a topological space, where $\mathcal{O}$ is the set of open sets in E. Then $\sigma(\mathcal{O})$ is called the Borel σ-algebra of the topological space. If $\mathcal{A} \subset \mathcal{B}$, then A is called a subalgebra of $\mathcal{B}$. A set B in $\mathcal{B}$ is also called a Borel set.

The last property is called σ-additivity.

Definition 2.3. A map X from a measure space $(\Omega, \mathcal{F})$ to another measure space $(\Delta, \mathcal{B})$ is called measurable if $X^{-1}(2) \in \mathcal{A}$ for all $B \in \mathcal{B}$. The set $X^{-1}(2)$ consists of all points $x \in \Omega$ for which $X(x) \in B$. This pullback set $X^{-1}(2)$ is defined even if X is noninvertible.

Definition 2.4. We say that $\mathbb{P}(.)$ is a probability measure on $(\Omega, \mathcal{F})$ or on $\mathcal{F}$ if:
1. $\mathbb{P}(1) \geq 0$ and $\mathbb{P}(\Omega) = 1$.

https://doi.org/10.1515/9783112251898-002

2. For every sequence of pairwise disjoint $A_1, \ldots, A_n, \ldots \in \mathcal{F}$, we have

$$\mathbb{P}\Big(\bigcup_n A_n\Big) = \sum_n \mathbb{P}(A_n).$$

If, on a measurable space $(\Omega, \mathcal{F})$, there is a defined probability measure $\mathbb{P}$, the triple $(\Omega, \mathcal{F}, \mathbb{P})$ is called a probability space.

Example 2.2. The triple, consisting of $[0,1](=\Omega)$, the σ-field $\mathcal{B}([0,1])$ of Borel subsets of $[0,1]$ (taken as $\mathcal{F}$) and Lebesgue measure ℓ as $(\mathbb{P})$ is a probability space.

Let $(\Omega, \mathcal{F}, \mathbb{P})$ be a complete probability space with a filtration $(\mathcal{F} = \mathcal{F}_t)_{t\geq 0}$ satisfying the usual conditions (i. e., right continuous and $\mathcal{F}_0$ containing all $\mathbb{P}$-null sets).

Definition 2.5. The algebra of all random variables is denoted by $\mathcal{L}$. It is a vector space over the field $\mathbb{R}$ of the real numbers in which one can multiply. An elementary function or step function is an element of $\mathcal{L}$ that is of the form

$$X = \sum_{i=1}^{n} \alpha_i \cdot 1_{A_i},$$

with $\alpha_i \in \mathbb{R}$ and where $A_i \in \mathcal{F}$ are disjoint sets. Denote by $\mathcal{S}$ the algebra of step functions. For $X \in \mathcal{S}$, we can define the integral

$$\begin{aligned}\mathbb{E}(X) &= \int_\Omega X d\mathbb{P} \\ &= \sum_{i=1}^{n} \alpha_i \mathbb{P}(A_i).\end{aligned}$$

Definition 2.6. Define $\mathcal{L}^1 \subset \mathcal{L}$ as the set of random variables X, for which

$$\sup_{Y\in\mathcal{S}, Y\leq X} \int Y d\mathbb{P},$$

is finite. For $X \in \mathcal{L}^1$, we can define the integral or expectation

$$\begin{aligned}\mathbb{E}(X) &= \int X d\mathbb{P} \\ &= \sup_{Y\in\mathcal{S}, Y\leq X^+} \int Y d\mathbb{P} - \sup_{Y\in\mathcal{S}, Y\leq X^-} \int Y d\mathbb{P},\end{aligned}$$

where

$$X^+ = X \vee 0 = \max(X, 0),$$

and

$$X^{-} = -X \vee 0 = \max(-X, 0).$$

The vector space $\mathcal{L}^1$ is called the space of integrable random variables. Similarly, for $p \geq 1$ write $\mathcal{L}^p$ for the set of random variables X for which

$$\mathbb{E}(|X|^P) < \infty.$$

Definition 2.7. For $X, Y \in \mathcal{L}^2$, define the covariance

$$\begin{aligned}\mathrm{Cov}(X, Y) &= \mathbb{E}[(X - \mathbb{E}(X))(Y - \mathbb{E}(Y))] \\ &= \mathbb{E}(XY) - \mathbb{E}(X)\mathbb{E}(Y).\end{aligned}$$

Two random variables in $\mathcal{L}^2$ are called uncorrelated if

$$\mathrm{Cov}(X, Y) = 0.$$

Definition 2.8. For $X \in \mathcal{L}^2$, we can define the variance

$$\mathrm{Var}(X) = \mathrm{Cov}(x, x) = \mathbb{E}((X - E(X))^2).$$

Definition 2.9. Write $J \subset I$ if J is a finite subset of I. A family $\{\mathcal{F}_i\}_{i \in I}$ of σ-sub-algebras of $\mathcal{F}$ is called independent if, for every $J \subset I$ and every choice

$$A_j \in \mathcal{F}_j, P\left[\bigcap_{j \in J} A_j\right] = \prod_{j \in J} \mathbb{P}(A_j).$$

A family $\{X_j\}_{j \in J}$ of random variables is called independent if $\{\sigma(X_j)\}_{j \in J}$ are independent σ-algebras. A family of sets $\{A_j\}_{j \in I}$ is called independent if the σ-algebras

$$\mathcal{F}_j = \{\emptyset, A_j, A_j^c, \Omega\},$$

are independent.

Probability distribution function

The probability distribution function

$$F_X : \mathbb{R} \longrightarrow [0, 1],$$

of a random variable X is defined as

$$F_X(x) = \mathbb{P}(X \leq x),$$

where $\mathbb{P}(X \le x)$ is a shorthand notation for

$$\mathbb{P}(\{\omega \in \Omega \mid X(\omega) < x\}).$$

2.1.1 Some inequalities

Let X and Y be random variables on $\mathcal{L}^p(\Omega, \mathcal{F}, \mathbb{P})$.

1. **Chebyshev's inequality:** For all $\lambda > 0$,

$$\mathbb{P}(|X| > \lambda) \le \frac{1}{\lambda^p}\mathbb{E}(|X|^p).$$

2. **Schwartz inequality:**

$$\mathbb{E}(XY) \le \sqrt{\mathbb{E}(X^2)\mathbb{E}(Y^2)}.$$

3. **Hölder inequality:**

$$\mathbb{E}(XY) \le \left[\mathbb{E}|X|^p\right]^{\frac{1}{p}} \left[\mathbb{E}|Y|^q\right]^{\frac{1}{q}},$$

   where $p, q > 1$ and

$$\frac{1}{p} + \frac{1}{q} = 1.$$

4. **Jensen inequality:** For any convex function $h : \mathbb{R} \longrightarrow \mathbb{R}$, we have

$$h(\mathbb{E}(X)) \le \mathbb{E}(hE(X)).$$

5. **Kolmogorov inequality:** Let $X_1, \dots, X_n$ be independent random variables with $E(X_i) = 0$ and $\operatorname{var}(X_i) < \infty$. If

$$S_n = X_1 + \cdots + X_n,$$

   then

$$\mathbb{P}\Big(\max_{1 \le k \le n} |S_k| \ge t\Big) \le \frac{1}{t^2} \operatorname{Var} S_k, \quad \forall t > 0.$$

2.1.2 Processes and filtration

Definition 2.10. A collection of sub σ-algebras

$$\{\mathcal{F}_t; t \ge 0\},$$

of the σ-algebra $\mathcal{F}$ is called a filtration if $s \le t$ implies that

$$\mathcal{F}_s \subset \mathcal{F}_t.$$

For a given stochastic process X, write $\mathcal{F}_t^X$ for the filtration

$$\sigma\{X_s; 0 \leq s \leq t\},$$

and

$$\{\mathcal{F}_t^X; t \geq 0\},$$

is referred to as the natural filtration associated with the process X.

Definition 2.11. A stochastic process is a parameterized collection of random variables $\{X_t\}_{t\in[0,T]}$ defined on a probability space $(\Omega, \mathcal{F}, \mathbb{P})$ and assuming values in $\mathbb{R}^n$.

The parameter space T is usually the half-line $[0, \infty)$, but it may also be an interval $[a, b]$, the nonnegative integers, and even subsets of $\mathbb{R}^n$ for $n \geq 1$. Note that for each $t \in T$ fixed, we have a random variable

$$\omega \longrightarrow X_t(\omega); \quad \omega \in \Omega.$$

On the other hand, fixing $\omega \in \Omega$, we can consider the function

$$t \longrightarrow X_t(\omega); \quad t \in T,$$

which is called a path of X_t.

Definition 2.12. The stochastic process X is adapted to the filtration $\{\mathcal{F}_t\}$ if, for each $t \geq 0, X_t$ is an $\mathcal{F}_t$-measurable random variable.

Definition 2.13. The parameter space T is usually the half-line $[0, \infty)$, but it may also be an interval $[a, b]$, the nonnegative integers, and even subsets of $\mathbb{R}^n$ for $n \geq 1$. Note that for each $t \in T$ fixed, we have a random variable

$$w \rightarrow X_t(w); \quad w \in \Omega.$$

On the other hand, fixing $w \in \Omega$, we can consider the function

$$t \rightarrow X_t(w); \quad t \in T,$$

which is called a path of X_t.

Definition 2.14. The stochastic process X is adapted to the filtration $\{\mathcal{F}_t\}$ for each $t \geq 0$. X_t is an $\{\mathcal{F}_t\}$-measurable random variable.

Definition 2.15. A filtration $\mathcal{F}_t$ is said to satisfy the usual conditions if it is right continuous and $\mathcal{F}_0$ contains all the $\mathbb{P}$-negligible events in $\mathcal{F}$.

2.1.3 Stopping times

Definition 2.16. A random variable $\tau : \Omega \longrightarrow [a,b]$ is called a stopping time with respect to a filtration

$$\{\mathcal{F}_t; a \le t \le b\},$$

if

$$\{\omega; \tau(\omega) \le t\} \in \mathcal{F}_t,$$

for all $t \in [a,b]$.

Properties:

1. If the filtration is right continuous, then τ is a stopping time if and only if, for all t,

$$\{\tau < t\} \in \mathcal{F}_t.$$

2. Let τ_1 and τ_1 be two stopping times. Then

$$\tau_1 + \tau_2, \quad \tau_1 \wedge \tau_2, \quad \tau_1 \vee \tau_2,$$

are stopping times.

Definition 2.17. The σ-algebra $\mathcal{F}_\tau$ for a stopping time τ is given by

$$\mathcal{F}_\tau = \{A \in \mathcal{F} : A \cap \{\tau \le t\} \in \mathcal{F}_t \text{ for all } t\}.$$

The following holds:

1. τ is measurable with respect to $\mathcal{F}_\tau$.
2. If $\tau(\omega) = t$ for almost all ω, then $\mathcal{F}_\tau = \mathcal{F}_t$.

2.1.4 Continuous time martingales

We shall consider exclusively real valued processes

$$X = \{X_t; 0 \le t < \infty\},$$

on a probability space $(\Omega, \mathcal{F}, \mathbb{P})$, adapted to a given filtration $\{\mathcal{F}_t\}$, and such that $\mathbb{E}|X_t| < \infty$ holds for every $t \ge 0$.

Definition 2.18. 1. A stochastic process $X = (X_t)_{t\ge 0}$ is called a martingale (with respect to $\mathbb{P}$ and $\mathcal{F}$) if:

a. X is adapted.

b. $\mathbb{E}(X_t) < \infty$ for all $t \geq 0$.
c. For $0 \leq s \leq t < \infty$,

$$\mathbb{E}(X_t \mid \mathcal{F}_s) = X_s \quad p.s.$$

2. X is called a submartingale

$$\mathbb{E}(X_t \mid \mathcal{F}_s) \geq X_s \quad p.s.$$

3. X is called a supermartingale

$$\mathbb{E}(X_t \mid \mathcal{F}_s) \leq X_s \quad p.s.$$

Definition 2.19. The stochastic process X is said to be adapted to $\mathcal{F}$ if, for each $t \geq 0$, X_t is $\mathcal{F}_t$-measurable.

Definition 2.20. A random variable $T : \Omega \longrightarrow \mathbb{R}^+ \cup \{\infty\}$ is called a stopping time with respect to $\mathcal{F}$ if

$$\{T \leq t\} \in \mathcal{F}_t \quad \text{for all } t \in \mathbb{R}^+.$$

Definition 2.21. The process X is continuous if all its paths $t \rightsquigarrow X_t(w)$ are continuous. The process X is cadlag if all its paths are right continuous with left-hand limits. If X is cadlag, we define the left-continuous process X_{t^-} as

$$X_{t^-} = \begin{cases} X_0, & t = 0, \\ \lim_{s \uparrow t} X_s, & t > 0, \end{cases}$$

and the jumps as

$$\Delta X_t = X_t - X_{t^-}, \quad t \geq 0.$$

For $T > 0$, we define the system of sets

$$\mathcal{G}_T = \{A \times \{0\} : A \in \mathcal{F}\} \cup \{A \times (s,t] : 0 \leq s \leq t \leq T \ A \in \mathcal{F}_s\},$$

and the predictable σ-algebra $\mathcal{P}_T = \sigma(\mathcal{G}_T)$. An $\mathcal{F}$-valued process $X = (X_t)_{t \in [0,T]}$ is called predictable if it is $\mathcal{P}_T$-measurable.

Definition 2.22. A modification Y of the stochastic process X is a stochastic process on the same probability space, with the same parameter set $\mathbb{R}^+$ such that

$$\mathbb{P}(X_t = Y_t) = 1 \quad \text{for all } t \in \mathbb{R}_+.$$

Definition 2.23. For a stochastic process $X = (X_t)_{t\geq 0}$ and a stopping time T, the stopped process

$$X^T = (X_t^T)_{t\geq 0},$$

is defined by

$$X_t^T = X_{T\wedge t} = \begin{cases} X_t, & \text{if } t \leq T, \\ X_T, & \text{if } t > T. \end{cases}$$

Definition 2.24. An adapted, right-continuous stochastic process $X = (X_t)_{t\geq 0}$ is called a local martingale if there exists a sequence of stopping times (T_n) with $T_n \to \infty$ $\mathbb{P}$-a. s., such that the stopped process

$$X^{T_n} I_{T_n>0} = (X_{T_n\wedge t} I_{T_n>0})_{t\geq 0},$$

is a (uniformly integrable) martingale with respect to $(\mathcal{F}_t)$.

The following two estimates are known as Doob's L^p-inequalities.

Theorem 2.1. *Let $M = (M_t)_{t\geq 0}$ be an $\mathcal{F}$-valued martingale. Then the following statements are valid:*

1. *For all $p \geq 1$ and $\lambda > 0$, we have*

$$\mathbb{P}\Big(\sup_{t\in[0,T]} \|M_t\| \geq \lambda\Big) \leq \frac{1}{\lambda^p} \mathbb{E}\|M_T\|^p.$$

2. *For every $p > 1$, we have*

$$\mathbb{E}\Big(\sup_{t\in[0,T]} \|M_t\|\Big) \leq \Big(\frac{p}{p-1}\Big)^p \mathbb{E}\|M_T\|^p.$$

Let us fix a number $T > 0$ and denote by $\mathcal{M}_T^2(E)$ the space of all E-valued continuous, square integrable martingales M. We will need the following.

Definition 2.25 (Square integrable martingales). A martingale (M_t) is called square integrable if it satisfies $\mathbb{E}(M_t^2) < \infty$, that is, $M_t \in L^2$, for all $t \geq 0$.

Proof. Let $0 \leq a \leq s \leq t$. Then

$$\mathbb{E}(M_s(M_t - M_s)/\mathcal{F}_s) = M_s\mathbb{E}((M_t - M_s)/\mathcal{F}_s) = 0,$$

and

$$\mathbb{E}(M_s(M_t - M_s)/\mathcal{F}_a) = 0.$$

Observing that

$$M_t^2 - M_s^2 - (M_t - M_s)^2 = M_s(M_t - M_s).$$

By (2.11), it follows by taking the conditional expectation. □

2.2 Brownian motion and the Wiener process

In 1828, Robert Brown observed that pollen grains suspended in water perform an unending chaotic motion. L. Bachelier (1900) derived the law governing the position W_t at time t of a single grain performing a one-dimensional Brownian motion starting at $a \in \mathbb{R}^+$ at time $t = 0$,

$$\mathbb{P}_a\{W_t \in dx\} = p(t, a, x)dx, \tag{2.1}$$

where

$$p(t, a, x) = \frac{1}{\sqrt{2\pi t}} e^{-(x-a)^2/2t},$$

is the fundamental solution of the heat equation

$$u_t = \frac{1}{2} u_{xx}.$$

Bachelier (1900) also pointed out the Markovian nature of the Brownian path and used it to establish the maximum displacement

$$\mathbb{P}_a\left\{\max_{s \le t} W_t \le b\right\} = \frac{2}{\sqrt{2\pi t}} \int_0^b e^{\frac{-x^2}{2t}} dx, \quad t > 0, b \ge 0.$$

Einstein (1905) also derived (2.1) from statistical mechanics considerations and applied it to the determination of molecular diameters. Bachelier was unable to obtain a clear picture of Brownian motion, and his ideas were unappreciated at the time. This is not surprising because the precise mathematical definition of Brownian motion involves a measure on the path space. Even after the ideas of Borel, Lebesgue, and Daniell appeared, N. Wiener (1923) only constructed a Daniell integral on the path space, which later was revealed to be the Lebesgue integral against a measure, the so-called Wiener measure.

The simplest model describing the movement of a particle subject to hits by much smaller particles is the following. Let η_k and $k = 1, 2, \dots$ be independent identically distributed random variables, which $\mathbb{E}\eta_k = 0$ and $E\eta_k^2 = 1$. Fix an integer n, and at times $1/n, 2/n, \dots$ let our particle experience instant displacements by $\eta_1 n^{\frac{-1}{2}}, \eta_1 n^{\frac{-1}{2}}, \dots$; at moment zero, let our particle be at zero. If

$$S_k = \eta_1 + \eta_2 + \cdots + \eta_k.$$

Then, at moment k/n, our particle will be at the point $S_k/\sqrt{n}$ and will stay there during the time interval

$$[k/n, (k+1)/n).$$

Since real Brownian motion has continuous paths, we replace our piecewise constant trajectory with a continuous piecewise linear one, preserving its positions at times k/n. Thus, we come to the process,

$$\xi_t^n = S_{[nt]}/\sqrt{n} + (nt - [nt])\eta_{[nt]+}/\sqrt{n}. \tag{2.2}$$

This process gives a very rough caricature of Brownian motion. Clearly, to get a better model, we have to let $n \to \infty$. By the way, precisely this necessity dictates the intervals of time between collisions to be $1/n$ and the displacements due to collisions to be $\eta_k/\sqrt{n}$, since then ξ_t^n is asymptotically normal with parameters $(0, 1)$.

Definition 2.26. The standard Brownian motion is a stochastic process $(W_t)_{t\in\mathbb{R}^+}$ such that:
1. $W_0 = 0$ almost surely.
2. The sample trajectories $t \to W_t$ are continuous, with a probability 1.
3. For any finite sequence of times $t_0 < t_1 < \cdots < t_n$, the increments

   $$W_{t_1} - W_{t_0}, W_{t_2} - W_{t_1}, \ldots, W_{t_n} - W_{t_{n-1}}.$$

   They are independent.
4. For any given times $0 \le s \le t$ and $W_t - W_s$, the Gaussian distribution $\mathcal{N}(0, t-s)$ has a mean of zero and a variance of $t - s$.

Theorem 2.2. *For every $t \in (0, 1]$, we have*

$$\max_{s\le t} W_s \sim |W_t|,$$

which is to say that for every $x \ge 0$,

$$\mathbb{P}\Big(\max_{s\le t} W_s \le x\Big) = \frac{2}{\sqrt{2\pi t}} \int_0^x e^{-y^2/2t}\,dy.$$

Proof. Take independent identically distributed random variables η_k such that

$$\mathbb{P}(\eta_k = 1) = \mathbb{P}(\eta_k = -1) = 1/2,$$

and define ξ_t^n by (2.2). First, we want to find the distribution of

$$\zeta^n = \max_{[0,1]} \xi_t^n = n^{\frac{-1}{2}} \max_{k \le n} S_k.$$

Observe that, for each n, the sequence $(S_1, \ldots, S_n)$ takes on each particular value with the same probability 2^{-n}. In addition, for each integer $i > 0$, the number of sequences favorable to the events

$$\begin{cases} \{\max_{k\le n} S_k \ge i, S_n < i\} \\ \text{and} \\ \{\max_{k\le n} S_k \ge i, S_n > i\}, \end{cases} \tag{2.3}$$

is the same. One proves this by using the reflection principle; that is, one takes each sequence favorable to the first event, keeps it until the moment it reaches the level i, and then reflects its remaining part about this level. This implies the equality of the probabilities of the events in (2.3). Furthermore, since i is an integer, we have

$$\{\zeta^n \ge in^{\frac{-1}{2}}, \xi_1^n < in^{\frac{-1}{2}}\} = \Big\{\max_{k\le n} S_k \ge i, S_n < i\Big\},$$

and

$$\{\zeta^n \ge in^{\frac{-1}{2}}, \xi_1^n > in^{\frac{-1}{2}}\} = \Big\{\max_{k\le n} S_k \ge i, S_n > i\Big\}.$$

Hence,

$$\mathbb{P}(\{\zeta^n \ge in^{\frac{-1}{2}}, \xi_1^n < in^{\frac{-1}{2}}\}) = \mathbb{P}\Big(\Big\{\max_{k\le n} S_k \ge i, S_n < i\Big\}\Big).$$

Moreover, obviously

$$\begin{aligned} \mathbb{P}(\{\zeta^n \ge in^{\frac{-1}{2}}, \xi_1^n > in^{\frac{-1}{2}}\}) &= \mathbb{P}\{\xi_1^n > in^{-1/2}\}, \\ \mathbb{P}(\{\zeta^n \ge in^{\frac{-1}{2}}\} &= \mathbb{P}(\{\zeta^n \ge in^{\frac{-1}{2}}, \xi_1^n > in^{\frac{-1}{2}}\}) \\ &\quad + \mathbb{P}\{\xi_1^n > in^{-1/2}, \xi_1^n < in^{\frac{-1}{2}}\} + \mathbb{P}\{\xi_1^n = in^{\frac{-1}{2}}\}. \end{aligned}$$

It follows that

$$\mathbb{P}(\{\zeta^n \ge in^{\frac{-1}{2}}\}) = \mathbb{P}(\{\xi_1^n > in^{\frac{-1}{2}}\}) + \mathbb{P}(\{\xi_1^n = in^{\frac{-1}{2}}\}), \tag{2.4}$$

for every integer $i > 0$. The last equality also obviously holds for $i = 0$. We see that for numbers a of type $in^{-1/2}$, where i is a nonnegative integer, we have

$$\mathbb{P}(\{\zeta^n \ge a\}) = \mathbb{P}(\{\xi_1^n > a\}) + \mathbb{P}(\{\xi_1^n = a\}). \tag{2.5}$$

Certainly, the last probability goes to zero as $n \to \infty$, since ξ_1^n is asymptotically normal with parameters $(0, 1)$.

Also, keeping in mind Donsker's theorem, it is natural to think that

$$\mathbb{P}\Big(\Big\{\max_{s\le 1}\xi_s^n \ge a\Big\}\Big) \longrightarrow \mathbb{P}\Big(\Big\{\max_{s\le 1} W_s \ge a\Big\}\Big),$$
$$2\mathbb{P}(\{\xi_1^n \ge a\}) \longrightarrow \mathbb{P}(\{W_1 \ge a\}).$$

Therefore, (2.5) naturally leads to the conclusion that

$$\mathbb{P}\Big(\Big\{\max_{s\le 1} W_s \ge a\Big\}\Big) = 2\mathbb{P}(\{W_1 \ge a\}) = \mathbb{P}(\{|W_1| \ge a\}) \quad \forall a \ge 0,$$

and this is our statement for $t = 1$:

$$\begin{aligned}\mathbb{P}(\{\zeta^n = in^{\frac{-1}{2}}\}) &= \mathbb{P}(\{\zeta^n \ge in^{\frac{-1}{2}}\}) - \mathbb{P}(\{\zeta^n \ge (i+1)n^{\frac{-1}{2}}\})\\ &= \mathbb{P}(\{\xi_1^n = (i+1)n^{\frac{-1}{2}}\}) + \mathbb{P}(\{\xi_1^n = (i)n^{\frac{-1}{2}}\}) - \mathbb{P}(\{\xi_1^n = (i+1)n^{\frac{-1}{2}}\})\\ &\mathbb{P}(\{\xi_1^n = (i+1)n^{\frac{-1}{2}}\}) + \mathbb{P}(\{\xi_1^n = in^{\frac{-1}{2}}\}), \quad i \ge 0.\end{aligned}$$

Now, for every bounded continuous function $f(x)$ that vanishes for $x < 0$, we get

$$\mathbb{E}(f(\zeta^n)) = \sum_{i=0}^{\infty} f(in^{\frac{-1}{2}})\mathbb{P}(\{\zeta^n = in^{\frac{-1}{2}}\}) = \mathbb{E}f(\xi_1^n - n^{-1/2}) + Ef(\xi_1^n).$$

By Donsker's theorem and the continuity of the function $x \longrightarrow \max_{[0,1]} x_t$, we have

$$\mathbb{E}f\Big(\max_{[0,1]} W_t\Big) = 2\mathbb{E}f(W_1) = \mathbb{E}f(|W_1|).$$

We have proved our statement for $t = 1$, saying that cW_{s/c^2} is a Wiener process for $s \in [0,1]$ if $c \ge 1$. The theorem is proved. □

The continuity question of Brownian motion can be answered using another famous theorem by Kolmogorov.

Theorem 2.3. *Suppose that the process ξ_t^n satisfies the following condition: For all $T > 0$, there exist positive constants α, β, N such that*

$$\mathbb{E}|\xi_t^n - \xi_s^n|^{\alpha} \le N|t-s|^{1+\beta} \quad \forall s,t \in [0,1].$$

Then there exists a continuous version of ξ_t^n.

Proof. For simplicity, we assume that $m_4 = \mathbb{E}(\eta_k)^4 < \infty$, referring the reader to[1] for the proof in the general situation. It suffices to prove that

---

1 B. G. Pachpatte, *Inequalities for Finite Difference Equations*, CRC Press, 2001.

$$\mathbb{E}|\xi_t^n - \xi_s^n|^4 \le N|t-s|^2 \quad \forall s,t \in [0,1], \tag{2.6}$$

where N is independent of n, t, s. Without loss of generality, assume that $s < t$. Denote

$$a_n = \mathbb{E}(S_n)^4.$$

By virtue of the independence of the η_k and the conditions $\mathbb{E}(\eta_k) = 0$ and $\mathbb{E}(\eta_k)^2 = 1$, we have

$$a_{n+1} = \mathbb{E}(\eta_{k+1} + S_n)^4 = a_n + 4\mathbb{E}(S_n^3\eta_{n+1}) + 6\mathbb{E}(S_n^2\eta_{n+1}^2) + \mathbb{E}(S_n\eta_{n+1}^3) + m_4 = a_n + 6n + m_4.$$

Hence (for instance, by induction),

$$a_n = 3n(n-1) + nm_4 \le 3n^2 + nm_4.$$

Furthermore, if s and t belong to the same interval $[\frac{k}{n}, \frac{k+1}{n}]$, then

$$\begin{aligned} |\xi_t^n - \xi_s^n| &= \sqrt{n}|\eta_{k+1}||t-s|, \\ \mathbb{E}|\xi_t^n - \xi_s^n|^4 &= n^2 m_4|t-s|^4 \le m_4|t-s|^2. \end{aligned} \tag{2.7}$$

Now, consider the following picture, where s and t belong to different intervals of type $[k/n, (k+1)/n)$, and by crosses, we denote points of type k/n and $s < s_1 < t_1 < t$.

Clearly,

$$s_1 - s \le t - s, \quad t - t_1 \le t - s, \quad t_1 - s_1 \le t - s \quad (t_1 - s_1)/n \le (t_1 - s_1)^2.$$

Put

$$\begin{aligned} s_1 &= ([ns]+1)/n, \\ t_1 &= ([nt])/n, \\ [nt] - ([ns]+1) &= n(t_1 - s_1). \end{aligned}$$

Hence, from (2.7) and the inequality

$$(a+b+c)^4 \le 81(a^4 + b^4 + c^4),$$

we conclude that

$$\begin{aligned} \mathbb{E}|\xi_t^n - \xi_s^n|^4 &\le 81(\mathbb{E}|\xi_t^n - \xi_{t_1}^n|^4 + \mathbb{E}|\xi_{t_1}^n - \xi_{s_1}^n|^4 + \mathbb{E}|\xi_{s_1}^n - \xi_s^n|^4) \\ &\le 162(t-s)^2 m_4 + 81\mathbb{E}|S_{[nt]}/\sqrt{n} - S_{[nt]+1/\sqrt{n}}| \\ &= 162(t-s)^2 m_4 + 81n^{-2}a_{[nt]-([ns]+1)} \\ &\le 162(t-s)^2 m_4 + 243(t-s)^2 + 81(t_1 - s_1)m_4/n \end{aligned}$$

$$\le 243(m_4 + 1)|t - s|^2.$$

Thus, for all positions of s and t, we have (2.6) with $N = 243(m_4 + 1)$. □

Remark 2.1. $\mathcal{M}^2$ = the collection of all cadlag martingales (M_t) with

$$\mathbb{E}\Big(\sup_{t\in[0,T]} \|M(t)\|^2\Big) \le \infty,$$

$$\mathcal{M}^2 = \{M \in \mathcal{M}^2 : M \text{ is continuous in } t\}.$$

Property 2.1. The space $\mathcal{M}_T^2(E)$, equipped with the norm

$$\|M\|_{\mathcal{M}_T^2} = \mathbb{E}\Big(\sup_{t\in[0,T]} \|M(t)\|^2\Big)^{1/2}, \tag{2.8}$$

is a Banach space.

Proof. Since $\|M(t)\|$ is a submartingale, Theorem 2.1 defines a norm. To prove completeness, assume that M_n is a Cauchy sequence, that is,

$$\mathbb{E}\Big(\sup_{t\in[0,T]} \|M_n(t) - M_m(t)\|^2\Big) \to 0 \quad \text{as } n, m \to \infty.$$

It follows that one can find a subsequence M_{n_k} such that

$$\mathbb{E}\Big(\sup_{t\in[0,T]} \|M_n(t) - M_m(t)\| \ge 2^{-k}\Big) \le 2^{-k}.$$

The Borel–Cantelli lemma implies that M_{n_k} converges $\mathbb{P}$-a. s. to a process $M(t)$, $t \in [0, T]$, uniformly on $[0, T]$. So M is a continuous process. It is clear that for arbitrary $t \in [0, T]$, the sequence M_{n_k} converges to $M(t)$ in the mean square. If $0 \le s \le t \le T$ and $k = 1, 2, \ldots,$ then

$$\mathbb{E}(M_{n_k}(t)/\mathcal{F}_s) = M_{n_k}(s) \quad \mathbb{P}\text{-a. s.}, \tag{2.9}$$

one can let k tend to infinity in (2.9) to get

$$\mathbb{E}(M(t)/\mathcal{F}_s) = M(s) \quad \mathbb{P}\text{-a. s.}$$

So, $M \in \mathcal{M}_T^2(E)$ and obviously

$$M_n \to M \in \mathcal{M}_T^2(E).$$ □

Definition 2.27. Let $M, N \in \mathcal{M}^2$ be continuous local martingales. Then the product MN is not, in general, a local martingale. However, using the polarization identity

$$MN = \frac{1}{4}((M+N)^2 - (M-N)^2), \tag{2.10}$$

we see that the process

$$\langle M,N\rangle = \frac{1}{4}(\langle M+N\rangle - \langle M-N\rangle),$$

is called a cross-variation (or quadratic covariation) of M and N. Note that

$$\langle M,M\rangle = \langle M\rangle,$$

is the quadratic variation process of M. Set

$$Q_\Delta(M,N) = \sum_{j=1}^{n}(M_{t_j} - M_{t_{j-1}})(N_{t_j} - N_{t_{j-1}}),$$

for each $t > 0$ and each partition

$$\Delta = \{0 = t_0 < \cdots < t_n = t\},$$

of the interval $[0,t]$. The polarization identity (2.10) yields

$$Q_\Delta(M,N) = \frac{1}{4}(Q_\Delta(M+N) - Q_\Delta(M-N)).$$

It now follows that

$$Q_\Delta(M,N) \longrightarrow \langle M,N\rangle_t, \quad \text{in probability, as } \Delta \to 0.$$

Here, the limit is taken over all partitions Δ of the interval $[0,t]$.

Lemma 2.1. *Let M_t be a square integrable martingale. Then*

$$\mathbb{E}((M_t - M_s)^2/\mathcal{F}_a) = \mathbb{E}((M_t^2 - M_s^2)/\mathcal{F}_a), \quad 0 \le a \le s \le t. \tag{2.11}$$

2.2.1 Functions of bounded variation

Definition 2.28. The variation of a function $f : [0,T] \longrightarrow \mathbb{R}$ is defined to be

$$\limsup_{\Delta t \to 0} \sum_{i=0}^{n-1} |f(t_{i+1}) - f(t_i)|,$$

where

$$t = (t_0, t_1, \ldots, t_n),$$

is a partition of $[0, T]$, that is,

$$0 = t_0 < t_1 < \cdots < t_n = T,$$

and where

$$\Delta t = \max_{i=0,1,\dots,n-1} |t_{i+1} - t_i|,$$

is bounded variation of f on $[0, T]$ is defined to be $V(f, [0, T])$.

Example 2.3. If f is constant on $[0, T]$, then f is of bounded variation on $[0, T]$.
Consider the constant function $f(x) = c$ on $[0, T]$. Notice that

$$\limsup_{\Delta t \to 0} \sum_{i=0}^{n-1} |f(t_{i+1}) - f(t_i)|,$$

is zero for every partition of $[0, T]$. Thus, $V(f; [0, T])$ is zero.

Another example of a function of bounded variation is a monotone function on $[0, T]$.

Theorem 2.4. *If f is increasing on $[0, T]$, then f is of bounded variation on $[0, T]$ and*

$$V(f, [0, T]) = f(T) - f(0).$$

Proof. Let

$$\{t_i : 0 \le i \le n-1\},$$

be a partition of $[0, T]$. Consider

$$\sum_{i=0}^{n-1} |f(t_{i+1}) - f(t_i)| = \sum_{i=0}^{n-1} (f(t_{i+1}) - f(t_i)).$$

Because of the telescoping nature of this sum, it is the same for every partition of $[0, T]$. Thus, we see that

$$V(f, [0, T]) = f(T) - f(0) < \infty.$$

Thus, f is of bounded variation on $[0, T]$. □

Similarly, if f is decreasing on $[0, T]$, then

$$V(f, [0, T]) = f(0) - f(T).$$

For the next example, we first recall a theorem involving rational and irrational numbers.

Theorem 2.5. *Let $f : [a,b] \to \mathbb{R}$ be a function, and let $c \in (a,b)$. If f is of bounded variation on $[a,c]$ and $[c,b]$, then f is of bounded variation on $[a,b]$ and*

$$V(f,[a,b]) = V(f,[a,c]) + V(f,[c,b]).$$

A proof of Theorem 2.5 is provided in the book.[2]

Theorem 2.6. *The variation of the paths of $W(t)$ is infinite a. s.*

Proof. Consider the sequence of partitions

$$t^n = (t_0^n, t_1^n, \dots, t_n^n),$$

of $[0,T]$ into n equal parts. Then

$$\sum_{i=0}^{n-1} (W_{t_{i+1}^n} - W_{t_i^n})^2 \le \max_{i=0,\dots,n-1} |W_{t_{i+1}^n} - W_{t_i^n}| \sum_{i=0}^{n-1} (W_{t_{i+1}^n} - W_{t_i^n}),$$

where

$$t_i^n = \frac{it}{n},$$

Since the paths of $W(t)$ are a. s. continuous on $[0,T]$,

$$\lim_{n\to\infty} \max_{i=0,\dots,n-1} |W_{t_{i+1}^n} - W_{t_i^n}| = 0 \quad a.s.,$$

there is a subsequence

$$t^{n_k} = (t_0^{n_k}, t_1^{n_k}, \dots, t_{n_k}^{n_k}),$$

of partitions such that

$$\lim_{k\to\infty} \sum_{i=0}^{n_k-1} |W_{t_{i+1}^n} - W_{t_i^n}|^2 = T \quad a.s.$$

This is because every sequence of random variables that is convergent in L^2 has a subsequence that is convergent a. s., which follows that

$$\lim_{k\to\infty} \sum_{i=0}^{n_k-1} |W_{t_{i+1}^n} - W_{t_i^n}| = \infty \quad a.s.,$$

while

2 Gordon A. Russell, *Real Analysis: A First Course*, 2nd edition, Boston: Pearson Education Inc., 2002.

$$\lim_{k\to\infty} \Delta t^{n_k} = \lim_{k\to\infty} \frac{T}{n_k} = 0,$$

which proves the theorem. □

Lemma 2.2. *Let $s \le t$.*

$$\operatorname{Cov}(W_s, W_t) = t \wedge s.$$

Proof.

$$\begin{aligned}\operatorname{Cov}(W_s, W_t) &= \mathbb{E}[W_s W_t] - \mathbb{E}[W_s]E[W_t] \\ &= \mathbb{E}[W_s W_t] - 0 \\ &= \mathbb{E}[W_s(W_t - W_s)] + \mathbb{E}[W_s^2] \\ &= \mathbb{E}[W_s]\mathbb{E}[(W_t - W_s)] + \mathbb{E}[W_s^2] \\ &= 0 + s \\ &= s.\end{aligned}$$

□

Lemma 2.3. *Let W Wiener process be a d-dimensional process with two constants $\lambda > 0$ and $r > 0$.*

1. *(Self-similar) The process*

$$X_t = \frac{1}{\sqrt{\lambda}} W_{\lambda t},$$

is a Wiener process.

2. *The process*

$$Y_t = W_{t+s} - W_s,$$

is a Wiener process.

Proof. The process X_t and Y_t p. s. are continuous, zero at 0, Gaussian centers. We have

$$\begin{aligned}\mathbb{E}(X_t^j X_s^k) &= \frac{1}{\lambda}\mathbb{E}(W_{\lambda t}^j W_{\lambda s}^k) \\ &= \frac{1}{\lambda}\delta_{jk}\min(\lambda t, \lambda s) \\ &= \delta_{jk}\min(t, s), \\ \mathbb{E}(Y_t^j Y_s^k) &= \mathbb{E}(W_{r+t}^j W_{r+s}^k + W_r^j W_r^k - W_r^j W_{r+s}^k + W_{r+t}^j W_r^k) \\ &= \delta_{jk}(r + \min(t, s) + r - r - r) \\ &= \delta_{jk}\min(t, s).\end{aligned}$$

□

Lemma 2.4. *For every*

$$0 \le t_1 < t_2 < t_3 \cdots < t_k \le 1,$$

the vectors

$$(\xi^n_{t_1}, \xi^n_{t_2} \dots \xi^n_{t_k}),$$

are asymptotically normal with parameters $(0, t_i \wedge t_j)$.

Proof. We only consider the case $k = 2$; other k are treated similarly. We have

$$\begin{aligned}\lambda_1\xi^n_{t_1} + \lambda_2\xi^n_{t_2} &= (\lambda_1 + \lambda_2)S_{[nt_1]}/\sqrt{n} + \lambda_2(S_{[nt_2]} - S_{[nt_1]+1})/\sqrt{n} \\ &+ \eta_{[nt_1]+1}\{(nt_1 - [nt_1])\}\lambda_1/\sqrt{n} + \lambda_2/\sqrt{n}\} \\ &+ \eta_{[nt_2+1]}(nt_2 - [nt_2])\lambda_2/\sqrt{n}.\end{aligned}$$

On the right, we have a sum of independent terms. In addition, the coefficients of $\eta_{[nt1]+1}$ and $\eta_{[nt2]+1}$ go to zero, and

$$\mathbb{E}\exp(ia_n\eta_{[nt+1]}) = \mathbb{E}\exp(ia_n\eta_1) \longrightarrow 1 \quad \text{as } a_n \longrightarrow 0.$$

Finally, by the central limit theorem, for $\varphi(\lambda) = \mathbb{E}\exp(i\lambda\eta_1)$,

$$\lim_{n\longrightarrow\infty} \varphi^n\left(\frac{\lambda}{\sqrt{n}}\right) = e^{-\lambda^2/2}.$$

Hence,

$$\begin{aligned}\lim_{n\longrightarrow\infty} E(e^{i(\lambda_1\xi^n_{t_1}+\lambda_2\xi^n_{t_2})}) &= \lim_{n\longrightarrow\infty} (\varphi(\lambda_1/\sqrt{n} + \lambda_2/\sqrt{n}))^{[nt_1]}(\varphi(\lambda_2/\sqrt{n}))^{[nt_2]-[nt_1]-1} \\ &= \exp\{-((\lambda_1 + \lambda_2)^2 t_1 + \lambda_2^2(t_2 - t_1))/2\} \\ &= \exp\{-(\lambda_1^2(t_1 \wedge t_2) + 2\lambda_1\lambda_2(t_1 \wedge t_2) + \lambda_2^2(t_1 \wedge t_2)\}.\end{aligned}$$

The lemma is proved. □

Theorem 2.7 (Quadratic variation). *Let*

$$0 = t_{0n} \le t_{1n} \le \dots \le t_{k_n n} = 1,$$

be a sequence of partitions of $[0,1]$ *such that* $\max_i(t_{i+1,n} - t_{i,n}) \to 0$ *a. s.* $n \to \infty$. *Also, let* $0 \le s \le t \le 1$. *Then, in probability, as* $n \to \infty$,

$$\sum_{s \le t_{i,n} \le t_{i+1,n} \le t} (W_{t_{i+1,n}} - W_{t_{i,n}})^2 \to t - s. \tag{2.12}$$

Proof. Let

$$\xi_n = \sum_{s \le t_{i,n} \le t_{i+1,n} \le t} (W_{t_{i+1,n}} - W_{t_{i,n}})^2,$$

and observe that ξ_n is a sum of independent random variables. Also, use that if

$$\eta \sim \mathcal{N}(0, \sigma^2),$$

then

$$\eta = \sigma\zeta,$$

where

$$\zeta \sim \mathcal{N}(0, 1),$$

and

$$\operatorname{Var}(\eta^2) = \sigma^4 \operatorname{Var}(\zeta).$$

Then

$$N = \operatorname{Var}(\zeta),$$

we obtain

$$\begin{aligned}\operatorname{Var}(\xi_n) &= \sum_{s \le t_{i,n} \le t_{i+1,n} \le t} \operatorname{Var}(W_{t_{i+1,n}} - W_{t_{i,n}})^2 = N \sum_{s \le t_{i,n} \le t_{i+1,n} \le t} (t_{i+1,n} - t_{i,n})^2 \\ &\le N \max_i (t_{i+1,n} - t_{i,n}) \sum_{s \le t_{i,n} \le t_{i+1,n} \le t} (t_{i+1,n} - t_{i,n}) = N \max_i (t_{i+1,n} - t_{i,n}) \to 0.\end{aligned}$$

In particular,

$$\xi_n - \mathbb{E}\xi_n \to 0,$$

in probability. In addition,

$$\mathbb{E}\xi_n = \sum_{s \le t_{i,n} \le t_{i+1,n} \le t} (t_{i+1,n} - t_{i,n}) \to t - s.$$

Hence,

$$\xi_n - (t - s) = \xi_n - \mathbb{E}\xi_n + \mathbb{E}\xi_n - (t - s) \to 0,$$

in probability, and the theorem is proved. □

Definition 2.29 (Riemann–Stieltjes integrals). Let $V, f : \mathbb{R}^+ \to \mathbb{R}$, and

$$0 = t_0 < t_1 < t_2 < \cdots < t_n = T,$$

be a partition of $[0, T]$.

1. A sum of the form

$$\sum_{i=1}^{n} f(s_i^n)(V(t_i^n) - V(t_{i-1}^n)),$$

is called a Riemann–Stieltjes sum of f with respect to V.

2. A function f is Riemann–Stieltjes integrable with respect to V on $[a,b]$ such that

$$\int_0^t f(s)dV = \lim_{n\to\infty} \sum_{i=1}^{n} f(s_i^n)(V(t_i^n) - V(t_{i-1}^n)).$$

2.2.2 Itô stochastic integrals

Assume that $W(t)$ is an infinite sequence of independent standard Brownian motions defined on $(\Omega, \mathcal{F}, \mathbb{P})$, that is,

$$W(t) = (W^1(t), W^2(t), \dots)^T.$$

An $\mathbb{R}$-valued random variable is an $\mathcal{F}$-measurable function $x(t) : \Omega \to \mathbb{R}$, and the collection of random variables

$$S = \{x(t,\omega) : \Omega \to \mathbb{R} \mid t \in J\},$$

is called a stochastic process. Generally, we just write $x(t)$ instead of $x(t,\omega)$.

Definition 2.30. An $\mathcal{F}$-adapted process X on $[0,b] \times \Omega$ is an elementary process if, for a partition

$$\phi = \{t = 0 < t_1 < \cdots < t_n = b\},$$

and $(\mathcal{F}_{t_i})$-measurable random variables $(X_{t_i})_{i<n}, X_t$, it satisfies

$$X_t(\omega) = \sum_{i=0}^{n-1} X_i(\omega)\chi_{[t_i,t_{i+1})}(t), \quad \text{for } 0 \le t \le b,\ \omega \in \Omega.$$

The Itô integral of the simple process X is defined as

$$\int_0^b X(s)dW^l(s) = \sum_{i=0}^{n-1} X_l(t_i)(W^l(t_{i+1}) - W^l(t_i)), \tag{2.13}$$

whenever

$$X_{t_i} \in L^2(\Omega, \mathcal{F}_t, \mathbb{P}; \mathbb{R}^n), \quad i \le n.$$

The following result is one of the elementary properties of square integrable stochastic processes.

Lemma 2.5 (Itô isometry for elementary processes). *Let $(X_l)_{l\in\mathbb{N}}$ be a sequence of elementary processes. Assume that*

$$\int_0^T \mathbb{E}|X(s)|^2 ds < \infty,$$

where

$$|X| = \left(\sum_{l=1}^{\infty} X_l^2\right).$$

Then

$$\mathbb{E}\left(\sum_{l=1}^{\infty}\int_0^T X_l(s)dW^l(s)\right)^2 = \mathbb{E}\left(\sum_{l=1}^{\infty}\int_0^T X_l^2(s)ds\right). \tag{2.14}$$

Proof. Set

$$\Delta W_i^l = W_{t_{i+1}}^l - W_{t_i}^l.$$

Let

$$M^k = \sum_{l=1}^{k}\int_0^T X_l(s)dW^l(s).$$

Observe that if $k = 1$, we have

$$\begin{aligned}
\mathbb{E}\left(\int_0^T X_1(s)dW(s)\right)^2 &= \mathbb{E}\left(\sum_{i=0}^{n-1} X_1(t_i)\Delta W_i^1\right)^2 \\
&= \sum_{i=0}^{n-1} E(X_1^2(t_i)(W_{t_{i+1}}^1 - W_{t_i}^1)^2) \\
&\quad + 2\sum_{i<j}\mathbb{E}(X_1(t_i)X_1(t_j)(W_{t_{i+1}}^1 - W_{t_i}^1)(W_{t_{j+1}}^1 - W_{t_j}^1)).
\end{aligned}$$

For $i < j$, since $t_i < t_j$ and $(W_{t_{j+1}} - W_{t_j})$ are independent of

$$X_i(t_i)X_j(t_j)(W_{t_{j+1}} - W_{t_j}).$$

Hence,

$$\mathbb{E}(X_1(t_i)X_1(t_j)(W^1_{t_{i+1}} - W^1_{t_i})(W^1_{t_{j+1}} - W^1_{t_j})) = \mathbb{E}(X_1(t_i)X_1(t_j)(W^1_{t_{i+1}} - W^1_{t_i})\mathbb{E}(W^1_{t_{j+1}} - W^1_{t_j}))$$
$$= 0.$$

This implies that

$$\begin{aligned}
\mathbb{E}\left(\int_0^T X_1(s)dW^1(s)\right)^2 &= \sum_{i=0}^{n-1} \mathbb{E}(X_1^2(t_i)(W^1_{t_{i+1}} - W^1_{t_i})^2) \\
&= \sum_{i=0}^{n-1} \mathbb{E}(X_1^2(t_i))\mathbb{E}(W^1_{t_{i+1}} - W^1_{t_i}))^2 \\
&= \sum_{i=0}^{n-1} \mathbb{E}(X_1^2(t_i))(t_{i+1} - t_i) \\
&= \mathbb{E}\left(\int_0^T X_1^2(s)ds\right).
\end{aligned}$$

Therefore,

$$\mathbb{E}\left(\int_0^T X_1(s)dW^1(s)\right)^2 = \left(\mathbb{E}\int_0^T X_1^2(s)ds\right).$$

When $k = 1$, the result is true. For $k = 2$, we use the estimate (2.14) to get

$$\begin{aligned}
M^2 &= \mathbb{E}\left(\sum_{l=1}^{2}\int_0^T X_l(s)dW^l(s)\right)^2 \\
&= \mathbb{E}\left(\sum_{l=1}^{2}\sum_{i=0}^{n-1} X_l(t_i)\Delta W_i^l\right)^2 \\
&= \mathbb{E}\left(\sum_{i=0}^{n-1} X_1(t_i)\Delta W_i^1\right)^2 + \mathbb{E}\left(\sum_{i=0}^{n-1} X_2(t_i)\Delta W_i^2\right)^2 + 2\mathbb{E}\left(\sum_{i=0}^{n-1} X_1(t_i)\Delta W_i^1\right)\mathbb{E}\left(\sum_{i=0}^{n-1} X_2(t_i)\Delta W_i^2\right) \\
&= \sum_{i=0}^{n-1} E(X_1^2(t_i))(t_{i+1} - t_i) + \sum_{i=0}^{n-1} \mathbb{E}(X_2^2(t_i))(t_{i+1} - t_i) \\
&= \mathbb{E}\int_0^T X_1^2(s)ds + \mathbb{E}\int_0^T X_2^2(s)ds \\
&= \mathbb{E}\left(\sum_{l=1}^{2}\int_0^T X_l^2(s)\right)ds.
\end{aligned}$$

Suppose that

$$\mathbb{E}\left(\sum_{l=1}^{k}\int_0^T X_l(s)dW^l(s)\right)^2 = \sum_{l=1}^{k}\mathbb{E}\left(\int_0^T X_l^2(s)ds\right),$$

is true for fixed $k \in \mathbb{N}$. We show that

$$\mathbb{E}\left(\sum_{l=1}^{k+1}\int_0^T X_l(s)dW^l(s)\right)^2 = \sum_{l=1}^{k+1}\int_0^T \mathbb{E}(X_l^2(s))ds.$$

Since $(X_l)_{l=1}^{k+1}$ is a set of elementary stochastic processes, then

$$\begin{aligned}
M^{k+1} &= \mathbb{E}\left(\sum_{l=1}^{k+1}\int_0^T X_l(s)dW^l(s)\right)^2 \\
&= \mathbb{E}\left(\sum_{l=1}^{k+1}\sum_{i=0}^{n-1} X_l(t_i)\Delta W_i^l\right)^2 \\
&= \mathbb{E}\left(\sum_{l=1}^{k}\sum_{i=0}^{n-1} X_l(t_i)\Delta W_i^l\right)^2 \\
&\quad + \mathbb{E}\left(\sum_{i=0}^{n-1} X_{k+1}(t_i)\Delta W_i^{k+1}\right)^2 \\
&\quad + 2\left(\mathbb{E}\sum_{i=0}^{n-1} X_{k+1}(t_i)\Delta W_i^{k+1}\sum_{l=1}^{k+1}\sum_{i=0}^{n-1} X_l(t_i)\Delta W_i^l\right).
\end{aligned}$$

Using the fact that $(W^l)_{l=1}^{k+1}$ is a set of independent standard Brownian motions, we have

$$\begin{aligned}
\mathbb{E}\left(\sum_{l=1}^{k+1}\int_0^T X_l(s)dW^l(s)\right)^2 &= \mathbb{E}\sum_{l=1}^{k}\int_0^T X_l^2(s)ds + \mathbb{E}\int_0^T X_{k+1}^2(s)ds \\
&= \mathbb{E}\left(\sum_{l=1}^{k+1}\int_0^T X_l^2(s)ds\right).
\end{aligned}$$

Hence, the formula is true for $k+1$. This implies that for every $k \in \mathbb{N}$, we have

$$M^k = \mathbb{E}\left(\sum_{l=1}^{k}\int_0^T X_l(s)dW^l(s)\right)^2 = \mathbb{E}\left(\sum_{l=1}^{k}\int_0^T X_l^2(s)ds\right).$$

Then M^k is convergent in $L^2(\mathcal{F}_t)$; then

$$\lim_{k\to\infty} M^k = \lim_{k\to\infty}\mathbb{E}\left(\sum_{l=1}^{k}\int_0^T X_l(s)dW^l(s)\right)^2 = \mathbb{E}\left(\sum_{l=1}^{\infty}\int_0^T X_l^2(s)ds\right).$$

Hence,

$$\mathbb{E}\left(\sum_{l=1}^{\infty}\int_0^T X_l(s)dW^l(s)\right)^2 = \mathbb{E}\left(\sum_{l=1}^{\infty}\int_0^T X_l^2(s)ds\right). \qquad \square$$

Remark 2.2. For a square integrable stochastic process X on $[0, b]$, its Itô integral is defined by

$$\int_0^b X(s)dW(s) = \lim_{n\to\infty}\int_0^b X_n(s)dW(s),$$

taking the limit in L^2, with X_n defined in Definition 2.30. Then the Itô isometry holds for all Itô integrable X; see Figure 2.1.

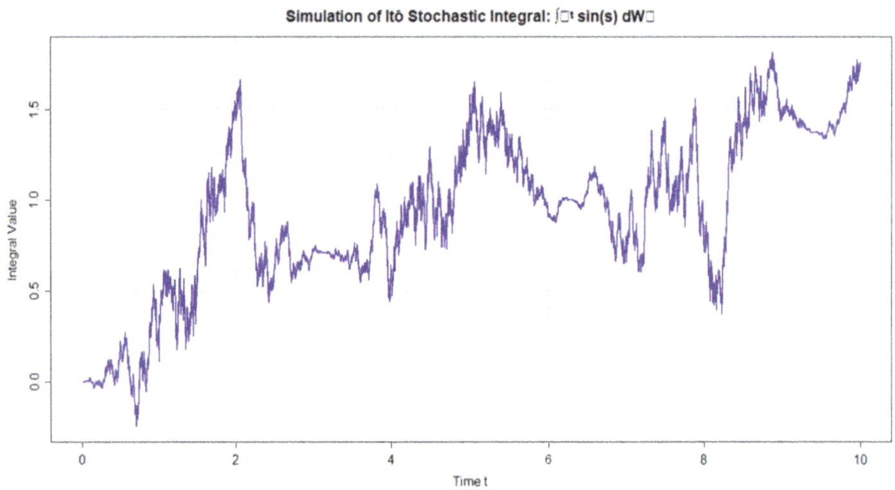

Figure 2.1: $\int_0^t \sin(t)dW(t)$ Itô stochastic integrals.

Theorem 2.8.[3] *Let $X, Y \in S$, $a, b \in \mathbb{R}$.*

Then:

1. *(Linearity) For all t at once, with probability one*

$$\mathbb{E}\int_0^t (aX(s) + bY(s)dW(s)) = a\mathbb{E}\int_0^t X(s)dW(s) + b\mathbb{E}\int_0^t Y(s)dW(s).$$

3 Introduction to the theory of random processes, *American Mathematical Society USA* **90** (1993) 11698–11702.

2. *We have*

$$\mathbb{E}\left(\int_0^t X(s)dW(s)\right) = 0.$$

3. *The process*

$$\int_0^t X(s)dW(s),$$

is a martingale relative to $\mathcal{F}_t$.
4. *Doob's inequality holds*

$$\mathbb{E}\sup_t\left(\int_0^t X(s)dW(s)\right)^2 \le 4\mathbb{E}\sup_t\left(\int_0^t X^2(s)ds.\right.$$

5. *If* $A \in \mathcal{F}$, $T \in [0,\infty]$, *and* $X_t(w) = Y_t(w)$ *are valid for all* $w \in A$ *and* $t \in [0,T)$, *then*

$$I_A\int_0^t X(s)dW(s) = I_A\int_0^t Y(s)dW(s),$$

for all $t \in [0,T]$ *at once with probability one.*

Theorem 2.9. *If* $f \in S$ *and* f *are continuous, then for any sequence* Π_i *of*

$$0 = t_{n,0} < t_{n,1} < t_{n,2} < \cdots < t_{n,n} = T,$$

it is a partition of $[0,T]$ *with* $|\Pi_i| \to 0$,

$$\mathbb{P}\left(\int_0^T f(s)dW(s) = \lim_{n\to\infty}\sum_{k=0}^{m_n-1} X_l(t_i)(W(t_{n,k+1}) - W(t_{n,k}))\right) = 1. \tag{2.15}$$

Proof. Introduce the step function g_n,

$$g_n(t) = f(t_{n,k}).$$

For $g_n(t) \to f(t)$, uniformly in $t \in [0,T)$ as $n \to \infty$.

Hence,

$$\int_0^T |g_n(t) - f(t)|^2 dt \to 0 \quad \text{a.s.}$$

We then have

$$\mathbb{P}\left(\lim_{n\to\infty}\int_0^T g_n(t)dW(t)=\int_0^T f(t)dW(t)\right)=1.$$

Since

$$\int_0^T g_n(t)dW(t)=\sum_{k=0}^{m_n-1} X_l(t_i)(W(t_{n,k+1})-W^l(n,t_n)),$$

the assertion (2.15) follows. □

Remark 2.3. For a square integrable stochastic process X on $[0,T]$, its Itô integral is defined by

$$\int_0^T X(s)dW(s)=\lim_{n\to\infty}\int_0^T X_n(s)dW(s),$$

taking the limit in L^2, with X_n defined in Definition 2.30. Then the Itô isometry holds for all Itô integrable X.

The next result is known as the Burholder–Davis–Gundy inequalities. It was first proved for discrete martingales and $p>0$ by Burkholder.[4] In 1968, Millar[5] extended the result to continuous martingales. In 1970, Davis[6] extended the result for discrete martingales to $p=1$. The extension to $p>0$ was obtained by Novikov,[7] in 1971.

Theorem 2.10 ([14]). *For each $p>0$, there exist constants $c_p, c_p \in (0,\infty)$, such that for any progressive process x with the property that for*

$$t\in[0,\infty),\qquad \int_0^t X_s^2 ds<\infty,$$

$\mathbb{P}$-a. s. we have

$$c_p\mathbb{E}\left(\int_0^t X_s^2 ds\right)^{\frac{p}{2}}\le \mathbb{E}\left(\sup_{s\in[0,t]}\int_0^t X_s^2 dW(s)\right)^p\le C_p\mathbb{E}\left(\int_0^t X_s^2 ds\right)^{\frac{p}{2}}.$$

4 D. L. Burkholder, Martingale transforms, *Ann. Math. Statist.* **37** (1966) 1494–1504.

5 P. Millar, Warwick Martingale integrals, *Trans. Amer. Math. Soc.* **133** (1968) 145–166.

6 B. Davis, On the integrability of the martingale square function, *Israel J. Math.* **8** (1970) 187–190.

7 A. A. Novikov, The moment inequalities for stochastic integrals (Russian), *Teor. Verojatnost. i Primenen.* **16** (1971) 548–551.

Example 2.4 (Mean and mean square of a stochastic integral). Let

$$I(f) = \int_0^1 W(s)dW(s).$$

Then, by the Theorem 2.9 of Itô integrals,

$$\mathbb{E}(I(f)) = 0,$$
$$\mathbb{E}(|I(f)|^2) = \int_0^1 \mathbb{E}|W(s)|^2 ds = \frac{1}{2}.$$

Lemma 2.6 (Stochastic product rule). *Let $X, Y \in S$. Then*

$$X_t Y_t = X_0 Y_0 + \int_0^t Y_s dX_s + \int_0^t X(s) dY_s + \langle X, Y\rangle_t, \tag{2.16}$$

for all $t \geq 0$.

Proof. Let

$$\Delta = 0 = t_0 < t_1 < \cdots < t_n = t,$$

be any partition of the interval $[0, t]$. Sum the equalities

$$X_{t_j} Y_{t_j} - X_{t_{j-1}} Y_{t_{j-1}} = (X_{t_j} - X_{t_{j-1}})(X_{t_j} - X_{t_{j-1}}) + X_{t_{j-1}}(Y_{t_j} - Y_{t_{j-1}}) + Y_{t_{j-1}}(X_{t_j} - X_{t_{j-1}}).$$

By Theorem 2.9 and $j = 1, 2, \ldots, n$, we obtain

$$X_t Y_t = X_0 Y_0 + \int_0^t X_s dY_s + \int_0^t Y_s dX_s + \langle X, Y\rangle_t.$$

The lemma is proved. □

2.2.3 Gaussian processes in Hilbert spaces

Let X be a Gaussian process in a Hilbert space H. Let

$$m(t) = \mathbb{E}(X(t)), \quad Q(t) = \mathbb{E}((X(t) - m(t)) \otimes (X(t) - m(t))), \quad t \geq 0,$$

and

$$\beta(t, s) = \mathbb{E}((X(t) - m(t)) \otimes (X(s) - m(s))), \quad t, s \geq 0.$$

Property 2.2 ([13, Proposition 2.1.4]). Consider a U-valued Gaussian random variable X with mean $m \in U$ and covariance operator $Q \in L(U)$, where Q is self-adjoint, positive semidefinite, and with finite trace, that is,

$$P^{-1} \circ X = \mathcal{N}(m, Q).$$

Then, for all $u \in U$, $(X, u)_U$ is a real-valued Gaussian random variable with:
1. $\mathbb{E}((X, u)_U) = (m, u)_U$ for all $u \in U$.
2. $\mathbb{E}((X - m, u)_U (X - m, v)_U) = (Qu, v)_U$ for all $u, v \in Q$.
3. $\mathbb{E}\|(X - m)\|_U^2 = \mathrm{Tr}(Q)$.

Definition 2.31. The process X is said to be stationary if

$$\mathbb{E}\left(\exp\left(i \sum_{k=1}^{n} \langle X(t_k + r), h_k \rangle \right) \right) = \mathbb{E}\left(\exp\left(i \sum_{k=1}^{n} \langle X(t_k), h_k \rangle \right) \right),$$

for all $n \in \mathbb{N}$, $t_1, \ldots, t_n \in [0, \infty[$, $h_1, \ldots, h_n \in H$, and $r \in [0, \infty)$.

Property 2.3 ([6]). A Gaussian process X, is stationary if and only if:
1. $m(t + r) = m(t)$, $t, r \geq 0$.
2. $\beta(t + r, s + r) = \beta(t, s)$, $t, s, r \geq 0$.

2.3 The stochastic integral in Hilbert spaces

Let H and U be two separable Hilbert spaces. This section is devoted to a construction of the stochastic Itô integral.

$$\int_0^t \Phi(s) dW(s), \quad t \in [0, T],$$

where $W(.)$ is a Wiener process on U and Φ is a process with values that are linear but not necessarily bounded operators from U into H. We will start by collecting basic facts on Hilbert space valued Wiener processes, including cylindrical ones. Then we define the stochastic integral in steps, starting from elementary processes and ending with the most general. We also establish basic properties of the stochastic integral, including the Itô formula.

2.3.1 Hilbert space valued Wiener processes

We consider two Hilbert spaces H and U, and a symmetric nonnegative operator $Q \in L(U)$. We will first consider the case when $\mathrm{Tr}\, Q < +\infty$. Then there exists a complete orthonormal system $\{e_k\}$ in U, and a bounded sequence of nonnegative real numbers λ_k such that

$$Qe_k = \lambda_k e_k, \quad k \in \mathbb{N}^*.$$

U-valued stochastic process $W(t), t > 0$ is called a Q-Wiener process if:
1. $W(0) = 0$.
2. W has continuous trajectories.
3. W has independent increments.
4. $\mathcal{L}(W(t) - W(s)) = \mathcal{N}((0, t-s)Q), t \geq s \geq 0$.

If a process $W(t), t \in [0,T]$ satisfies (1)–(2) and (4) for $t, s \in [0,T]$, then we say that $W(t)$ is a Q-Wiener process on $[0,T]$.

Property 2.4. Assume that W is a Q-Wiener process, with $\operatorname{Tr} Q < +\infty$. Then the following statements hold:
1. W is a Gaussian process on U and

$$\mathbb{E}\big(W(t)\big) = 0, \quad \operatorname{Cov}\big(W(t)\big) = tQ, \quad t \geq 0, \tag{2.17}$$

2. For arbitrary t, W has the expansion

$$W(t) = \sum_{j=1}^{\infty} \sqrt{\lambda_j}\beta_j e_j, \tag{2.18}$$

where

$$\beta_j(t) = \frac{1}{\sqrt{\lambda_j}}\langle W(t), e_j\rangle, \quad j = 1, 2, \ldots, \tag{2.19}$$

are real valued Brownian motions mutually independent on $(\Omega, \mathcal{F}, \mathbb{P})$ and the series in (2.18) is convergent in $L^2(\Omega, \mathcal{F}, \mathbb{P})$.

Proof. Let

$$0 < t_1 < \cdots < t_n,$$

and let

$$u_1, \ldots, u_n \in U.$$

Consider the random variable Z:

$$Z = \sum_{j=1}^{n}\langle W(t_j), u_j\rangle = \left\langle W(t_1), \sum_{k=1}^{n} u_k \right\rangle$$

$$+\left\langle W(t_2)-W(t_1),\sum_{k=2}^{n}u_k\right\rangle+\cdots$$
$$+\left\langle W(t_n)-W(t_{n-1}),\sum_{k=2}^{n}u_k\right\rangle.$$

Since W has independent increments, Z is Gaussian for any choice of $u_1,\dots,u_n$, and (1) follows.

We now prove (2). Let $t>s>0$; by (2.19), it follows that

$$\begin{aligned}\mathbb{E}(\beta_i(t)\beta_j(s)) &= \frac{1}{\sqrt{\lambda_i\lambda_j}}\mathbb{E}((\langle W(t),e_i\rangle)(\langle W(t),e_j\rangle))\\ &= \frac{1}{\sqrt{\lambda_i\lambda_j}}\mathbb{E}((\langle W(t)-W(s),e_i\rangle)(\langle W(s),e_j\rangle))\\ &\quad+\frac{1}{\sqrt{\lambda_i\lambda_j}}\mathbb{E}((\langle W(s),e_i\rangle)(\langle W(s),e_j\rangle))\\ &= \frac{1}{\sqrt{\lambda_i\lambda_j}}s\langle Qe_i,e_j\rangle = s\delta_{ij}.\end{aligned}$$

Therefore, the independence of β_i and $i=1,2,\dots,$ follows. To prove representation (2.18), it is enough to notice that, for $m\ge n\ge 1$,

$$\mathbb{E}\left\|\sum_{j=n}^{m}\sqrt{\lambda_j}\beta_j e_j\right\|^2 = t\sum_{j=n}^{m}\sqrt{\lambda_j}. \tag{2.20}$$

Recall that

$$\sum_{j=n}^{m}\sqrt{\lambda_j}<\infty. \qquad \square$$

2.3.2 Definition of the stochastic integral

We are given here a Q-Wiener process in $(\Omega,\mathcal{F},\mathbb{P})$ having values in U. By Proposition 2.4, $W(t)$ is given by (2.20). For the sake of simplicity of notation, we require that $\lambda_k>0$ for all $k=1,2,\dots$. We are also given a normal filtration $\{\mathcal{F}_t\}_{t\ge0}$ in $\mathcal{F}$ and we assume that:

1. $W(t)$ is $\mathcal{F}_t$-measurable.
2. $W(t+h)-W(t)$ is independent of $\mathcal{F}_t$, $\forall h,t\ge0$.

If a Q-Wiener process W satisfies (1) and (2), we say that W is a Q-Wiener process with respect to $\{\mathcal{F}_t\}_{t\ge0}$. However, to shorten the formulation, we usually avoid stressing the dependence on the filtration.

Definition 2.32. Let us fix $T < \infty$. An $L = L(U,H)$-valued process $\Phi(t)$, $t \in [0,T]$ taking only a finite number of values is said to be elementary if there exists a sequence

$$0 = t_0 < t_1 < \cdots < t_k = T,$$

and a sequence

$$\Phi_0, \Phi_1, \dots, \Phi_{k-1},$$

of L-valued random variables taking only a finite number of values such that Φ_m are $\mathcal{F}_{t_m}$ measurable and

$$\Phi(t) = \Phi_m, \quad \text{for } t \in (t_m, t_{m+1}], \quad m = 0, 1, \dots, k-1.$$

For elementary processes Φ, one defines the stochastic integral by the formula

$$\int_0^t \Phi(s)dW(s) = \sum_{m=0}^{k-1} \Phi_m W_{(t_{m+1}\wedge t)} - W_{(t_m \wedge t)},$$

and denotes it by $\Phi.W(t)$, $t \in [0,T]$.

It is useful, at this moment, to introduce the subspace $U_0 = Q^{1/2}(U)$ of U, which is endowed with the inner product

$$\begin{aligned} \langle u, v\rangle_0 &= \sum_{k=1}^{\infty} \frac{1}{\lambda_k} \langle u, e_k\rangle \langle v, e_k\rangle \\ &= \langle Q^{1/2}u, Q^{1/2}v\rangle. \end{aligned} \tag{2.21}$$

It is a Hilbert space.

In the construction of the stochastic integral for more general processes, an important role is played by the space of all Hilbert spaces.

Definition 2.33 (Hilbert–Schmidt operator). Let e_j, f_j, $j \in \mathbb{N}$ be an orthonormal basis of U_0. An operator $\Psi \in L_2^0$ is called a Schmidt operator

$$L_2^0 = L_2(U_0, H),$$

from U_0 into H. The space L_2^0 is also a separable Hilbert space, equipped with the norm

$$\|\Psi\|^2_{L_2^0} = \sum_{h,k=1}^{\infty} |\langle \Psi g_h, f_k\rangle|^2 = \sum_{h,k=1}^{\infty} \lambda_h |\langle \Psi e_h, f_k\rangle|^2 = \|\Psi Q^{1/2}\|^2 = \mathrm{Tr}(\Psi Q \Psi^*),$$

where $\{g_j\}$, with

$$g_j = \sqrt{\lambda_j} e_j, \quad j = 1, 2, \ldots,$$

e_j and f_j are complete orthonormal bases in U_0, U, and H, respectively. Clearly, $L \subset L_2^0$, but not all operators from L_2^0 can be regarded as restrictions of operators from L. The space L_2^0 contains genuinely unbounded operators on U.

Let $\Phi(t)$, $t \in [0, T]$, be a measurable L_2^0-valued process; we define the norms

$$\begin{aligned}
\|\Phi\|_t &= \left\{ \mathbb{E} \int_0^t \|\Phi\|_{L_2^0}^2 ds \right\}^{\frac{1}{2}} \\
&= \left\{ \mathbb{E} \int_0^t \operatorname{Tr}(\Phi Q^{1/2})(\Phi Q^{1/2})^* ds \right\}^{\frac{1}{2}}, \quad t \in [0, T].
\end{aligned}$$

Property 2.5. If a process Φ is elementary and

$$\mathbb{E} \int_0^t \|\Phi\|_{L_2^0}^2 < \infty,$$

then the process

$$\int_0^t \Phi(s) dW(s),$$

is a continuous, square integrable H-valued martingale on $[0, T]$ and

$$\mathbb{E}\left(\int_0^t \Phi(s) dW(s) \right)^2 = \mathbb{E} \int_0^t \|\Phi\|_{L_2^0}^2 ds, \quad t \in [0, T]. \tag{2.22}$$

Proof. The proof is straightforward. We will check, for instance, that (2.22) holds for $t = t_m < T$. Define

$$\Delta_j = W(t_{j+1}) - W_{t_j}, \quad j = 1, \ldots, m - 1.$$

Then

$$\begin{aligned}
\mathbb{E}\left(\left| \int_0^t \Phi(s) dW(s) \right| \right)^2 &= \mathbb{E} \left| \sum_{j=}^{m-1} \Phi(t_j) \Delta_j \right|^2 \\
&= \sum_{j=}^{m-1} E |\Phi(t_j) \Delta_j|^2 + 2 \sum_{i<j=}^{n} \langle \Phi(t_j) \Delta_j, \Phi(t_i) \Delta_i \rangle.
\end{aligned}$$

We will first show that

$$\sum_{j=1}^{m-1} \mathbb{E}\left|\Phi(t_j)\Delta_j\right|^2 = \sum_{j=}^{m-1} (t_{j+1} - t_j) E\left\|\Phi(t_j)\right\|_{L_2^0}^2. \tag{2.23}$$

To this purpose, note that the random variable $\Phi^*(t_j)$ is $\mathcal{F}_{t_j}$ measurable, and Δ_j is a random variable independent of $\mathcal{F}_{t_j}$.

$$\begin{aligned}
\mathbb{E}\left|\Phi(t_j)\Delta_j\right|^2 &= \sum_{l=1}^{\infty} \mathbb{E}\Big(\left|\langle \Phi(t_j)\Delta_j, f_l\rangle\right|^2\Big) \\
&= \sum_{l=1}^{\infty} \mathbb{E}\Big(\mathbb{E}(\langle \Delta_j, \Phi^*(t_j) f_l\rangle^2 \mid \mathcal{F}_{t_j})\Big) \\
&= (t_{j+1} - t_j) \sum_{l=1}^{\infty} \mathbb{E}\Big(\langle Q\Phi^*(t_j) f_l, \Phi^*(t_j) f_l\rangle\Big) \\
&= (t_{j+1} - t_j)\mathbb{E}\left\|\Phi(t_j)\right\|_{L_2^0}^2.
\end{aligned}$$

This shows (2.23). Similarly, one has

$$\mathbb{E}\big\langle \Phi(t_i)\Delta_i, \Phi(t_j)\Delta_j\big\rangle = 0, \quad i \neq j,$$

and the conclusion follows. □

Remark 2.4. Note that the stochastic integral is an isometric transformation from the space of all elementary processes equipped with the norm $\|\cdot\|_T$ into the space $\mathcal{M}_T^2(H)$ of H-valued martingales.

Remark 2.5. The mapping $h \longrightarrow W(h)$ is linear. Indeed, for any $\lambda, \mu \in \mathbb{R}$, and $h, g \in H$, we have

$$\begin{aligned}
&\mathbb{E}\big(W(\lambda h + \mu g) - \lambda W(h) - \mu g(h)\big)^2 \\
&\quad = \|\lambda h + \mu g\|_t^2 + \lambda\|h\|_t^2 + \lambda\|g\|_t^2 - 2\lambda\langle \lambda h + \mu g, h\rangle - 2\mu\langle \lambda h + \mu g, g\rangle + 2\lambda\mu\langle h, g\rangle = 0.
\end{aligned}$$

We are now able to extend the definition of the stochastic integral to all L_2^0 predictable processes $\|\Phi\|_T$. Note that they form a Hilbert space denoted by $\mathcal{N}_W^2(0,T;L_2^0)$, more simply $\mathcal{N}_W^2(0,T)$ or $\mathcal{N}_W^2$, and by the previous proposition, elementary processes form a dense set in $\mathcal{N}_W^2$. By Proposition 2.5, the stochastic integral

$$\int_0^t \Phi(s)dW(s),$$

is an isometric transformation from that dense set into $\mathcal{M}_T^2(H)$; therefore, the definition of the integral can be immediately extended to all elements of $\mathcal{N}_W^2$. Moreover, (2.22) holds, and

$$\int_0^t \Phi(s)dW(s)$$

is a continuous square integrable martingale. As a final step, we extend the definition of the stochastic integral to L_2^0-predictable processes satisfying an even weaker condition:

$$\mathbb{P}\left(\int_0^t \|\Phi\|_{L_2^0}^2 ds,\ t \in [0,T]\right) = 1. \tag{2.24}$$

All such processes are called stochastically integrable on $[0, T]$. They form a linear space denoted by $\mathcal{N}_W^2(0, T; L_2^0)$, more simply $\mathcal{N}_W^2(0, T)$, or even $\mathcal{N}_W^2$. The extension can be accomplished by the so-called localization procedure. To do so, we need the following.

Lemma 2.7. *Assume that $\Phi \in \mathcal{N}_W^2(0, T)$ and τ are $\mathcal{F}_t$-stopping times such that $\mathbb{P}(\tau < T) = 1$. Then $\mathbb{P}$-a. s.*

Proof. Assume that Φ is elementary and that τ is a simple stopping time. If Φ is elementary and τ general, then there exists a sequence of simple stopping times $\{\tau_n\}$ such that $\tau_n \downarrow \tau$ and $\mathbb{P}$-a. s.

$$\int_0^t I_{[0,\tau_n]}\Phi(s)dW(s) \to \int_0^t I_{[0,\tau]}\Phi(s)dW(s).$$

On the other hand,

$$\|I_{[0,\tau]}\Phi - I_{[0,\tau_n]}\Phi\|_T^2 = \mathbb{E}\int_0^t I_{[0,\tau_n]}\|\Phi\|_{L_2^0}^2 ds \to 0$$

and, therefore, for a subsequence, still denoted by $\{\tau_n\}$,

$$\int_0^t I_{[0,\tau_n]}\Phi(s)dW(s) \to \int_0^t I_{[0,\tau]}\Phi(s)dW(s), \quad \mathbb{P}\text{-a. s. and uniformly in } [0,T].$$

If Φ is general and $\|\Phi - \Phi_m\|_T$ for a sequence of elementary processes, we have

$$\int_0^t \Phi_m(s)dW(s) \to \int_0^t \Phi(s)dW(s),$$

and, for an appropriate subsequence,

$$\int_0^t I_{[0,\tau_n]}\Phi_{m_k}(s)dW(s) \to \int_0^t I_{[0,\tau]}\Phi(s)dW(s). \qquad \square$$

Lemma 2.8. *Assume that condition (2.24) holds and define*

$$\tau_n = \inf\left\{ t \in [0,T] : \int_0^t \|\Phi\|^2_{L_2^0} ds \geq n \right\},$$

with the convention that the infimum of an empty set is T. Then $\{\tau_n\}$ is a sequence such that

$$\mathbb{E}\left(\int_0^t \|I_{[0,\tau_n]}\Phi\|^2_{L_2^0} ds \right) < \infty. \tag{2.25}$$

Proof. Consequently, stochastic integrals

$$\int_0^t I_{[0,\tau_n]}\Phi(s)dW(s),$$

are well-defined for all $n = 1, 2, \ldots$. Moreover, if $n < m$, then $\mathbb{P}$-a. s.

$$\begin{aligned} \int_0^t I_{[0,\tau_n]}\Phi(s)dW(s) &= \int_0^t (I_{[0,\tau_n]}I_{[0,\tau_m]})\Phi(s)dW(s) \\ &= \int_0^{t\wedge\tau_n} (I_{[0,\tau_m]}I_{[0,\tau_m]})\Phi(s)dW(s), \quad t \in [0,T]. \end{aligned} \tag{2.26}$$

Therefore, one can assume that (2.25) holds for all $w \in \Omega$, $n < m$, and $t \in [0,T]$. For arbitrary $t \in [0,T]$, define

$$\int_0^t \Phi(s)dW(s) = \int_0^t I_{[0,\tau_n]}\Phi(s)dW(s), \tag{2.27}$$

where n is an arbitrary natural number such that $\tau_n > t$. Note that if $\tau_m > t$ and $m > n$ also hold, then

$$\int_0^t I_{[0,\tau_m]}\Phi(s)dW(s) = \int_0^{t\wedge\tau_n} I_{[0,\tau_m]}\Phi(s)dW(s) = I_{[0,\tau_n]}\int_0^t \Phi(s)dW(s)$$

and, therefore, Definition (2.27) is consistent. By analogous reasoning, if $\tau_n' \to T$ is another sequence satisfying (2.25), then Definition (2.27) leads to a stochastic process identical to $\mathbb{P}$-a. s. for all $t \in [0,T]$. Note that for arbitrary $n = 1, 2, \ldots$, $w \in \Omega$, $t \in [0,T]$:

$$\int_0^t I_{[0,\tau_n]}\Phi(s)dW(s) = \sum_{m=0}^{k-1} \Phi_m W_{\tau_n\wedge(t_{m+1}\wedge t)} - W_{\tau_n\wedge(t_m\wedge t)} = M_n(\tau_n \wedge t),$$

where M_n is a square integrable continuous H-valued martingale. This property will be referred to as the local martingale property of the stochastic integral. □

2.3.3 Stochastic integral for cylindrical Wiener processes

The construction of the stochastic integral required the assumption that Q was a nuclear operator; only then did the Q-Wiener process have values in U. We can, however, extend the definition of the integral to the case of general bounded, self-adjoint, nonnegative operators Q on U. To avoid trivial complications, we will assume that Q is strictly positive: $Qx \neq 0$ for $x \neq 0$. Let $U_0 = |Q^{1/2}U|$ with the induced norm $\|u\|\|_0 = \|Q^{-1/2}u\|$, and let U_1 be an arbitrary Hilbert space such that U is embedded continuously into U_1, and the embedding of U_0 into U_1 is Hilbert–Schmidt. Let $\{g_j\}$ be an orthonormal and complete basis in U_0, and $\{\beta_j\}$ a family of independent real valued standard Wiener processes.

Proposition 2.1. *The formula*

$$W(t) = \sum_{j=1}^{\infty} g_j\beta_j(t), \quad t \geq 0 \tag{2.28}$$

defines the aQ_1-Wiener process on U_1 with $\operatorname{Tr} Q_1 < \infty$*. For arbitrary $a \in U$, the process*

$$\langle a, W(t)\rangle = \sum_{j=1}^{\infty}\langle a, g_j\rangle\beta_j(t)$$

is a real-valued Wiener process and

$$\mathbb{E}(\langle a, W(t)\rangle\langle b, W(s)\rangle) = (t \wedge s)\langle Qa, b\rangle.$$

Moreover, we have $\operatorname{Im} Q_1^{1/2} = U_0$ *and*

$$\|u\|_0 = \left\|Q^{-1/2}u\right\|_1. \tag{2.29}$$

Proof. The series defining $W(t)$ is convergent in $L^2(\Omega, \mathcal{F}, \mathbb{P}, U_1)$ because

$$\mathbb{E}\left(\left\|\sum_{j=n}^{m} g_j\beta_j(t)\right\|_1^2\right) = \sum_{j=n}^{m}\|g_j\|_1^2, \quad m \geq n \geq 1,$$

and the embedding $J : U_0 \to U_1$ is Hilbert–Schmidt, which means

$$\sum_{j=1}^{\infty} \|Jg_j\|_1^2 < \infty.$$

One can also construct U_1. The series defining $\langle a, W(t)\rangle$ does converge in

$$L^2(\Omega, \mathcal{F}, \mathbb{P}, U_1),$$

and, therefore, $\mathbb{P}$-a. s. since

$$\begin{aligned}\sum_{j=1}^{\infty} |\langle a, g_j\rangle|^2 &\leq |a|^2 \sum_{j=1}^{\infty} \|g_j\|^2 \\ &\leq C\|J\|^2\|a\|^2 \sum_{j=1}^{\infty} \|g_j\|^2 < \infty.\end{aligned}$$

In addition, if $t \geq s \geq 0$,

$$\begin{aligned}\mathbb{E}(\langle a, W(t)\rangle\langle b, W(s)\rangle) &= \mathbb{E}(\langle a, W(s)\rangle\langle b, W(s)\rangle) \\ &= s\sum_{j=1}^{\infty}\langle a, g_j\rangle\langle b, g_j\rangle \\ &= s\sum_{j=1}^{\infty}\langle aQ^{1/2}, Q^{-1/2}g_j\rangle\langle Q^{1/2}b, Q^{-1/2}g_j\rangle \\ &= s\sum_{j=1}^{\infty}\langle aQ, g_j\rangle_0\langle Qb, g_j\rangle_0 \\ &= s\langle aQ, Qb\rangle_0 \\ &= s\langle aQ, b\rangle.\end{aligned}$$

To prove the last part of the proposition, note that, by the definition,

$$\begin{aligned}\langle aQ_1, b\rangle_1 &= \mathbb{E}(\langle a, W(1)\rangle\langle b, W(1)\rangle_1) \\ &= \sum_{j=1}^{\infty}\langle a, g_j\rangle_1\langle b, g_j\rangle_1 \\ &= \sum_{j=1}^{\infty}\langle a, Jg_j\rangle_1\langle b, Jg_j\rangle_1 \\ &= \sum_{j=1}^{\infty}\langle aJ^*, g_j\rangle_0\langle J^*b, g_j\rangle_0 \\ &= \langle J^*a, J^*b\rangle_0 = \langle J^*Ja, b\rangle_1.\end{aligned}$$

Consequently, $Q_1 = JJ^*$. In particular,

$$\|Q_1^{1/2}\|_1^2 = \langle J^*Ja, a\rangle_1 = \|J^*a\|_0^2, \quad a \in U_1. \tag{2.30}$$

Thus,

$$\operatorname{Im} Q_1^{1/2} = \operatorname{Im} J = U_0,$$

and the operator $G = Q^{-1/2}J$ is bounded from U_0 onto U_1. It follows from (2.30) that $G^* = J^* Q^{-1/2}$ is an isometry; therefore, G itself is an isometry. So,

$$\|Q_1^{-1/2}u\|_1 = \|Q_1^{-1/2}Ju\|_1 = \|u\|_0.$$

This complete the proof. □

Proposition 2.2 ([6]). *For*

$$\Phi \in \mathcal{N}_W^2(0, T, H),$$

and

$$A \in L(H, H_1),$$

where H_1 *denotes a further Hilbert space* $(H_1, \langle .,. \rangle_{H_1}, \| \cdot \|)$, *it holds that*

$$A \circ \Phi \in \mathcal{N}_W^2(0, T, H),$$

and

$$A \int_0^t \Phi(s) dW(s) = \int_0^t A\Phi(s) dW(s) \quad \mathbb{P}\text{-a. s.}$$

2.3.4 Stochastic integration with respect to continuous semimartingales

Definition 2.34. A stochastic process X is called a continuous semimartingale if there exist processes M_t that are local martingales, and A_t denotes the family of all continuous and adapted processes with paths of finite variation that vanish at $t = 0$:

$$X_t = X_0 + M_t + A_t \quad \text{for all } t \geq 0.$$

Using the fact that there are no nontrivial continuous local martingales of finite variation, one can show that for a continuous semimartingale, the decomposition $X = X_0 + M + A$ into the initial value, a continuous local martingale, and a continuous adapted process of finite variation is unique.

This decomposition is called the semimartingale decomposition of X.

Definition 2.35 (Quadratic variation and covariation). If X and Y are continuous semimartingales, we define the covariation process $\langle X, Y\rangle$ as

$$\langle X, Y\rangle = \langle M, N\rangle,$$

where M and N are the local martingale parts of X and Y, respectively. In particular, if A is a continuous bounded variation process, then A is a continuous semimartingale with a constant martingale part. It follows easily that $\langle X, a\rangle = 0$ holds for each continuous semimartingale X. The following justifies our definition of $\langle X, Y\rangle$.

Proposition 2.3. *Let X, Y be continuous semimartingales, A a continuous bounded variation process, and $t \geq 0$. Then:*
1. $Q_\Delta(X, a) \to 0$, $\mathbb{P}$-*a. s.* $\|\Delta\| \to 0$,
2. $Q_\Delta(X, Y) \to \langle X, Y\rangle_t$, $\mathbb{P}$-*a. s.* $\|\Delta\| \to 0$.

Here, the limits are taken over all partitions Δ of the interval $[0, t]$.

Proof. (1) Let $w \in \Omega$ be such that the path $s \to A_s(w)$ is of bounded variation, and the path $s \to X_s(w)$ is continuous, and hence uniformly continuous on the interval $[0, t]$. This is the case for $\mathbb{P}$, a. s. $w \in \Omega$. Let $|A|_t(w) < \infty$ denote the total variation of the path $s \to A_s(w)$ on the interval $[0, t]$ and set, for any partition

$$\Delta = \{0 = t_0 < t_1 < \cdots < t_n = t\},$$

of the interval $[0, t]$,

$$C_\Delta(w) = \sup_{1\leq j\leq n} \left|X_{t_j}(w) - X_{t_{j-1}}(w)\right|.$$

The uniform continuity of the path $s \to X_s(w)$ on the interval $[0, t]$ implies that

$$\lim_{\Delta\to 0} C_\Delta(w) = 0.$$

Thus,

$$\begin{aligned}\left|Q_\Delta(X, a)(w)\right| &\leq \sum \left|X_{t_j}(w) - X_{t_{j-1}}(w)\right|\left|A_{t_j}(w) - A_{t_{j-1}}(w)\right| \\ &\leq C_\Delta(w) \sum \left|A_{t_j}(w) - A_{t_{j-1}}(w)\right| \leq C_\Delta(w)|A|_t(w) \to 0,\end{aligned}$$

as $\|\Delta\| \to 0$. This shows (1).

(2) Let

$$X = M + \widetilde{A},$$

and

$$Y = N + B,$$

be the semimartingale decompositions of X and Y. Fix $t \geq 0$ and let

$$\Delta = \{0 = t_0 < t_1 < \cdots < t_n = t\},$$

be a partition of the interval $[0, t]$. By elementary algebra,

$$\begin{aligned} Q_\Delta(X, Y)(w) &= \sum((M_{t_j}(w) - M_{t_{j-1}}(w)) + (\widetilde{A}_{t_j}(w) - \widetilde{A}_{t_{j-1}}(w))) \\ &\quad \times ((N_{t_j}(w) - N_{t_{j-1}}(w)) + (B_{t_j}(w) - B_{t_{j-1}}(w))) \\ &= Q_\Delta(Y, \widetilde{A})(w) + Q_\Delta(M, N)(w) + Q_\Delta(M, b)(w). \end{aligned}$$

Now let

$$\|\Delta\| \to 0.$$

Then we have

$$\begin{aligned} Q_\Delta(Y, \widetilde{A})(w) &\to 0, \\ Q_\Delta(M, b)(w) &\to 0, \end{aligned}$$

according to (1).

$$Q_\Delta(M, N)(w) \to \langle M, N \rangle_t.$$

It follows that

$$Q_\Delta(X, Y)(w) \to \langle M, N \rangle_t = \langle X, Y \rangle_t,$$

in probability. □

Lemma 2.9 (Kunita–Watanabe inequality). *Let M, N be continuous local martingales, and U, V be $\mathcal{B} \times \mathcal{F}$-measurable processes. Then*

$$\int_0^t |U_s V_s| d\langle M, N \rangle_s \leq \left(\int_0^t |U_s|^2 |d\langle M \rangle_s| \right)^{\frac{1}{2}} \left(\int_0^t |V_s|^2 d\langle M \rangle_s \right)^{\frac{1}{2}}, \quad \mathbb{P}\text{-a. s.} \tag{2.31}$$

Proof. Let $w \in \Omega$ be such that Lemma 2.9 holds for all nonnegative Borel measurable functions $f, g : [0, t] \to \overline{\mathbb{R}}$, and let $f(s) = U_s(w)$ and $g(s) = V_s(w)$. Recalling that

$$\begin{aligned} |\mu_w|(ds) &= |d\langle M \rangle_s|, \\ \sigma_w &= d\langle M \rangle_s, \end{aligned}$$

and

$$\nu_w = d\langle N\rangle_s.$$

Lemma 2.9 shows that (2.31) holds at the point w. □

2.3.5 Itô's formula

Definition 2.36. Let W_t be 1-dimensional Brownian motion on $(\Omega, \mathcal{F}, \mathbb{P})$ (i. e., 1-dimensional). An Itô process (or stochastic integral) is a stochastic process X_t on $(\Omega, \mathcal{F}, \mathbb{P})$ of the form

$$X_t = X_0 + \int_0^t u(s,w)ds + \int_0^t v(s,w)dW(s), \tag{2.32}$$

where $u(t,w)$, $v(t,w)$ is $\mathcal{F}_t$-adapted.

$$\mathbb{P}\left(\int_0^t v^2(s,w)ds < \infty \text{ for all } t \geq 0\right) = 1,$$

and

$$\mathbb{P}\left(\int_0^t u(s,w)ds < \infty \text{ for all } t \geq 0\right) = 1.$$

If X_t is an Itô process of the form (2.32), it is sometimes written in the shorter differential form

$$dX_t = udt + vdW(t).$$

2.3.6 The multidimensional Itô formula

Let $X = (X^1, X^2, \dots, X^d)$ be an $\mathbb{R}^d$-valued process with continuously differentiable paths and consider the process $Y_t = f(X_t)$, where $f \in C^2(\mathbb{R}^d)$. Let us write

$$D_j f = \frac{\partial f}{\partial X_j} \quad \text{and} \quad D_{i,j} f = \frac{\partial^2 f}{\partial X_i \partial X_j}.$$

The process Y has continuously differentiable paths with

$$\frac{d}{dt} f(X_t) = \sum_{j=1}^d D_j f(X_t(w)) \frac{d}{dt} X_t^j(w).$$

Fixing $w \in \Omega$ and integrating yield

$$f(X_t(w)) - f(X_0(w)) = \sum_{j=1}^{d} \int_0^t D_j f(X_s(w)) \frac{d}{ds} X_s^j(w),$$

where this integral is to be interpreted pathwise. Written as

$$f(X_t(w)) - f(X_0(w)) = \sum_{j=1}^{d} \int_0^t D_j f(X_s(w)) dX_s^j(w). \tag{2.33}$$

This equation remains true if X is a continuous bounded variation process. The situation becomes more complicated if the process X is a continuous semimartingale, and hence no longer has paths that are of bounded variation on finite intervals in general. Then a new term appears on the right-hand side of (2.33) (Itô's formula). We will provide a very explicit derivation that clearly shows where the new term comes from.

Theorem 2.11. *Let $G \subset \mathbb{R}^d$ be an open set,*

$$X = (X^1, X^2, \dots, X^d),$$

a continuous semimartingale with values in G and $f \in C^2(G)$. Then

$$\begin{aligned} &f(X_t(w)) - f(X_0(w)) \\ &\quad = \sum_{j=1}^{d} \int_0^t D_j f(X_s(w)) dX_s^j(w) + \frac{1}{2} \sum_{i,j=1}^{d} \int_0^t D_{i,j} f(X_s(w)) d\langle X_s^i(w), X_s^j(w)\rangle \quad \mathbb{P}\text{-a. s.}, \end{aligned} \tag{2.34}$$

for each $t \geq 0$.

Proof. We may assume that the path $t \to X_t(w)$ is continuous for every $w \in \Omega$. Assume first that F, K are compact sets such that

$$F \subseteq K^\circ \subseteq K \subseteq G,$$

and the range of X is contained in F. Fix $t \geq 0$. Let (Δ_n) be a sequence of partitions of the interval $[0, t]$ such that

$$\|\Delta_n\| \to 0,$$

as $n \to \infty$. For $n \geq 1$, write

$$\Delta_n = \{0 = t_0^n < t_1^n \dots t_{k_n}^n = t\}.$$

Set

$$\epsilon = \text{dist}(F, K^\circ) > 0,$$

and

$$\Omega_m = \{w \in \Omega \; \|X_{t_k^n} - X_{t_{k-1}^n}\| < \epsilon, \forall n \geq m, 1 \leq k \leq k_n\}.$$

If $w \in \Omega$, then the path

$$s \in [0, t] \to X_s(w),$$

is uniformly continuous and so $w \in \Omega_m$, for some $m \geq 1$. Thus, $\Omega_m \uparrow \Omega$, as $m \uparrow \infty$. It will thus suffice to show that (2.34) holds $\mathbb{P}$-as. on the set Ω_m, for each $m \geq 1$.

Fix $m \geq 1$. If $w \in \Omega$, $X_{t_k^n}(w) \in B_\epsilon(X_{t_k^{n-1}})$, and hence the line segment from $X_{t_{k-1}^n}(w)$ to $X_{t_k^n}(w)$ is contained in the ball $B_\epsilon(X_{t_k^{n-1}}) \subseteq K$ for all $n \geq m$ and all $1 \leq k \leq k_n$. Let $n \geq m$ and write

$$f(X_t - f(X_0) = \sum_{k=1}^{k_n} (f(X_{t_k^n}) - f(X_{t_{k-1}^n})). \tag{2.35}$$

Consider $k = \{1, 2, \ldots, k_n\}$ and $w \in \Omega$. A second-degree Taylor expansion for $f(x)$ centered at $x = X_{t_{k-1}^n}(w)$ yields

$$f(X_{t_k^n}) - f(X_{t_{k-1}^n}) = \sum_{j=1}^{d} D_j f(X_{t_{k-1}^n})(X^j_{t_k^n} - X^j_{t_{k1}^n}) + \frac{1}{2} \sum_{i,j=1}^{d} D_{i,j} f(\xi_{n,k})(X^j_{t_k^n} - X^i_{t_{k-1}^n})(X^j_{t_k^n} - X^i_{t_{k-1}^n}),$$

where the point $\xi_{n,k} = \xi_{n,k}(w)$ is on the line segment from $X_{t_{k-1}^n}(w)$ to $X_{t_k^n}(w)$. Note that this line segment is contained in K and that $D_{i,j}f$ is uniformly continuous on K. Entering the above expansion into (2.35) and commuting the order of summation, we can write

$$f(X_{t_k^n}) - f(X_{t_{k-1}^n}) = \sum_{j=1}^{d} A_j^n + \frac{1}{2} \sum_{i,j=1}^{d} B_{i,j}^n,$$

where

$$A_j^n = \sum_{k=1}^{k_n} D_j f(X_{t_{k-1}^n})(X^j_{t_k^n} - X^j_{t_{k-1}^n}),$$

and

$$B_{i,j}^n = \sum_{k=1}^{k_n} D_{i,j} f(\xi_{n,k})(X^i_{t_k^n} - X^i_{t_{k-1}^n})(X^i_{t_k^n} - X^i_{t_{k-1}^n}),$$

at all points $w \in \Omega$. By Theorem 2.9, we have

$$A_j^n \to \int_0^t D_j f(X_s(w))dX_s^j(w),$$

in probability, as $n \uparrow \infty$. Since limits in probability are uniquely determined $\mathbb{P}$-a. s., it will now suffice to show that

$$B_{i,j}^n \to \int_0^t D_{i,j} f(X_s(w))d\langle X^i(w), X^j(w)\rangle_s,$$

in probability on the set Ω_m as $n \uparrow \infty$. To see this, we will compare $B_{i,j}^n$ to the similar term

$$\widetilde{B}_{i,j}^n = \sum_{k=1}^{k_n} D_{i,j} f(X_{t_{k-1}^n})(X_{t_k^n}^i - X_{t_{k-1}^n}^i)(X_{t_k^n}^j - X_{t_{k-1}^n}^j),$$

which is known to converge to

$$\int_0^t D_{i,j} f(X_s(w))d\langle X^i(w), X^j(w)\rangle_s,$$

in probability.

It will thus suffice to show that

$$|B_{i,j}^n - \widetilde{B}_{i,j}^n| \to 0,$$

in probability on the set Ω_m as $n \uparrow \infty$. Indeed, using the Cauchy–Schwartz-inequality,

$$\begin{aligned} |B_{i,j}^n - \widetilde{B}_{i,j}^n| &= \left|\sum_{k=1}^{k_n} (D_{i,j} f(\xi_{n,k}) - D_{i,j} f(X_{t_{k-1}^n}))(X_{t_k^n}^i - X_{t_{k-1}^n}^i)(X_{t_k^n}^j - X_{t_{k1}^n}^j)\right| \\ &\leq C_{i,j}^n \sum_{k=1}^{k_n} |X_{t_k^n}^i - X_{t_{k-1}^n}^i||X_{t_k^n}^j - X_{t_{k1}^n}^i| \leq C_{i,j}^n B_i^n B_j^n, \end{aligned}$$

where

$$(B_i^n)^2 = \sum_{k=1}^{k_n} |X_{t_k^n}^i - X_{t_{k-1}^n}^i|^2, \quad (B_j^n)^2 = \sum_{k=1}^{k_n} |X_{t_k^n}^j - X_{t_{k-1}^n}^j|^2,$$

and

$$C_{i,j}^n = \sup_{1\leq k\leq k_n} |D_{i,j} f(\xi_{n,k}) - D_{i,j} f(X_{t_{k-1}^n})|.$$

From the uniform continuity of $D_{ij}f$ on K, it follows that $C^n_{i,j} \to 0$, $\mathbb{P}$-a. s., and hence in probability on the set Ω_m, $n \uparrow \infty$. Moreover,

$$B^n_i \to \sqrt{\langle X^i\rangle_t} < \infty,$$

and

$$B^n_j \to \sqrt{\langle X^j\rangle_t} < \infty,$$

in probability. It follows that

$$|B^n_{i,j} - \widetilde{B}^n_{i,j}| \to 0,$$

in probability on Ω_m as $n \uparrow \infty$ as desired. Let us now deal with the general case. Choose a sequence (K_m) of compact sets such that $K_m \subseteq K$ and $G = \bigcup_m K_m$ and set

$$T_m = \{t \geq 0 : X_t \in K_m\}.$$

By path continuity of X, (T_m) is a sequence of optional times such that $T_m \uparrow 0$ on all of Π as $m \uparrow \infty$. Since X_0 is constant, we can choose m_0 such that $X_0 \in K^0_{m_0}$, and hence $X^{T_m}_t \in K_m$ for all $m \geq m_0$ and $t \geq 0$. Consider such m. The form of the equation (2.34) relates to the process X^{T_m}, and observing that

$$g(X^{T_m}) = g(X)^{T_m},$$
$$\langle Y^{T_m}, Z^{T_m}\rangle = \langle Y, Z\rangle^{T_m},$$

we have

$$f(X^{T_m}_t(w)) - f(X_0(w)) = \sum_{j=1}^{d} \int_0^t D_j f(X_s(w))dX^{j,T_m}_s(w)$$
$$+ \frac{1}{2}\sum_{i,j=1}^{d} \int_0^t D_{i,j} f(X_s(w))d\langle X^i_s(w), X^j_s(w)\rangle^{T_m}.$$

Let $m \uparrow \infty$, we obtain

$$f(X_t(w)) - f(X_0(w)) = \sum_{j=1}^{d} \int_0^t D_j f(X_s(w))dX^j_s(w)$$
$$+ \frac{1}{2}\sum_{i,j=1}^{d} \int_0^t D_{i,j} f(X_s(w))d\langle X^i_s(w), X^j_s(w)\rangle, \quad \mathbb{P}\text{-a. s.},$$

for each $t \geq 0$. □

Theorem 2.12. *Let $G \subset \mathbb{R}^d$ be an open set,*

$$X = (X^1, X^2, \dots, X^d),$$

a continuous semimartingale with values in G and

$$f \in C^{1,2}([0,T] \times G).$$

Then

$$\begin{aligned} f(t, X_t(w)) - f(0, X_0(w)) &= \int_0^t \frac{\partial f(s, X_s(w))}{\partial s} ds + \sum_{j=1}^d \int_0^t D_j f(s, X_s(w)) dX_s^j(w) \\ &\quad + \frac{1}{2} \sum_{i,j=1}^d \int_0^t D_{i,j} f(s, X_s(w)) d\langle X_s^i(w), X_s^j(w)\rangle, \quad \mathbb{P}\text{-a. s.}, \end{aligned}$$

for each $t \geq 0$.

Proof. The notation

$$f \in C^{1,2}([0,T] \times G),$$

is to be interpreted as follows: the partial derivative $\frac{\partial f}{\partial s}$ exists on $(0,T) \times G$ and has a continuous extension to $[0,T] \times G$. Continuous partial derivatives $D_j f$, $D_{ij} f$ with respect to the remaining variables are assumed to exist on $[0,T] \times G$. This ensures that all partial derivatives are uniformly continuous on $[0,T] \times K$ for each compact subset $K \subseteq G$. Consider a partition

$$\Delta_n = \{0 = t_0^n < t_1^n \dots t_{k_n}^n = t\},$$

of $[0,t]$ and write

$$\begin{aligned} f(t, X_t) - f(t, X_0) &= \sum_{k=1}^{k_n} (f(t_k^n, X_{t_k^n}) - f(t_{k-1}^n, X_{t_{k-1}^n})). \\ &= \sum_{k=1}^{k_n} (f(t_k^n, X_{t_k^n}) - f(t_{k-1}^n, X_{t_{k-1}^n}) + f(t_k^n, X_{t_k^n}) - f(t_{k-1}^n, X_{t_{k-1}^n})), \end{aligned}$$

and

$$f(t_k^n, X_{t_k^n}) - f(t_{k-1}^n, X_{t_{k-1}^n}) = \frac{\partial f}{\partial t}(\eta_{n,k}, X_{t_k^n})(t_k^n - t_{k-1}^n),$$

for some

$$\eta_{n,k} = \eta_{n,k}(w) \in (t_k^n, t_{k-1}^n).$$

The summand

$$f(t_k^n, X_{t_k^n}) - f(t_{k-1}^n, X_{t_{k-1}^n}),$$

is dealt with exactly as in the proof of Theorem 2.12 (second-degree Taylor expansion of $g(x) = f(t_{k-1}^n, x)$ around the point $x = X_{t_{k-1}^n}$). □

Let us write down the special case where $X \in H_{SP}$ is a scalar semimartingale ($d = 1$).

Theorem 2.13. *Let $G \subset \mathbb{R}$ be an open set, $X = (X^1)$ (i. e., $d = 1$) a continuous semimartingale with values in G and $f \in C^2(G)$. Then*

$$f(X_t(w)) - f(X_0(w)) = \int_0^t \frac{\partial f(X_s(w))}{\partial X_s} dX_s(w) + \frac{1}{2}\int_0^t \frac{\partial^2 f(X_s(w))}{\partial X_s^2} d\langle X_s(w), X_s(w)\rangle, \quad \mathbb{P}\text{-a. s.},$$

for each $t \geq 0$.

Proof. The same evidence is dealt with exactly as in the proof of Theorem 2.13 for $j = d = 1$. □

Theorem 2.14. *Let $G \subset \mathbb{R}$ be an open set, X a continuous semimartingale with values in G and $f \in C^{1,2}([0,T] \times G)$. Then*

$$f(t, X_t(w)) - f(0, X_0(w)) = \int_0^t \frac{\partial f(s, X_s(w))}{\partial s} ds + \int_0^t \frac{\partial f(X_s(w))}{\partial X_s} dX_s(w) + \frac{1}{2}\int_0^t \frac{\partial^2 f(X_s(w))}{\partial X_s^2} d\langle X_s(w), X_s(w)\rangle \quad \mathbb{P}\text{-a. s.},$$

for each $t \geq 0$.

Proof. The same evidence is dealt with exactly as in the proof of Theorem 2.14 for $j = d = 1$. □

As a first consequence of Itô's formula, we show that the family S of continuous semimartingales is not only a real algebra but is, in fact, closed under the application of twice continuously differentiable functions:

Theorem 2.15. *Let G be an open subset of $\mathbb{R}^d$; $X \in S^d$. Let S^d denote the family of all $\mathbb{R}^d$-valued functions with values in G; $f \in C^2(G)$. For each $i = 1, \ldots, d$, let $X^i = M^i + A^i$ be the semimartingale decomposition of X^i, especially $A^i = u_{X^i}$. Then $Z = f(X)$ is again a continuous semimartingale, and its local martingale part M_t and compensator $u_Z(t)$ are given by*

$$M = Z_0 + \sum_{j=1}^{d} \int_0^t D_j f(X_s(w)) dM_s^i(w),$$

and

$$u_Z = \sum_{j=1}^{d} \int_0^t D_j f(X_s(w)) dA_s^i(w) + \frac{1}{2} \sum_{i,j=1}^{d} \int_0^t D_{i,j} f(X_s(w)) d\langle X_s^i(w), X_s^j(w)\rangle.$$

Proof. Writing $f(X)$ to denote the process $f(X_t)$ and using the formula:

$$\begin{aligned} Z = f(X) = Z_0 &+ \int_0^t \frac{\partial f(X_s(w))}{\partial s} \langle X_s^i, X_s^j\rangle_s + \sum_{i=1}^{d} \int_0^t D_j f(s, X_s(w)) dX_s^i(w) \\ &= Z_0 + \sum_{i=1}^{d} \int_0^t D_j f(s, X_s(w)) dM_s^i(w) + \sum_{i=1}^{d} \int_0^t D_j f(s, X_s(w)) dA_s^i(w) \\ &+ \frac{1}{2} \sum_{i,j=1}^{d} \int_0^t D_{i,j} f(X_s(w)) d\langle X_s^i(w), X_s^j(w)\rangle = M + A, \end{aligned}$$

where

$$M = Z_0 + \sum_{i=1}^{d} \int_0^t D_j f(s, X_s(w)) dM_s^i(w),$$

and

$$A = \frac{1}{2} \sum_{i,j=1}^{d} \int_0^t D_{i,j} f(X_s(w)) d\langle X_s^i(w), X_s^j(w)\rangle + \sum_{i=1}^{d} \int_0^t D_j f(s, X_s(w)) dA_s^i(w).$$

Here, M is a continuous local martingale, since A is a continuous bounded variation process that vanishes at zero. This shows that Z is a continuous semimartingale with a semimartingale decomposition $Z = M + A$, as desired. □

We can save this result in the table below.

$[d., d.]$	ds	$dW^i(s)$	$dW^j(s)$
ds	0	0	0
$dW^i(s)$	0	ds	0
$dW^j(s)$	0	0	ds

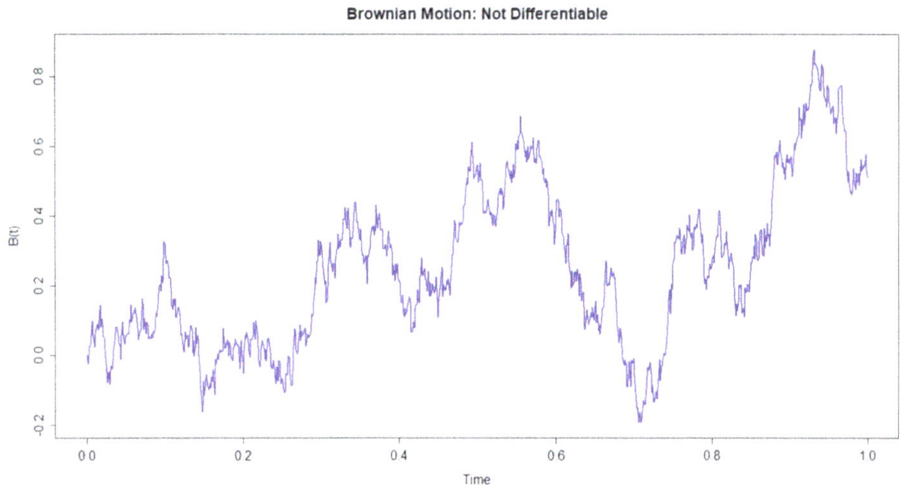

Figure 2.2: Brownian motion (Wiener process) is not differentiable.

2.4 Fractional Brownian motion

The fractional Brownian motion was first introduced within a Hilbert space framework by Kolmogorov in 1940 in,[8] where it was called the Wiener helix. The name, fractional Brownian motion, is due to Mandelbrot and Van Ness, who in 1968 provided in.[9] A stochastic integral representation of this process in terms of standard Brownian motion; see Figure 2.2.

Definition 2.37. Given $H \in (0,1)$, a continuous centered Gaussian process B^H with the covariance function

$$\begin{aligned} R_H(t,s) &= \mathbb{E}[B_l^H(t))B_k^H(s)], \\ R_{H_{lk}}(t,s) &= \frac{1}{2}(|t|^{2H} + |s|^{2H} - |t-s|^{2H})\delta_{lk}, \quad t,s \in [0,T], \end{aligned}$$

where

$$\delta_{lk} = \begin{cases} 1, & l = k, \\ 0, & l \neq k, \end{cases}$$

is called a two-sided one-dimensional fractional Brownian motion (*fBm*), and H is the Hurst parameter.

8 D. Nualart, *The Malliavin Calculus and Related Topics*, 2nd ed. Berlin: Springer-Verlag, 2006.

9 P. Embrechts, Maejima, Makoto, *Self-similar Processes*. Princeton, NJ: Princeton University Press, 2002.

For $H = 1/2$, the fBm is then a standard Brownian motion. By Definition 2.37, we obtain that a standard fBm $B^{(H)}$ has the following properties:

1. $B^{(H)}(0) = 0$ and $E(B^{(H)}(t)) = 0$ for all $t \geq 0$.
2. $B^{(H)}$ has homogeneous increments, that is, $B^{(H)}(t+s) - B^{(H)}(s)$ has the same law of $B^{(H)}(t)$ for $s, t \geq 0$.
3. $B^{(H)}$ is a Gaussian process and $E(B^{(H)}(t)^2) = t^{2H}$, $t \geq 0$, for all $H \in (0,1)$.
4. $B^{(H)}$ has continuous trajectories.

2.4.1 Correlation between two increments

For $H = 1/2$, $B^{(H)}$ is a standard Brownian motion; hence, in this case, the increments of the process are independent. On the contrary, for $H \neq 1/2$, the increments are not independent. More precisely, by Definition 2.37, we know that the covariance between

$$B^{(H)}(t+h) - B^{(H)}(t),$$

and

$$B^{(H)}(s+h) - B^{(H)}(s),$$

with $s + h \leq t$ and $t - s = nh$,

$$\rho_H(n) = \frac{1}{2} h^{2H} \left((n+1)^{2H} + (n-1)^{2H} - 2n^{2H} \right).$$

In particular, we obtain that the two increments of the form

$$B^{(H)}(t+h) - B^{(H)}(t),$$

and

$$B^{(H)}(t+2h) B^{(H)}(t+h),$$

are positively correlated for $H > 1/2$, while they are negatively correlated for $H < 1/2$. In the first case, the process demonstrates an aggregation behavior, and this property can be used to describe "cluster" phenomena (systems with memory and persistence). In the second case, it can be used to model sequences with intermittency and antipersistence.

2.4.2 Long-range dependence

Definition 2.38. A stationary sequence $(X_n)_{n \in N}$ exhibits long-range dependence if the autocovariance functions

$$\rho(n) = \mathrm{Cov}(X_k, X_{k+n}),$$

satisfy

$$\lim_{n\to\infty} \frac{\rho(n)}{cn^{-\alpha}} = 1,$$

for some constant c and $\alpha \in (0,1)$. In this case, the dependence between X_k and X_{k+n} decays slowly as n tends to infinity and

$$\sum_{n=1}^{\infty} \rho(n) = \infty.$$

Hence, we immediately obtain that the increments

$$X_k = B^{(H)}(k) - B^{(H)}(k-1),$$

of $B^{(H)}$ and

$$X_{k+n} = B^{(H)}(k+n) - B^{(H)}(k+n-1),$$

of $B^{(H)}$ have the long-range dependence property for $H > 1/2$ since

$$\rho_H(n) = \frac{1}{2} h^{2H}\left((n+1)^{2H} + (n-1)^{2H} - 2n^{2H}\right) \sim H(2H-1)n^{2H-2},$$

as n goes to infinity. In particular,

$$\lim_{n\to\infty} \frac{\rho(n)}{H(2H-1)n^{2H-2}} = 1.$$

Summarizing, we obtain:
1. For $H > 1/2$, we have

$$\sum_{n=1}^{\infty} \rho_H(n) = \infty.$$

2. For $H < 1/2$, we have

$$\sum_{n=1}^{\infty} |\rho_H(n)| < \infty.$$

Definition 2.39. We say that an $\mathbb{R}^d$-valued random process $X = (X_t)_{t\geq 0}$ is self-similar or satisfies the property of self-similarity if, for every $a > 0$, there exists $b > 0$ such that

$$\mathrm{Law}(X_{at}, t \geq 0) = \mathrm{Law}(bX_t, t \geq 0). \tag{2.36}$$

Note that (2.36) means that the two processes X_{at} and bX_t have the same finite-dimensional distribution functions, that is, for every choice $t_0, \dots, t_n$ in $\mathbb{R}$,

$$\mathbb{P}(X_{at_0} \le x_0, \dots, X_{at_n} \le x_n) = \mathbb{P}(bX_{t_0} \le x_0, \dots, bX_{t_n} \le x_n),$$

for every $x_0, \dots, x_n$ in $\mathbb{R}$.

Definition 2.40. If $b = a^{-H}$ in Definition 2.39, then we say that $X = (X_t)_{t \ge 0}$ is a self-similar process with a Hurst index H or that it satisfies the property of (statistical) self-similarity with a Hurst index H. The quantity $D = 1/H$ is called the statistical fractal dimension of X.

2.4.3 Hölder continuity

We recall that according to the Kolmogorov criterion, a process $X = (X_t)_{t \in \mathbb{R}}$ admits a continuous modification if there exist constants $0 < \alpha \le 1$ and $k > 0$ such that

$$\mathbb{E}(|X(t) - X(s)|) \le k|t - s|^\alpha, \quad \text{for all } s, t \in \mathbb{R}.$$

Theorem 2.16. *Let $H \in (0, 1)$. The fBm $B^{(H)}$ admits a version whose sample paths are almost surely Hölder continuous of order strictly less than H.*

Proof. We recall that a function $f : \mathbb{R} \to \mathbb{R}$ is Holder continuous of order $0 < \alpha \le 1$, and we write $f \in C^\alpha(\mathbb{R})$ if there exists $M > 0$ such that

$$|f(t) - f(s)| \le M|t - s|^\alpha,$$

for every $s, t \in \mathbb{R}$. For any $\alpha > 0$, we have

$$\mathbb{E}(|B^{(H)}(t) - B^{(H)}(s))|^\alpha) = \mathbb{E}(|B^{(H)}(1)|^\alpha)|t - s|^{\alpha H}.$$

Hence, by the Kolmogorov criterion, we conclude that the sample paths of $B^{(H)}$ are almost everywhere Hölder continuous of order strictly less than H. Moreover, we have

$$\limsup_{t \to 0^+} \frac{|B^{(H)}(t)|}{t^h \sqrt{\log \log t^{-1}}} = c_H,$$

with probability one, where c_H is a suitable constant. Hence, $B^{(H)}$ cannot have sample paths with Hölder continuity order greater than H. □

2.4.4 Path differentiability

We also obtain that the process $B^{(H)}$ is not mean square differentiable and does not have differentiable sample paths.

Property 2.6. Let $H \in (0,1)$. The fBm sample path $B^{(H)}(\cdot)$ is not differentiable.
In fact, for every $t_0 \in [0,\infty)$,

$$\limsup_{t\to t_0}\left|\frac{B^{(H)}(t)-B^{(H)}(t_0)}{t-t_0}\right| = \infty,$$

with probability one.

Proof. Note that we assume $B^{(H)}(0) = 0$. The result is proved by exploiting the self-similarity of $B^{(H)}$. Consider the random variable

$$\mathcal{R}_{t,t_0} = \frac{B^{(H)}(t)-B^{(H)}(t_0)}{t-t_0},$$

that represents the incremental ratio of $B^{(H)}$. Since $B^{(H)}$ is self-similar, we have that the law of $\mathcal{R}_{t,t_0}$ is the same as that of $(t-t_0)^{H-1}B^{(H)}(1)$. If one considers the event,

$$A(t,w) = \left\{\sup_{0\le s\le t}\left|\frac{B^H(s)}{s}\right| > d\right\}.$$

For any sequence $t \to 0$, we have

$$A(t_n,w) \supseteq A(t_{n+1},w).$$

Thus,

$$\mathbb{P}\Big(\lim_{n\to\infty} A(t_n)\Big) = \lim_{n\to\infty} \mathbb{P}(A(t_n)),$$

and

$$\mathbb{P}(A(t_n)) \ge \mathbb{P}\left(\frac{B^H(t_n)}{(t_n)}\right) = \mathbb{P}(B^H(1) > t_n^{1-H}d),$$

which tends to one as $n \to \infty$. Note that the proof goes through under the assumption of self-similarity; see Figure 2.3. □

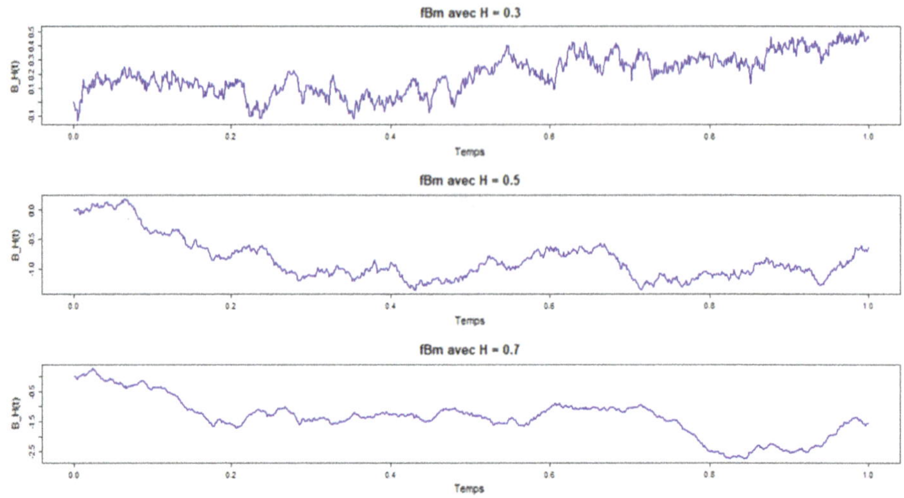

Figure 2.3: Path differentiability.

2.4.5 The fBm is not a semimartingale for $H \neq 1/2$

The definition of the Itô integral is a direct consequence of the martingale property of Brownian motion. But fBm does not exhibit this property; in fact, *fBm* is not even a semimartingale. We state the theorem and present a simple proof here. But first, we need to find the p-variation of $B^H(t)$. In order to verify that $B^{(H)}$ is not a semimartingale for $H \neq 1/2$, it is sufficient to compute the p-variation of $B^{(H)}$.

Definition 2.41. Let $(X(t))_{t\in[0,T]}$ be a stochastic process and consider a partition $\pi = \{0 = t_0 < t_1 < \cdots < t_n = T\}$. Put

$$S_p(X,\pi) = \sum_{k=1}^{n} \left|X(t_k) - X(t_{k-1})\right|^p.$$

The p-variation of X over the interval $[0,T]$ is defined as

$$V_p(X,[0,T]) = \sup_p S_p(X,\pi),$$

where π is a finite partition of $[0,T]$. The index of p-variation of a process is defined as

$$I(X,[0,T]) = \inf\{p > 0; V_p(X,[0,T]) < \infty\}.$$

Lemma 2.10 ([3]). *Let*

$$I(B^H(t),[0,T]) = \frac{1}{H}.$$

Moreover,

$$V_p(B^H(t), [0,T]) = 0 \quad \textit{when } pH > 1,$$

and

$$V_p(B^H(t), [0,T]) = \infty \quad \textit{when } pH < 1.$$

Theorem 2.17. *Let*

$$\{B^H(t) : t \geq 0\},$$

for $H \neq 1/2$, not be a semimartingale.

Proof. A process

$$\{X(t), \mathcal{F}_t, t \geq 0\},$$

is called a semimartingale if it admits the Doob–Meyer decomposition

$$X(t) = X(0) + M(t) + A(t),$$

where $M(t)$ is an $\mathcal{F}_t$ local martingale with $M(0) = 0$, $A(t)$ is a cadlag adapted process of locally bounded variation, and $X(0)$ is $\mathcal{F}_0$-measurable. Moreover, any semimartingale has locally bounded quadratic variation. Now, let $X(t) = B^H(t)$.

For $p > 0$, set

$$Y_{n,p} = n^{pH-1} \sum_{j=1}^{n} |B_{j/n} - B_{(j-1)/n}|^p.$$

By the self-similar property of fBm, the sequence $\{Y_{n,p}, n \geq 1\}$ has the same distribution as

$$\{\widetilde{Y}_{n,p}, n \geq 1\},$$

where

$$\widetilde{Y}_{n,p} = n^{-1} \sum_{j=1}^{n} |B_j - B_{j-1}|^p.$$

The stationary sequence

$$\{B_j - B_{j-1} j \geq 1\},$$

is mixing. Hence, by the Ergodic theorem $\widetilde{Y}_{n,p}$ converges almost surely and in $L^1(\Omega)$ to $\mathbb{E}(|B_1|^p)$ as n tends to infinity. As a consequence, $Y_{n,p}$ converges in probability as n tends to infinity to $\mathbb{E}(|B_1|^p)$. Therefore,

$$V_p(X,[0,T]) = \sup_p \sum_{j=1}^{n} |B_{j/n} - B_{(j-1)/n}|^p.$$

1. If $H < 1/2$, we can choose $p > 2$ such that $pH < 1$, and we obtain that the p-variation of *fBm* (defined as the limit in probability $\lim_{n\to\infty} V_{n,p}$) is infinite. Hence, the quadratic variation ($p = 2$) is also infinite.
2. If $H > 1/2$, we can choose p such that $\frac{1}{H} < p < 2$. Then the p variation is zero, and, as a consequence, the quadratic variation is also zero. On the other hand, if we choose p such that $1 < p < \frac{1}{H}$, we deduce that the total variation is infinite.

Therefore, we have proved that for $H \neq 1/2$ the fractional Brownian motion cannot be a semimartingale. □

Cheridito[10] has introduced the notion of weak semimartingale as a stochastic process $\{X_t, t \geq 0\}$ such that for each $T > 0$, the set of random variables

$$\left\{\sum_{j=1}^{n} f_j(X_{t_j} - X_{t_{j-1}}), n \geq 1, 0 = t_0 < t_1 < \cdots < t_n = T, |f_j| \leq 1, f_j \in \mathcal{F}^X_{t_{j-1}}\right\},$$

is bounded. Here, $\mathcal{F}^X$ represents the natural filtration associated with the process X. Moreover, it is shown that if $B(t)$ is a standard Brownian motion independent of $B^{(H)}$, then the process:

$$M(t)^H = B^H(t) + B(t).$$

Then

$$\{M(t)^H, t \geq 0\},$$

is not a weak semimartingale if

$$H \in (0,1/2) \cup (1/2,3/4],$$

and it is a semimartingale, equivalent in law to Brownian motion on any finite time interval $[0,T]$, if $H \in (3/4,1)$.

10 A. N. Kolmogorov, Wienersche Spiralen und einige andere interessante Kurven im Hilbertschen Raum, C. R. (Doklady) Acad. URSS (N. S.) **26** (1940), 115–118.

2.4.6 Fractional integrals and derivatives

Let $a, b \in \mathbb{R}, a < b$. Let $f \in L^1(a, b)$ and $\alpha > 0$. The left- and right-sided fractional integrals of f of order α are defined for almost all $x \in (a, b)$ by

$$I_{a^+}^{\alpha} f(x) = \frac{1}{\Gamma(\alpha)} \int_a^x (x-y)^{\alpha-1} f(y) dy, \tag{2.37}$$

and

$$I_{b^-}^{\alpha} f(x) = \frac{1}{\Gamma(\alpha)} \int_x^b (y-x)^{\alpha-1} f(y) dy, \tag{2.38}$$

respectively. Let $I_{a^+}^{\alpha}(L^p)$ (resp., $I_{b^-}^{\alpha}(L^p)$) the image of $L^p(a, b)$ by the operator $I_{a^+}^{\alpha}$ (resp. $I_{b^-}^{\alpha}$).

If $f \in I_{a^+}^{\alpha}(L^p)$ (resp., $f \in I_{b^-}^{\alpha}(L^p)$) and $0 < \alpha < 1$, then the left- and right-sided fractional derivatives are defined by

$$D_{a^+}^{\alpha} f(x) = \frac{1}{\Gamma(1-\alpha)} \left(\frac{f(x)}{(x-a)^{\alpha}} + \alpha \int_a^x \frac{f(x) - f(y)}{(x-y)^{\alpha+1}} dy \right), \tag{2.39}$$

and

$$D_{b^-}^{\alpha} f(x) = \frac{1}{\Gamma(1-\alpha)} \left(\frac{f(x)}{(x-a)^{\alpha}} + \alpha \int_x^b \frac{f(x) - f(y)}{(y-x)^{\alpha+1}} dy \right), \tag{2.40}$$

for almost all $x \in (a, b)$, the convergence of the integrals at the singularity $y = x$ holds pointwise for almost all $x \in (a, b)$ if $p = 1$ and, moreover, in L^p-sense if $1 < p < \infty$.

Recall the following properties of these operators:

1. If $\alpha < \frac{1}{p}$ and $q = \frac{p}{1-\alpha p}$. Then

$$I_{a^+}^{\alpha}(L^p) = I_{b^-}^{\alpha}(L^p) \subset L^q(a, b).$$

2. If $\alpha > \frac{1}{p}$. Then

$$I_{a^+}^{\alpha}(L^p) \cup I_{b^-}^{\alpha}(L^p) \subset C^{\alpha - \frac{1}{p}}(a, b),$$

   where $C^{\alpha - \frac{1}{p}}(a, b)$ denotes the space of $(\alpha - \frac{1}{p})$-Hölder continuous functions of order $\alpha - \frac{1}{p}$ in the interval $[a, b]$.

The following inversion formulas hold:

$$I_{a^+}^{\alpha}(D_{a^+}^{\alpha} f) = f,$$

for all $f \in I^{\alpha}_{a^+}(L^p)$, and

$$D^{\alpha}_{a^+}(I^{\alpha}_{a^+}f) = f,$$

for all $f \in L^1(a,b)$. Similar inversion formulas hold for the operators $I^{\alpha}_{b^-}$ and $D^{\alpha}_{b^-}$.

The following integration by parts formula holds:

$$\int_a^b (D^{\alpha}_{a^+}f)(s)g(s)ds = \int_a^b f(s)(D^{\alpha}_{b^-}g)(s)ds, \tag{2.41}$$

for any

$$f \in I^{\alpha}_{a^+}(L^p),$$
$$g \in I^{\alpha}_{b^-}(L^p), \quad \frac{1}{p} + \frac{1}{q} = 1.$$

We start our tour through the different definitions of stochastic integration for fBm with Hurst index $H \in (0,1)$ using the Wiener integrals since they deal with the simplest case of deterministic integrands. We show how they can be expressed in terms of an integral with respect to standard Brownian motion, extend their definition also to the case of stochastic integrands, and then proceed to define the stochastic integral by using the divergence operator. In both cases, we need to distinguish between $H > 1/2$ and $H < 1/2$.

2.4.7 Integrals for fractional Brownian motion

Fix an interval $[0,T]$ and let $B^{(H)}(t)$, $t \in [0,T]$ be a fBm with a Hurst index $H \in (0,1)$ on the probability space $(\Omega, \mathcal{F}^{(H)}, \mathcal{F}^{(H)}_t, \mathbb{P}^H)$ endowed with the natural filtration $\mathcal{F}^{(H)}_t$ and the law $\mathbb{P}^H$ of $B^{(H)}$ (for a construction of the measure $\mathbb{P}^H$, we refer to Section 2.1)

Definition 2.42. An $\mathcal{F}$-adapted process ϕ on $[0,T] \times \Omega$ is an elementary process if a partition

$$\psi = \{\bar{t} = 0 < \bar{t}_1 < \cdots < \bar{t}_n = T\},$$

and $(\mathcal{F}^H_{\bar{t}_i})$-measurable random variables $(\phi_{\bar{t}_i})_{i<n}$, ϕ_t satisfy

$$\phi_t(\omega) = \sum_{i=1}^{n} \phi_i(\omega)\chi_{(\bar{t}_{i-1},\bar{t}_i]}(t), \quad \text{for } 0 \le t \le T,\ \omega \in \Omega.$$

The Itô integral of the simple process ϕ is defined as

$$I_H(\phi) = \int_0^T \phi_l(s)dB^H_l(s) = \sum_{i=1}^{n} \phi_l(\bar{t}_i)(B^H_l(\bar{t}_i) - B^H_l(\bar{t}_{i-1})), \tag{2.42}$$

whenever $\phi_{\bar{t}_i} \in L^2(\mathcal{F}^H_{\bar{t}_i})$ for all $i \le n$.

If $H > \frac{1}{2}$, it is easy to see that the covariance of fBm can be written as

$$R_H(t,s) = \alpha_H \int_0^t \int_0^s |r-u|^{2H-2} du dr, \tag{2.43}$$

where $\alpha_H = H(2H-1)$. The formula (2.43) implies that

$$\langle \varphi, \psi \rangle_{\mathcal{H}} = \alpha_H \int_0^T \int_0^T |r-u|^{2H-2} \varphi_r \psi_u du dr, \tag{2.44}$$

for any pair of step functions φ and ψ in $\mathcal{E}$.

We can write

$$|r-u|^{2H-2} = \frac{(ru)^{H-\frac{1}{2}}}{\beta(2-2H, H-\frac{1}{2})} \int_0^{r \wedge u} v^{1-2H}(r-v)^{H-\frac{3}{2}}(u-v)^{H-\frac{3}{2}} dv, \tag{2.45}$$

where β denotes the Beta function. Let us show Equation (2.45). Suppose $r > u$. By means of the change of variables $z = \frac{r-v}{u-v}$ and $x = \frac{r}{uz}$, we obtain

$$\begin{aligned}\int_0^u v^{1-2H}(r-v)^{H-\frac{3}{2}}(u-v)^{H-\frac{3}{2}} dv &= (r-u)^{2H-2} \int_{\frac{r}{u}}^{\infty} (zu-r)^{1-2H} z^{H-\frac{3}{2}} dz \\ &= (ru)^{\frac{1}{2}-H}(r-u)^{2H-2} \int_0^1 (1-x)^{1-2H} x^{H-\frac{3}{2}} dx \\ &= \beta\left(2-2H, H-\frac{1}{2}\right)(ru)^{\frac{1}{2}-H}(r-u)^{2H-2}.\end{aligned}$$

Consider the square integrable kernel

$$K_H(t,s) = c_H s^{\frac{1}{2}-H} \int_s^t (u-s)^{H-\frac{3}{2}} u^{H-\frac{1}{2}} du, \tag{2.46}$$

where

$$c_H = \left[\frac{H(2H-1)}{\beta(2-2H, H-\frac{1}{2})}\right]^{1/2},$$

and $t > s$.

Taking into account the formulas (2.43) and (2.45), we deduce that this kernel verifies

$$\int_0^{t\wedge s} K_H(t,u)K_H(s,u)du = c_H^2 \int_0^{t\wedge s}\left(\left(\int_u^t (y-u)^{H-\frac{3}{2}} y^{H-\frac{1}{2}} dy\right)\left(\int_u^s (z-u)^{H-\frac{3}{2}} z^{H-\frac{1}{2}} dz\right) u^{1-2H} du\right.$$
$$= c_H^2 \beta\left(2-2H, H-\frac{1}{2}\right)\int_0^t\int_0^s |y-z|^{2H-2} dz dy$$
$$= R_H(t,s), \tag{2.47}$$

and the latter sum coincides with the usual Riemann–Stieltjes sum. A question arises, in which sense, can we consider Formula (2.42) as the extension of the sum (2.42). Note that this is true for a step function. It is known that $B^H(t)$ with $H > \frac{1}{2}$ admits the following Volterra representation:

$$B^H(t) = \int_0^t K_H(t,s)dB(s),$$

where B is a standard Brownian motion, and the Volterra kernel $K(t,s)$ is given by

$$K_H(t,s) = c_H \int_s^t (u-s)^{H-\frac{3}{2}} \left(\frac{u}{s}\right)^{H-\frac{1}{2}} du, \quad t \geq s,$$

where

$$c_H = \sqrt{\frac{H(2H-1)}{\beta(2H-2, H-\frac{1}{2})}},$$

and $\beta(.,.)$ denote the Beta function. We put $K(t,s) = 0$ if $t \leq s$:

$$\frac{\partial K_H}{\partial t}(t,s) = c_H\left(\frac{t}{s}\right)^{H-\frac{1}{2}} (t-s)^{H-\frac{3}{2}}.$$

We denote by $\mathcal{E}$ the set of step functions on $[0,T]$. Let $\mathcal{H}$ be the Hilbert space defined as the closure of $\mathcal{E}$ with respect to the scalar product

$$\langle \chi_{[0,t]}, \chi_{[0,s]}\rangle_{\mathcal{H}} = R_H(t,s).$$

Consider the linear operator K_H^* from $\mathcal{E}$ to $L^2([0,T])$ defined by

$$(K_H^*\phi)(t) = \int_s^T \phi(t)\frac{\partial K_H}{\partial t}(t,s)dt.$$

Notice that

$$(K_H^*\chi_{[0,t]})(s) = K_H(t,s)\chi_{[0,t]}(s).$$

The operator K_H^* is an isometry between $\mathcal{E}$ and $L^2([0,T])$ that can be extended to the Hilbert space $\mathcal{H}$. In fact, for any $s,t \in [0,T]$, then

$$\langle K_H^*\chi_{[0,t]}, K_H^*\chi_{[0,t]}\rangle_{L^2([0,T])} = \langle \chi_{[0,t]}, \chi_{[0,s]}\rangle_{\mathcal{H}} = R_H(t,s).$$

In addition, for any $\phi \in \mathcal{H}$,

$$\int_0^T \phi_l(s)dB_l^H(s) = \int_0^T (K_H^*\phi_l)(s)dB_l(s),$$

if and only if $K_H^*\phi_l \in L^2([0,T])$. Next, we are interested in considering an *fBm* with values in a Hilbert space.

Consider the process $B_l = \{B_l(t),\ t \in [0,T]\}$ defined by

$$B_l(t) = B_l^H\big((K_H^*)^{-1}\chi_{[0,t]}\big).$$

Indeed, for any $s,t \in [0,T]$, we have

$$\begin{aligned}
\mathbb{E}(B_l(t)B_l(s)) &= E\big(B_l^H((K_H^*)^{-1}\chi_{[0,t]})B_l^H((K_H^*)^{-1}\chi_{[0,s]})\big) \\
&= \big\langle (K_H^*)^{-1}\chi_{[0,t]}, (K_H^*)^{-1}\chi_{[0,s]}\big\rangle_{\mathcal{H}} \\
&= \langle \chi_{[0,t]}, \chi_{[0,s]}\rangle_{L^2[0,T]} \\
&= s \wedge t.
\end{aligned}$$

Let

$$(X, \langle\cdot,\cdot\rangle, |\cdot|_X),\ (Y, \langle\cdot,\cdot\rangle, |\cdot|_Y),$$

be separable Hilbert spaces. Let $\mathcal{L}(Y,X)$ denote the space of all linear bounded operators from Y into X. Let $e_n, n = 1,2,\dots$ be a complete orthonormal basis in Y, and $Q \in \mathcal{L}(Y,X)$ be an operator defined by $Qe_n = \lambda_n e_n$ with finite trace

$$\operatorname{tr} Q = \sum_{n=1}^{\infty} \lambda_n < \infty,$$

where λ_n, $n = 1,2,\dots$ are nonnegative real numbers. Let $(\beta_n^H)_{n\in N}$ be a sequence of two-sided, one-dimensional standard fractional Brownian motions that are mutually independent on $(\Omega, \mathcal{F}, \mathbb{P})$. If we define the infinite-dimensional *fBm* on Y with covariance Q as

$$B^H(t) = \sum_{n=1}^{\infty} \sqrt{\lambda_n}\beta_n^H(t)e_n. \tag{2.48}$$

Then it is well-defined as an Y-valued Q-cylindrical fractional Brownian motion. In order to define Wiener integrals with respect to a *Q-fBm*, we introduce the space $L_Q^0 = L_Q^0(Y,X)$ of all Q-Hilbert–Schmidt operators $\varphi : Y \longrightarrow X$. We recall that $\varphi \in L(Y,X)$ is called a Q-Hilbert–Schmidt operator if

$$\|\varphi\|_{L_Q^0}^2 = \|\varphi Q^{1/2}\|_{HS}^2 = \operatorname{tr}(\varphi Q \varphi^*) < \infty,$$

the following useful result holds.

Lemma 2.11 ([12]). *There exists a positive constant $c_1(H)$ such that for any $\phi \in L^{1/H}([0,T])$, it holds*

$$H(2H-1)\int_0^T\int_0^T |\phi(y)||\phi(z)||y-z|^{2H-2}dydz \le c_1(H)\|\phi\|_{L^{1/H}([0,T])}^2. \tag{2.49}$$

Definition 2.43. Let $\phi(s), s \in [0,T]$ be a function with values in $L_Q^0(Y,X)$. The Wiener integral of ϕ with respect to *fBm* given by (2.48) is defined by

$$\int_0^T \phi(s)dB^H(s) = \sum_{n=1}^{\infty}\int_0^t \sqrt{\lambda_n}\phi(s)e_n d\beta_n^H = \sum_{n=1}^{\infty}\int_0^T \sqrt{\lambda_n}K_H^*(\phi e_n)(s)d\beta_n.$$

Notice that if

$$\sum_{n=1}^{\infty} \|\phi Q^{1/2}e_n\|_{L^{1/H}([0,T];X)} < \infty, \tag{2.50}$$

the next result ensures the convergence of the series in the previous definition. It can be proved by similar arguments to those used to prove Lemma 2 in Caraballo[11] and Lemma 2.1 in Guilan.[12]

Lemma 2.12. *For any $\phi : [0,T] \to L_Q^0(Y,X)$, such that (2.50) holds, and for any $\alpha, \beta \in [0,T]$ with $\alpha > \beta$,*

$$E\left|\int_\alpha^\beta \phi(s)dB^H(s)\right|_X^2 \le \overline{C}\sum_{n=1}^{\infty}\int_\alpha^\beta |\phi(s)Q^{1/2}e_n|_X^2 ds, \tag{2.51}$$

---

11 T. Caraballo, M. Garrido-Atienza, T. Taniguchi, The existence and exponential behavior of solutions to stochastic delay evolution equations with a fractional Brownian motion, *Nonlinear Anal.* **74** (2011), 3671–3684.

12 C. Guilan and H. Kai, On a type of stochastic differential equations driven by countably many Brownian motions, *J. Funct. Anal.* **203** (2003), 262–285.

where $\overline{C} = c_2(H)H(2H-1)(\alpha-\beta)^{2H-1}$ *and* $c_2(H)$ *are constants depending on* H. *If, in addition,*

$$\sum_{n=1}^{\infty} |\phi Q^{1/2} e_n|_X \quad \text{is uniformly convergent for } t \in [0,T],$$

then

$$\mathbb{E}\left|\int_{\alpha}^{\beta} \phi(s)dB^H(s)\right|_X^2 \leq \overline{C}\int_{\alpha}^{\beta} \|\phi(s)\|_{L_Q^0}^2 ds. \tag{2.52}$$

Corollary 2.1. *Let* $(\mathcal{F}_t)_{t\geq 0}$ *be a right-continuous complete σ-algebras filtration generated by a sequence of fBm* B_l^H, $l = 1,2,\ldots$, *mutually independent, and consider another sequence of* $(\mathcal{F}_t)$*-adapted processes* ϕ_l. *Denote*

$$M^k(t) = \sum_{l=1}^{k} \int_0^t \phi_l(y)dB_l^H(y), \quad k \in \mathbb{N},$$

and assume that

$$\sum_{l=1}^{\infty} \int_0^T \int_0^T \|\phi_l(y)\|_{L_Q^0} \|\phi_l(z)\|_{L_Q^0} |y-z|^{2H-2} dydz < \infty,$$

for all $t \geq 0$. *Then* M^k *is convergent in* $L^2(\Omega, \mathcal{F}_t, \mathbb{P}; X)$.

Proof. Let $M^k(t) = (\sum_{l=1}^k \int_0^t \phi_l(y)dB_l^H(y))$ for any $t \in [0,T]$. By Lemma 2.12, we have that

$$\begin{aligned}\mathbb{E}|M^n(t) - M^m(t)|_X^2 &= \mathbb{E}\left|\sum_{l=1}^{n} \int_0^t \phi_l(y)dB_l^H(y) - \sum_{l=1}^{m} \int_0^t \phi_l(y)dB_l^H(y)\right|_X^2 \\ &= \mathbb{E}\left|\sum_{l=n\wedge m+1}^{n\vee m} \int_0^t \phi_l(y)dB_l^H(y)\right|_X^2 \\ &\leq c_3(H)H(2H-1)T^{2H-1} \sum_{l=n\wedge m+1}^{n\vee m} \int_0^T \|\phi_l(y)\|_{L_Q^0}^2 dy \to 0,\end{aligned}$$

as $n, m \to \infty$ where $c_3(H) > 0$. Consequently, $M^n(\cdot)$ is a Cauchy sequence with respect to the norm

$$\sup_{0\leq t\leq T} E|\cdot|_X^2,$$

and the limit is M. Then we can conclude

$$\lim_{n\to\infty}\sup_{0\le t\le T} E|M^n(t)-M(t)|_X^2=0,$$

and observe that

$$\sum_{l=1}^{\infty}\int_0^T\int_0^T \|\phi_l(y)\|_{L_Q^0}\|\phi_l(z)\|_{L_Q^0}|y-z|^{2H-2}dydz<\infty.$$

Then M in $L^2(\Omega,\mathcal{F}_t,\mathbb{P};X)$. □

The following result is one of the elementary properties of square-integrable stochastic processes.

Lemma 2.13 (Itô isometry for elementary processes). *Let $(\phi_l)_{l\in\mathbb{N}}$ be a sequence of elementary processes. If*

$$\sum_{l=1}^{\infty}\int_0^T\int_0^T \|\phi_l(y)\|_{L_Q^0}\|\phi_l(z)\|_{L_Q^0}|y-z|^{2H-2}dydz<\infty,$$

then

$$\mathbb{E}\left|\sum_{l=1}^{\infty}\int_0^T\phi_l(y)dB_l^H(y)\right|_X^2\le H(2H-1)\sum_{l=1}^{\infty}\int_0^T\int_0^T \|\phi_l(y)\|_{L_Q^0}\|\phi_l(z)\|_{L_Q^0}|y-z|^{2H-2}dydz. \tag{2.53}$$

For $H<1/2$, the kernel

$$K_H(t,s)=b_H\left(\left(\frac{t}{s}\right)^{H-1/2}(t-s)^{H-1/2}-(H-1/2)s^{1/2-H}\int_s^t(u-s)^{H-1/2}u^{H-3/2}du\right), \tag{2.54}$$

with

$$b_H=\sqrt{\frac{2H}{(1-2H)\beta(1-2H,H+1/2)}},$$

and $t>s$ satisfy

$$R_H(t,s)=\int_0^{t\wedge s}K_H(t,u)K_H(s,u)du.$$

Consider the linear operator K_H^* from the space $\mathcal{E}$ of step functions on $[0,T]$ to $L^2([0,T])$ defined by

$$(K_H^*\phi)(s)=K(T,s)\phi(s)+\int_s^T(\phi(t)-\phi(s))\frac{\partial K_H}{\partial t}(t,s)dt. \tag{2.55}$$

Then (2.55) evaluated for $\phi = \chi_{[0,t]}$ gives

$$(K_H^* \chi_{[0,t]})(s) = K_H(t,s)\chi_{[0,t]}(s),$$

and for $H < 1/2$, we have

$$\mathcal{H} = (K_H^*)^{-1}(L^2([0,T]) = I_-^{1/2-H}(L^2([0,T])).$$

This representation of $\mathcal{H}$ guaranties, in addition, that the inner product space $\mathcal{H}$ is complete when endowed with the inner product:

$$\langle f, g\rangle = \int_0^T K_H^* f(s) K_H^* g(s) ds.$$

For $H = 1/2$, we have

$$K_{1/2}(t,s) = \chi_{[0,t]}(s).$$

Lemma 2.14. *Let*

$$I_-^{1/2-H} f(u) = c_H \int_u^\infty (t-u)^{H-3/2} f(t) dt,$$

where

$$c_H = \sqrt{H(2H-1)\Gamma(3/2-H)/\Gamma(H-1/2)\Gamma(2-2H)},$$

and Γ *denotes the gamma function. Then* $I_-^{1/2-H}$ *is an isometry from*

$$\overline{L([0,T],\langle,\rangle)} = \mathcal{H} = L_\phi^2(\mathbb{R}) \subseteq L_\phi^2(\mathbb{R}),$$

to $L^2(\mathbb{R})$.

Proof. By a limiting argument, we may assume that f and g are continuous with compact support. By definition,

$$\langle I_-^{1/2-H}(f), I_-^{1/2-H}(g)\rangle_{L^2(\mathbb{R})}.$$

Thus, we have

$$\begin{aligned} &= \langle I_-^{1/2-H}(f), I_-^{1/2-H}(g)\rangle_{L^2(\mathbb{R})} \\ &= c_H^2 \int_{\mathbb{R}} \left\{ \int_u^\infty (s-u)^{H-3/2} f(s) ds \int_u^\infty (s-u)^{H-3/2} g(t) dt \right\} du \end{aligned}$$

$$= c_H^2 \int_{\mathbb{R}^2} f(s)g(t) \left\{ \int_{-\infty}^{s \wedge t} (s-u)^{H-3/2}(t-u)^{H-3/2} du \right\} dsdt$$

$$= H(2H-1) \int_{\mathbb{R}^2} f(s)g(t)|s-t|^{2H-2},$$

where we have used the identity,

$$c_H^2 \int_{-\infty}^{s \wedge t} (s-u)^{H-3/2}(t-u)^{H-3/2} du = H(2H-1)|s-t|^{2H-2}.$$

□

3 Elements of functional analysis

Let $J = [a, b]$ be an interval of $\mathbb{R}$. Let $(E, |\cdot|)$ be a real Banach space. $C(J, E)$ is the Banach space of all continuous functions from J into E with the norm

$$\|y\|_\infty = \sup_{t\in J} |y(t)|.$$

$L^1([a, b], E)$ denotes the Banach space of measurable functions, and $y : [a, b] \longrightarrow E$ is Bochner integrable, normed by

$$\|y\|_{L^1} = \int_a^b |y(t)| dt,$$

if and only if $|y|$ is Lebesgue integrable.

Definition 3.1. A map f is said to be compact if the image is relatively compact. f is said to be completely continuous if it is continuous and the image of every bounded set is relatively compact.

Lemma 3.1.[1] *Let u, g: $J \to \mathbb{R}$ be positive real continuous functions. Assume there exist $c > 0$ and a continuous nondecreasing function h: $\mathbb{R} \to (0, +\infty)$ such that*

$$u(t) \leq c + \int_a^t g(s)h(u(s))\, ds, \quad \forall t \in J.$$

Then

$$u(t) \leq H^{-1}\left(\int_a^t g(s)\, ds\right), \quad \forall t \in J,$$

provided

$$\int_c^{+\infty} \frac{dy}{h(y)} > \int_a^b g(s)\, ds.$$

Here, H^{-1} refers to the inverse of the function

$$H(u) = \int_c^u \frac{dy}{h(y)} \quad \text{for } u > c.$$

1 I. Bihari, A generalization of a lemma of Bellman and its application to uniqueness problems of differential equations, *Acta Math. Acad. Sci. Hungar.* 7 (1956), 81–94.

https://doi.org/10.1515/9783112251898-003

3.1 Generalized metric and Banach spaces

If

$$x, y \in \mathbb{R}^n, \quad x = (x_1, \dots, x_n)$$
$$y = (y_1, \dots, y_n),$$

by $x \leq y$ we mean

$$x_i \leq y_i, \quad i = 1, \dots, n.$$

Also, we have

$$|x| = (|x_1|, \dots, |x_n|),$$
$$\max(x, y) = (\max(x_1, y_1), \dots, \max(x_n, y_n)),$$

and

$$\mathbb{R}_+^n = \{x \in \mathbb{R}^n : x_i > 0\}.$$

If $c \in \mathbb{R}$, then $x \leq c$ means $x_i \leq c$ for each $i = 1, \dots, n$.

Definition 3.2. Let X be a nonempty set. By a vector-valued metric on X, we mean a map $d : X \times X \to \mathbb{R}^n$ with the following properties:

1. $d(u, v) \geq 0$ for all $u, v \in X$; if $d(u, v) = 0$, then $u = v$.
2. $d(u, v) = d(v, u)$ for all $u, v \in X$.
3. We have

$$d(u, v) \leq d(u, w) + d(w, v),$$

for all $u, v, w \in X$.

We call the pair (X, d) a generalized metric space with

$$d(x, y) = \begin{pmatrix} d_1(x, y) \\ \cdots \\ d_n(x, y) \end{pmatrix}.$$

Notice that d is a generalized metric space on X if and only if d_i, $i = 1, \dots, n$ are metrics on X.

For

$$r = (r_1, \dots, r_n) \in \mathbb{R}_+^n,$$

we will denote by

$$\begin{aligned} B(x_0, r) &= \{x \in X : d(x_0, x) < r\} \\ &= \{x \in X : d_i(x_0, x) < r_i,\ i = 1, \ldots, n\}, \end{aligned}$$

the open ball centered in x_0 with radius r and

$$\begin{aligned} \overline{B(x_0, r)} &= \{x \in X : d(x_0, x) \le r\} \\ &= \{x \in X : d_i(x_0, x) \le r_i,\ i = 1, \ldots, n\}, \end{aligned}$$

the closed ball centered in x_0 with radius r. We mention that for generalized metric spaces, the notions of open subsets, closed sets, convergence, Cauchy sequences, and completeness are similar to those in usual metric spaces.

Let (X, d) be a generalized metric space; we define the following metric spaces: Let $X_i = X, i = 1, \ldots, n$. Consider $\prod_{i=1}^{n} X_i$ with $\bar{d}$:

$$\bar{d}((x_1, \ldots, x_n), (y_1, \ldots, y_n)) = \sum_{i=1}^{n} d_i(x_i, y_i).$$

The diagonal space of $\prod_{i=1}^{n} X_i$ is defined by

$$\widetilde{X} = \left\{(x, \ldots, x) \in \prod_{i=1}^{n} X_i : x \in X,\ i = 1, \ldots, n\right\}.$$

Thus, it is a metric space with the following distance:

$$d_*((x, \ldots, x), (y, \ldots, y)) = \sum_{i=1}^{n} d_i(x, y), \quad \text{for each } x, y \in X.$$

It is clear that $\widetilde{X}$ is a closed set in $\prod_{i=1}^{n} X_i$.

Intuitively, X and $\widetilde{X}$ are equivalent. This is shown in the following result.

Lemma 3.2. *Let (X, d) be a generalized metric space. Then there exists a $h : X \to \widetilde{X}$ homeomorphism map.*

Proof. Consider $h : X \to \widetilde{X}$ defined by

$$h(x) = (x, \ldots, x) \quad \text{for all } x \in X.$$

Obviously h is bijective.

– To prove that h is a continuous map.
  Let $x, y \in X$. Thus,

$$d_*(h(x), h(y)) \le \sum_{i=1}^{n} d_i(x, y).$$

For $\epsilon > 0$, we take $\delta = (\frac{\epsilon}{n}, \dots, \frac{\epsilon}{n})$; let fixed $x_0 \in X$ and

$$B(x_0, \delta) = \{x \in X : d(x_0, x) < \delta\},$$

then for every $x \in B(x_0, \delta)$, we have

$$d_*(h(x_0), h(x)) \leq \epsilon.$$

- Now, $h^{-1} : \widetilde{X} \to X$ is a continuous map defined by

$$h^{-1}(x, \dots, x) = x, \quad (x, \dots, x) \in \widetilde{X}.$$

To show that h^{-1} is continuous, let

$$(x, \dots, x), (y, \dots, y) \in \widetilde{X},$$

then

$$d(h^{-1}(x, \dots, x), h^{-1}(y, \dots, y)) = d(x, y).$$

Let

$$\epsilon = (\epsilon_1, \dots, \epsilon_n) > 0.$$

We take

$$\delta = \frac{\min_{1 \leq i \leq n} \epsilon_i}{n},$$

and fix

$$(x_0, \dots, x_0) \in \widetilde{X}.$$

Set

$$B((x_0, \dots, x_0), \delta) = \{(x, \dots, x) \in \widetilde{X} : d_*((x_0, \dots, x_0), (x, \dots, x)) < \delta\}.$$

For

$$(x, \dots, x) \in B((x_0, \dots, x_0), \delta),$$

we have

$$d_*((x_0, \dots, x_0), (x, \dots, x)) < \delta,$$

$$\Rightarrow$$

$$\sum_{i=1}^{n} d_i(x_0, x) < \frac{\min_{1\le i\le n} \epsilon_i}{n}.$$

Then

$$d_i(x_0, x) < \frac{\min_{1\le i\le n} \epsilon_i}{n}, \quad i = 1, \dots, n$$
$$\Rightarrow$$
$$d(x_0, x) < \epsilon.$$

Hence, h^{-1} is continuous. □

Definition 3.3. A square matrix of real numbers is said to be convergent to zero if and only if its spectral radius $\rho(M)$ is strictly less than 1. In other words, this means that all the eigenvalues of M are in the open unit disk, that is, $|\lambda| < 1$, for every $\lambda \in \mathbb{C}$ with

$$\det(M - \lambda I) = 0,$$

where I denotes the unit matrix of $\mathcal{M}_{n\times n}(\mathbb{R})$.

Definition 3.4. Let (X, d) be a generalized metric space. An operator $N : X \to X$ is said to be contractive if there exists a convergent matrix M that approaches zero such that

$$d(N(x), N(y)) \le Md(x, y) \quad \text{for all } x, y \in X.$$

For $n = 1$, we recover the classical Banach's contraction fixed-point result.

Definition 3.5. We say that a nonsingular matrix

$$A = (a_{ij})_{1\le i,j\le n} \in \mathcal{M}_{n\times n}\mathbb{R},$$

has the absolute value property if

$$A^{-1}|A| \le I,$$

where

$$|A| = (|a_{ij}|)_{1\le i,j\le n} \in \mathcal{M}_{n\times n}(\mathbb{R}_+).$$

Some examples of matrices converging to zero $A \in \mathcal{M}_{n\times n}(\mathbb{R})$, which also satisfy the property

$$(I - A)^{-1}|I - A| \le I,$$

are:

(1) $A = \begin{pmatrix} a & 0 \\ 0 & b \end{pmatrix}$, where $a, b \in \mathbb{R}_+$ and $\max(a, b) < 1$.
(2) $A = \begin{pmatrix} a & -c \\ 0 & b \end{pmatrix}$, where $a, b, c \in \mathbb{R}_+$ and $a + b < 1$, $c < 1$.
(3) $A = \begin{pmatrix} a & -a \\ b & -b \end{pmatrix}$, where $a, b, c \in \mathbb{R}_+$ and $|a - b| < 1$, $a > 1$, $b > 0$.

Definition 3.6. Let $Q \in \mathcal{M}_{2\times 2}(\mathbb{R})$ be said to be order preserving (or positive) if $p_1 \geq p_0$, $q_1 \geq q_0$ implies

$$Q\begin{pmatrix} p_0 \\ q_0 \end{pmatrix} \leq Q\begin{pmatrix} p_1 \\ q_1 \end{pmatrix}$$

in the sense of components.

Lemma 3.3.[2] *Let*

$$Q = \begin{pmatrix} a & -b \\ -c & d \end{pmatrix},$$

where $a, b, c, d \geq 0$ and $\det Q > 0$. Then Q^{-1} is order preserving.

Theorem 3.1.[3] *Let (X, d) be a complete generalized metric space and $N : X \to X$ a contractive operator with Lipschitz matrix M. Then N has a unique fixed point x_*, and for each $x_0 \in X$ we have*

$$d(N^k(x_0), x_*) \leq M^k(I - M)^{-1} d(x_0, N(x_0)) \quad \textit{for all } k \in \mathbb{N}.$$

As a consequence of Perov's theorem, we have the following result.

Theorem 3.2. *Let (X, d) be a generalized Banach space and $N : X \to X$ an M-contraction. Then $I_X - N$ is a homeomorphism, that is, $I_X - N$ bijective, and its inverse is continuous; $(I_X - N)^{-1}$ is continuous, too.*

Theorem 3.3 ([5]). *Let E be a generalized Banach space, $C \subset E$ be a nonempty closed convex subset of E, and $N : C \to C$ be a continuous operator with a relatively compact range. Then N has at least one fixed point in C.*

Theorem 3.4. *Let (X, d) be a generalized metric space. For any compact set $A \subset X$ and for any closed set $B \subset X$ that is disjoint from A, there exists a continuous function*

$$f : X \to [0, 1],$$
$$g : X \to [0, 1] \times [0, 1] \times \cdots [0, 1] = [0, 1]^n,$$

---

2 A. I. Perov, On the Cauchy problem for a system of ordinary differential equations, *Pviblizhen. Met. Reshen. Differ. Uvavn.* **2** (1964), 115–134.

3 A. I. Perov, On the Cauchy problem for a system of ordinary differential equations, *Pviblizhen. Met. Reshen. Differ. Uvavn.* **2** (1964), 115–134.

such that

- $f(x) = 0$ *for all* $x \in B$.
- $f(x) = 1$ *for all* $x \in A$.
- $g(x) = (1, \dots, 1)$ *for all* $x \in B$.
- $g(x) = (0, \dots, 0)$ *for all* $x \in A$.

Proof. Note that $d_i(x, b) = 0$ for any $x \in B$ and $d_i(x, a) = 0$ and $d_i(x, a) > 0$ for any $x \in A$. Thus, we obtain (1) and (2). Let $f : X \to [0, 1]$ be defined by

$$f(x) = \frac{\sum_{i=1}^n d_i(x, b)}{\sum_{i=1}^n d_i(x, a) + \sum_{i=1}^n d_i(x, b)}, \qquad x \in X.$$

To prove that f is continuous, let $(x_m)_{m \in \mathbb{N}}$ be a sequence converging to $x \in X$. Then

$$\begin{aligned}
|f(x_m) - f(x)| &= \left| \frac{\sum_{i=1}^n d_i(x_m, b)}{\sum_{i=1}^n d_i(x_m, a) + \sum_{i=1}^n d_i(x_m, b)} - \frac{\sum_{i=1}^n d_i(x, b)}{\sum_{i=1}^n d_i(x, a) + \sum_{i=1}^n d_i(x, b)} \right| \\
&= \left| \frac{\sum_{i=1}^n d_i(x_m, b) \sum_{i=1}^n d_i(x, a) - \sum_{i=1}^n d_i(x_m, a) \sum_{i=1}^n d_i(x, b)}{(\sum_{i=1}^n d_i(x, a) + \sum_{i=1}^n d_i(x, b))(\sum_{i=1}^n d_i(x_m, a) + \sum_{i=1}^n d_i(x_m, b))} \right| \\
&\leq \frac{\sum_{i=1}^n d_i(x, a) \sum_{i=1}^n |d_i(x_m, b) - d_i(x, b)|}{(\sum_{i=1}^n d_i(x, a) + \sum_{i=1}^n d_i(x, b))(\sum_{i=1}^n d_i(x_m, a) + \sum_{i=1}^n d_i(x_m, b))} \\
&\quad + \frac{\sum_{i=1}^n d_i(x, b) \sum_{i=1}^n |d_i(x_m, a) - d_i(x, a)|}{(\sum_{i=1}^n d_i(x, a) + \sum_{i=1}^n d_i(x, b))(\sum_{i=1}^n d_i(x_m, a) + \sum_{i=1}^n d_i(x_m, b))}.
\end{aligned}$$

Since for each $i = 1, \dots, m$, we have

$$|d_i(x_m, b) - d_i(x, b)| \to 0, \ |d_i(x_m, a) - d_i(x, a)| \to 0 \quad \text{as } m \to \infty.$$

Therefore,

$$\frac{\sum_{i=1}^n d_i(x, a) \sum_{i=1}^n |d_i(x_m, b) - d_i(x, b)|}{(\sum_{i=1}^n d_i(x, a) + \sum_{i=1}^n d_i(x, b))(\sum_{i=1}^n d_i(x_m, a) + \sum_{i=1}^n d_i(x_m, b))} \to 0 \quad \text{as } m \to \infty,$$

and

$$\frac{\sum_{i=1}^n d_i(x, a) \sum_{i=1}^n |d_i(x_m, a) - d_i(x, a)|}{(\sum_{i=1}^n d_i(x, a) + \sum_{i=1}^n d_i(x, b))(\sum_{i=1}^n d_i(x_m, a) + \sum_{i=1}^n d_i(x_m, b))} \to 0 \quad \text{as } m \to \infty.$$

Thus, we get

$$|f(x_m) - f(x)| \to 0 \quad \text{as } m \to \infty.$$

We can easily prove that the following function $g : X \to [0, 1]^n$ is defined by

$$g(x) = \begin{pmatrix} \frac{d_1(x,a)}{d_1(x,b)+d_1(x,b)} \\ \dots \\ \frac{d_n(x,a)}{d_n(x,b)+d_n(x,b)} \end{pmatrix}, \quad x \in X,$$

is a continuous function and satisfies (3) and (4). □

Definition 3.7. Let E be a vector space over $\mathbb{K} = \mathbb{R}$ or $\mathbb{C}$. By a vector-valued norm on E, we mean a map $\|\cdot\| : E \to \mathbb{R}^n_+$ with the following properties:
1. $\|x\| \geq 0$ for all $x \in E$; if $\|x\| = 0$, then $x = 0$.
2. $\|\lambda x\| = |\lambda|\|x\|$ for all $x \in E$ and $\lambda \in \mathbb{K}$.
3. $\|x + y\| \leq \|x\| + \|y\|$ for all $x, y \in E$.

The pair $(E, \|\cdot\|)$ is called a generalized normed space. If the generalized metric generated by $\|\cdot\|$ (i. e., $d(x,y) = \|x - y\|$) is complete, then the space $(E, \|\cdot\|)$ is called a generalized Banach space, where

$$\|x - y\| = \begin{pmatrix} \|x - y\|_1 \\ \dots \\ \|x - y\|_n \end{pmatrix}.$$

Notice that $\|\cdot\|$ is a generalized Banach space on E if and only if $\|\cdot\|_i$, $i = 1, \dots, n$ are norms on E.

Definition 3.8. Let $(E, \|\cdot\|)$ be a generalized Banach space and $U \subset E$ an open subset such that $0 \in U$. The function $p_U : E \to \mathbb{R}_+$ is defined by

$$p_U(x) = \inf\{a > 0 : x \in aU\},$$

is called the Minkowski functional of U.

Lemma 3.4. *Let $(E, \|\cdot\|)$ be a generalized Banach space and $U \subset E$ an open subset such that $0 \in U$. Then:*
1. *If $\lambda \geq 0$, then*

$$p_U(\lambda x) = \lambda p_U(x).$$

2. *If U is convex, we have*
   a) $p_U(x + y) \leq p_U(x) + p_U(y)$, *for every $x, y \in U$.*
   b) $\{x \in E : p_U(x) < 1\} \subset U \subset \{x \in E : p_U(x) \leq 1\}$.
   c) *If U is symmetric; then $p_U(x) = p_U(-x)$.*
3. *p_U is continuous.*

Proof. 1. Let $x \in E$ be arbitrary, and $\lambda \geq 0$. We have

$$p_U(\lambda x) = \inf\{a > 0 : \lambda x \in aU\}$$

$$
\begin{aligned}
&= \inf\{\alpha > 0 : x \in \lambda^{-1}\alpha U\} \\
&= \inf\{\lambda\beta > 0 : x \in \beta U\} \\
&= \lambda \inf\{\beta > 0 : x \in \beta U\} \\
&= \lambda p_U(x).
\end{aligned}
$$

2. (a) Let $\alpha_1 > 0$ and $\alpha_2 > 0$ be such that

$$x \in \alpha_1 U \quad \text{and} \quad y \in \alpha_2 U,$$

then

$$x + y \in \alpha_1 U + \alpha_2 U \quad \Rightarrow \quad \frac{x+y}{\alpha_1 + \alpha_2} \in \frac{\alpha_1}{\alpha_1 + \alpha_2} U + \frac{\alpha_2}{\alpha_1 + \alpha_2} U.$$

Hence

$$x + y \in (\alpha_1 + \alpha_2) U. \tag{3.1}$$

For every $\epsilon > 0$, there exist $\alpha_\epsilon > 0$, $\beta_\epsilon > 0$ such that

$$\alpha_\epsilon \le p_U(x) + \epsilon,$$

and

$$\beta_\epsilon \le p_U(y) + \epsilon.$$

From (3.1), we have

$$
\begin{gathered}
p_U(x+y) \le p_U(x) + p_U(y) + 2\epsilon \\
\Rightarrow \\
p_U(x+y) \le p_U(x) + p_U(y) + 2\epsilon.
\end{gathered}
$$

Letting $\epsilon \to 0$, we obtain

$$p_U(x+y) \le p_U(x) + p_U(y) \quad \text{for every } x, y \in U.$$

(b) Let $x \in E$ be such that $p_U(x) < 1$. Then there exists $\alpha \in (0,1)$ such that

$$p_U(x) \le \alpha < 1 \quad \text{and} \quad x \in \alpha U \quad \Rightarrow \quad x = \alpha a + (1-\alpha)0 \in U.$$

Therefore,

$$\{x \in E : p_U(x) < 1\} \subset U.$$

For $x \in U$, we have

$$x = \alpha x \in U, \quad \alpha = 1 \quad \Rightarrow \quad p_U(x) \le 1.$$

Then

$$\{x \in E : p_U(x) < 1\} \subset U \subset \{x \in E : p_U(x) \le 1\}.$$

3. Since $0 \in U$, there exist $r > 0$ such that

$$B(0, r) = \{x \in E : \|x\| < r_*\} \subset U,$$

where

$$\|x\| = \begin{pmatrix} \|x\|_1 \\ \dots \\ \|x\|_n \end{pmatrix},$$

and

$$r_* = \begin{pmatrix} r \\ \dots \\ r \end{pmatrix}.$$

Given $\epsilon > 0$, then $x + \epsilon B(0, r_*)$ is a neighborhood of x. For every $y \in x + \epsilon B(0, r_*)$, we have

$$\frac{x - y}{\epsilon} \in B(0, r_*) \quad \Rightarrow \quad p_U\left(\frac{x - y}{\epsilon}\right) \le 1.$$

It is clear that

$$|p_U(x) - p_U(y)| \le p_U(x - y) = \epsilon p_U\left(\frac{x - y}{\epsilon}\right) \le \epsilon.$$

Hence, p_U is continuous. □

Remark 3.1. In a generalized metric space in the sense of Perov's, the notation of convergence sequence, Cauchy sequence, completeness, open subset, and closed subset are similar to those for usual metric spaces.

Definition 3.9. Let E be a real generalized normed (or normed) space. A mapping $N : E \to E$ is called compact if T maps every bounded subset of E to a relatively compact subset in E. N is said to be completely continuous if N is continuous and compact.

Definition 3.10. A square matrix of real numbers is said to be convergent to zero if and only if its spectral radius $\rho(M)$ is strictly less than 1. In other words, this means that all the eigenvalues of M are in the open unit disk.

3.2 Compactness criteria

Let

$$C([a,b],E),$$
$$C_b([a,\infty),E),$$

and E be a real Banach space, and $[a,b]$ be an interval.

Definition 3.11. A family $\mathcal{A}$ in $C([a,b],E)$ is equicontinuous at t in $[a,b]$ if for each $\epsilon > 0$ there exists $\delta(\epsilon,t) > 0$ such that, for each $s \in [a,b]$ with $|t-s| < \delta(\epsilon,t)$, we have

$$\|f(t) - f(s)\| < \epsilon,$$

uniformly with respect to $f \in \mathcal{A}$.

The family $\mathcal{A}$ is equicontinuous on $[a,b]$ if it is equicontinuous at each point $t \in [a,b]$ in the sense mentioned above.

The family $\mathcal{A}$ is uniformly equicontinuous on $[a,b]$ if it is equicontinuous on $[a,b]$, and $\delta(\epsilon,t)$ can be chosen independently of t in $[a,b]$.

Theorem 3.5 (Arzelà–Ascoli).[4] *A bounded subset $\mathcal{A}$ in $C([a,b],E)$ is relatively compact if and only if item $\mathcal{A}$ is equicontinuous on $[a,b]$ and there exists a dense subset D in $[a,b]$ such that, for each $t \in D$,*

$$\mathcal{A}(t) = \{f(t) \mid f \in \mathcal{A}\},$$

is relatively compact in E.

The following compactness criterion for subsets of C_b is a consequence of the well-known Arzelà–Ascoli theorem.

Theorem 3.6. *Let $B \subset C([a,b],\mathbb{R}^n)$ be a subset; assume conditions are satisfied:*
For every $t \in \mathbb{R}^+$, the set

$$\{x(t) \mid x \in B\}$$

is relatively compact.

- *For every $a > 0$, the set B is equicontinuous on the interval $[0,a]$.*
- *For every $\epsilon > 0$, there exist $T = T(\epsilon)$ and $\delta = \delta(\epsilon)$ such that $x, y \in B$ with*

$$\|x(T) - y(T)\| \le \delta,$$

4 T. Yoshizawa, *Stability Theory by Liapunov's Second Method*, The Mathematical Society of Japan, Tokyo, 1966. **12** (2011), 2851–2860.

implies that

$$\|x(t) - y(t)\| \leq \epsilon,$$

for all $t \in [T, \infty)$. *Then the set* B *is compact in* $C_b([a, \infty), \mathbb{R}^n)$.

As a consequence, we have the following.

Corollary 3.1. *Let* $M \subset C_b$ *be the space of functions, which have limits at positive infinity. Then* M *is relatively compact in* C_b *if the following conditions hold:*
- M *is uniformly bounded in* C_b.
- *The functions belonging to* M *are almost equicontinuous on* $\mathbb{R}^+$, *that is, equicontinuous on every compact interval of* $\mathbb{R}^+$.
- *The functions from* M *are equiconvergent at* ∞, *that is, given* $\epsilon > 0$, *there corresponds* $T = T(\epsilon) > 0$ *such that*

$$\|x(t) - x(\infty)\| < \epsilon,$$

for any $t \geq T(\epsilon)$ *and* $x \in M$.

Lemma 3.5.[5] *Let* M *be a square matrix of nonnegative numbers. The following assertions are equivalent:*
1. M *is convergent towards zero;*
2. *The matrix* $I - M$ *is nonsingular and*

$$(I - M)^{-1} = I + M + M^2 + \cdots + M^k + \cdots;$$

3. $\|\lambda\| < 1$ *for every* $\lambda \in \mathbb{C}$ *with* $\det(M - \lambda I) = 0$;
4. $(I - M)$ *is nonsingular and* $(I - M)^{-1}$ *has nonnegative elements.*

Definition 3.12. We say that a nonsingular matrix

$$A = (a_{ij})_{1 \leq i,j \leq n} \in \mathcal{M}_{n \times n}(\mathbb{R})$$

has the absolute value property if

$$A^{-1}|A| \leq I,$$

where

$$|A| = (|a_{ij}|)_{1 \leq i,j \leq n} \in \mathcal{M}_{n \times n}(\mathbb{R}_+).$$

5 I. A. Rus, The theory of a metrical fixed-point theorem: Theoretical and applicative relevance, *Fixed-Point Theory* **9** (2008), 541–559.

Some examples of matrices that converge to zero are as follows:

1) $A = \begin{pmatrix} a & 0 \\ 0 & b \end{pmatrix}$, where $a, b \in \mathbb{R}_+$ and $\max(a, b) < 1$.
2) $A = \begin{pmatrix} a & -c \\ 0 & b \end{pmatrix}$, where $a, b, c \in \mathbb{R}_+$ and $a + b < 1$, $c < 1$.
3) $A = \begin{pmatrix} a & -a \\ b & -b \end{pmatrix}$, where $a, b, c \in \mathbb{R}_+$ and $|a - b| < 1$, $a > 1$, $b > 0$.

3.3 Measures of noncompactness

First, we collect some definitions and properties of measures of noncompactness in Banach spaces.

Definition 3.13. Let X be a generalized Banach space and $(\mathcal{A}, \leq)$ a partially ordered set. A map

$$\beta : \mathcal{P}(X) \to \mathcal{A} \times \mathcal{A} \cdots \times \mathcal{A},$$

is called a generalized measure of noncompactness on X if

$$\beta(\overline{co}\Omega) = \beta(\Omega) \quad \text{for every } \Omega \in \mathcal{P}(X),$$

where

$$\beta(\Omega) = \begin{pmatrix} \beta_1(\Omega) \\ \vdots \\ \beta_n(\Omega) \end{pmatrix}.$$

Definition 3.14. A measure of noncompactness β is called:

1. Monotone if $\Omega_0, \Omega_1 \in \mathcal{P}(X)$, $\Omega_0 \subset \Omega_1$ implies
$$\beta(\Omega_0) \leq \beta(\Omega_1).$$
2. Nonsingular if
$$\beta(\{a\} \cup \Omega) = \beta(\Omega),$$
for every $a \in X$, $\Omega \in \mathcal{P}(X)$.
3. Invariant with respect to the union with compact sets if
$$\beta(K \cup \Omega) = \beta(\Omega),$$
for every relatively compact set $K \subset X$, and $\Omega \in \mathcal{P}(X)$.
4. Real if $\mathcal{A} = \overline{R}_+$ and $\beta(\Omega) < \infty$ for every $i = 1, \ldots, n$ and every bounded Ω.
5. Semiadditive if
$$\beta(\Omega_0 \cup \Omega_1) = \max(\beta(\Omega_0), \beta(\Omega_1)),$$

for every $\Omega_0, \Omega_1 \in \mathcal{P}(X)$.

6. Lower-additive if β is real and

$$\beta(\Omega_0 + \Omega_1) \leq \beta(\Omega_0) + \beta(\Omega_1),$$

for every $\Omega_0, \Omega_1 \in \mathcal{P}(X)$.

7. Regular if the condition $\beta(\Omega) = 0$ is equivalent to the relative compactness of Ω.

Definition 3.15. Let X, Y be two generalized normed spaces and $F : X \to \mathcal{P}(Y)$ be a multivalued map. F is called a M-contraction (with respect to β) if there exists $M \in M_{n\times n}(\mathbb{R})$ converging to zero such that for every $D \in \mathcal{P}(X)$, we have

$$\beta(F(D)) \leq M\beta(D).$$

The next result is concerned with β-condensing or M-contractive.

Theorem 3.7. *Let $V \subset X$ be a bounded, closed, convex subset and $N : V \to V$ a generalized β-condensing continuous mapping, where β is a nonsingular measure of noncompactness defined on subsets of X. Then the set*

$$Fix(N) = \{x \in V : x \in N(x)\},$$

is nonempty.

Proof. Let

$$\mathcal{M}_1 = V,$$
$$\mathcal{M}_{k+1} = \overline{co}N(\mathcal{M}_k), \quad k \in \mathbb{N}.$$

It is clear that the sequence $(\mathcal{M}_k)_{k\in\mathbb{N}}$ consists of nonempty closed convex decreasing subsets of V. Since N is β-condensing, then

$$\beta(\overline{co}N(\mathcal{M}_1)) = \beta(N(\mathcal{M}_1)) \leq M\beta(V).$$

We continue this process and we obtain

$$\beta(\mathcal{M}_{k+1}) \leq M^{k+1}\beta(V).$$

Therefore,

$$\lim_{k\to\infty} \beta(\mathcal{M}_k) = 0.$$

Thus,

$$C = \bigcap_{k=1}^{\infty} \mathcal{M}_k \neq \emptyset$$

is convex and compact. Furthermore, by the convexity of $\{\mathcal{M}_k\}_{k\in\mathbb{N}}$ and $N(\mathcal{M}_1) \subseteq \mathcal{M}_1$, we have

$$N(\mathcal{M}_2) \subseteq \mathcal{M}_2 \quad \Rightarrow \quad N(\mathcal{M}_3) \subseteq \mathcal{M}_3.$$

Proceeding by induction, we obtain

$$N(\mathcal{M}_k) \subseteq \mathcal{M}_k \quad \text{for every } k \in \mathbb{N} \Rightarrow N(C) \subseteq C.$$

Hence, we have N has at least one fixed point. □

3.4 Fixed-point theory

In this section, we present the random versions of the Perov, Schauder, and Krasnosel'skii fixed-point theorems in generalized metric spaces.

Theorem 3.8. *Let $(\Omega, \mathcal{F})$ be a measurable space, let X be a real separable generalized Banach space, and let $F : \Omega \times X \to X$ be a continuous random operator. Let*

$$M(\omega) \in \mathcal{M}_{n\times n}(\mathbb{R}^+),$$

be a random variable matrix such that for every $\omega \in \Omega$, the matrix $M(\omega)$ converges to 0 and

$$d(F(\omega, x_1), F(\omega, x_2)) \leq M(\omega)d(x_1, x_2) \quad \text{for each } x_1, x_2 \in X, \omega \in \Omega.$$

Then there exists a random variable $x : \Omega \to X$ that is the unique random fixed point of F.

Theorem 3.9. *Let $(\Omega, \mathcal{F}, \mu)$ be a probability space, let X be a real separable generalized Banach space, and let $F : \Omega \times X \to X$ be a continuous random operator. Let*

$$M(\omega) \in \mathcal{M}_{n\times n}(\mathbb{R}^+),$$

be a nonnegative real matrix random variable such that $\rho(M(\omega)) < 1$ and

$$\|F(\omega, x_1) - F(\omega, x_2)\| \leq M(\omega)\|x_1 - x_2\| \quad \text{for each } x_1, x_2 \in X, \omega \in \Omega.$$

Then there exists a random variable $x : \Omega \times \mathbb{N}(0, b) \to X$, which is the unique random fixed point of F.

3.5 Some properties of set-valued maps

Let (X, d) be a metric space and Y be a subset of X. We denote

$$\begin{aligned}\mathcal{P}_{cl}(X) &= \{y \in \mathcal{P}(X) : y \text{ closed}\},\\ \mathcal{P}_b(X) &= \{y \in \mathcal{P}(X) : y \text{ bounded}\},\\ \mathcal{P}_c(X) &= \{y \in \mathcal{P}(X) : y \text{ convex}\},\\ \mathcal{P}_{cp}(X) &= \{y \in \mathcal{P}(X) : y \text{ compact}\}.\end{aligned}$$

Consider

$$H_d : \mathcal{P}(X) \times \mathcal{P}(X) \longrightarrow \mathbb{R}_+^n \cup \{\infty\},$$

defined by

$$H_d(A, B) = \begin{pmatrix} H_{d_1}(A, B) \\ \dots \\ H_{d_n}(A, B) \end{pmatrix}.$$

Let (X, d) be a generalized metric space with

$$d(x, y) = \begin{pmatrix} d_1(x, y) \\ \dots \\ d_n(x, y) \end{pmatrix}.$$

Notice that d is a generalized metric space on X if and only if d_i, $i = 1, \dots, n$ are metrics on X,

$$H_d(A, B) = \max\Big\{\sup_{a \in A} d(a, B), \sup_{b \in B} d(A, b)\Big\},$$

where

$$d(A, b) = \inf_{a \in A} d(a, b), d(a, B) = \inf_{b \in B} d(a, b).$$

Then $(\mathcal{P}_{b,cl}(X), H_d)$ is a metric space and $(\mathcal{P}_{cl}(X), H_d)$ is a generalized metric space.

A multivalued map $F : X \longrightarrow \mathcal{P}(X)$ is convex (closed) valued if $F(y)$ is convex (closed) for all $y \in X$, F is bounded on bounded sets if

$$F(2) = \bigcup_{y \in B} F(y),$$

is bounded in X for all $B \in \mathcal{P}_b(X)$. F is called upper semicontinuous (u. s. c. for short) on X if, for each $y_0 \in X$, the set $F(y_0)$ is a nonempty subset of X, and for each open set $\mathcal{U}$ of X containing $F(y_0)$, there exists an open neighborhood $\mathcal{V}$ of y_0 such that $F(\mathcal{V}) \in \mathcal{U}$. F is said to be completely continuous if $F(2)$ is relatively compact for every $B \in \mathcal{P}_b(X)$. F is quasicompact if, for each subset $A \subset X$, $F(1)$ is relatively compact.

If the multivalued map F is completely continuous with nonempty compact values, then F is $u.s.c.$ if and only if F has a closed graph, that is, $x_n \longrightarrow x_*, y_n \longrightarrow y_*, y_n \in F(x_n)$ imply $y_* \in F(x_*)$.

A A multivalued map

$$F : J \longrightarrow \mathcal{P}_{cp,c}(X)$$

is said to be measurable if, for each $y \in X$, the mean-square distance between y and $F(t)$ is measurable.

Definition 3.16. The set-valued map

$$F_1, F_2 : J \times X \times X \to \mathcal{P}(X)$$

is said to be L^2-Carathéodory if:
1. $t \mapsto F(t, v)$ is measurable for each $v \in X \times X$;
2. $v \mapsto F(t, v)$ is $u.s.c.$ for almost all $t \in J$;
3. For each $q > 0$, there exists $h_q \in L^1(J, \mathbb{R}^+)$ such that

$$\|F(t,v)\|^2 = \sup_{f \in F(t,v)} \|f\|^2 \le h_q(t), \quad \text{for all } \|v\|^2 \le q \text{ and for a. e. } t \in J.$$

Remark 3.2. 1. For each $x \in C(J, X)$, the set $S_{F,x}$ is closed whenever F has closed values. It is convex if and only if $F(t, x(t))$ is convex for a. e., $t \in J$.
2. From Theorem 5.10,[6] we know that $S_{F,x}$ is nonempty if and only if the mapping

$$t \longmapsto \inf\{\|u\| : u \in F(t, x(t))\},$$

belongs to $L^2(J)$.

Lemma 3.6 ([11]). *Let J be a compact interval and X be a Hilbert space. Let F be an L^2-Carathéodory multivalued map with $S_{F,y} \neq \emptyset$, and let Γ be a linear continuous mapping from $L^2(J, X)$ to $C(J, X)$. Then the operator*

$$\begin{aligned} \Gamma \circ S_F : C(J,X) &\longrightarrow \mathcal{P}_{cp,c}(L^2(J,X)), \\ y &\longmapsto (\Gamma \circ S_F)(y) = \Gamma(S_F, y), \end{aligned}$$

is a closed graph operator in

6 D. Wagner, Survey of measurable selection theorems, *SIAM J. Control Optim.* **15** (1977), 859–903.

$$C(J,X) \times C(J,X),$$

where $S_{F,y}$ is known as the selectors set from F and is given by

$$f \in S_{F,y} = \{f \in L^2([0,T],X) : f(t) \in F(t,y) \text{ for a. e. } t \in J\}.$$

Definition 3.17. The map

$$f^i : J \times \mathcal{D}_{\mathcal{F}_0} \times \mathcal{D}_{\mathcal{F}_0} \to X,$$

is said to be L^2-Carathéodory if:
1. $t \mapsto f^i(t,x,y)$ is measurable for each $x, y \in \mathcal{D}_{\mathcal{F}_0}$;
2. $(x,y) \mapsto f^i(t,x,y)$ is continuous for almost all $t \in J$;
3. For each $q_1, q_2 > 0$, there exists $\alpha_q \in L^1(J, \mathbb{R}^+)$ such that

$$E|f^i(t,x,y)|^2 \le \alpha_q(t), \quad \text{for all } \|x\|^2_{\mathcal{D}_{\mathcal{F}_0}} \le q_1,\ \|y\|^2_{\mathcal{D}_{\mathcal{F}_0}} \le q_2, \text{ and for a. e. } t \in J.$$

Definition 3.18. A map $f : J \times E \to E$ is said to be L^p-Carathéodory if:
1. $t \mapsto f(t,v)$ is measurable for each $v \in E$;
2. $v \mapsto f(t,v)$ is continuous for almost all $t \in J$;
3. For each $q > 0$, there exists $\alpha_q \in L^p(J, \mathbb{R}^+)$ such that

$$\|f(t,v)\|^p \le \alpha_q(t), \quad \text{for all } \|v\|^p_E \le q \text{ and for a. e. } t \in J.$$

We denote the graph of G to be the set

$$gr(G) = \{(x,y) \in X \times Y,\ y \in G(x)\}.$$

Lemma 3.7 ([7]). *If*

$$G : X \to \mathcal{P}_{cl}(Y)$$

is u. s. c., then $gr(G)$ is a closed subset of $X \times Y$. Conversely, if G is locally compact and has nonempty compact values and a closed graph, then it is u. s. c.

Lemma 3.8. *If*

$$G : X \to \mathcal{P}_{cp}(Y),$$

is quasicompact and has a closed graph, then G is u. s. c.

The following two results are easily deduced from the limit properties

Lemma 3.9 ([1, Theorem 1.4.13]). *If*

$$G : X \to \mathcal{P}_{cp}(X),$$

is u. s. c., then for any $x_0 \in X$,

$$\limsup_{x \to x_0} G(x) = G(x_0).$$

Lemma 3.10.[7] *Let*

$$(K_n)_{n \in N} \subset K \subset X,$$

be a sequence of subsets where K is compact in the separable Banach space X. Then

$$\overline{co}\Big(\limsup_{n \to \infty} K_n\Big) = \bigcap_{N>0} \overline{co}\big(\bigcup_{n \geq N} K_n\big),$$

where $\overline{co}A$ *refers to the closure of the convex hull of A.*

Lemma 3.11.[8] *Let X be a normed space and*

$$\{x_k\}_{k \in N} \subset X,$$

be a sequence that weakly converges to a limit $x \in X$. *Then there exists a sequence of convex combinations*

$$y_m = \sum_{k=1}^{m} \alpha_{mk} x_k,$$

with $\alpha_{mk} > 0$ *for* $k = 1, 2, \ldots, m$ *and*

$$\sum_{k=1}^{m} \alpha_{mk} = 1,$$

that converges strongly to x.

Definition 3.19. A sequence $(v_n)_{n \in \mathbb{N}}$ is said to be semicompact if:

1. It is integrable and bounded, that is, there exists $q \in L^1(J, \mathbb{R})$ such that

$$|v_n|_X \leq q(t),$$

for a. e. $t \in J$ and every $n \in \mathbb{N}$,

---

7 J. P. Aubin, H. Frankowska, *Set-Valued Analysis*, Birkhauser, Boston, 1990.

8 L. Górniewicz, *Topological Fixed-Point Theory of Multivalued Mappings*, Mathematics and its Applications, 495, Kluwer Academic Publishers, Dordrecht, 1999.

2. The image sequence $(v_n)_{n\in\mathbb{N}}$ is relatively compact in X for a. e. $t \in J$.

Our next result describes a basic theorem of reflexive spaces.

Theorem 3.10. *E is reflexive if and only if*

$$B_E = \{x \in E; \|x\| \leq 1\},$$

is compact in the weak topology.

This result is of particular importance if X is reflexive, in which case (1) implies (2) in Definition 3.19.

Lemma 3.12. *Every semicompact sequence $L^1(J,X)$ is weakly compact in $L^1(J,X)$.*

Definition 3.20. A multivalued operator

$$N : X \to \mathcal{P}_{cl}(X),$$

is called:
(a) γ-Lipschitz if and only if there exists $\gamma > 0$ such that

$$H_d(N(x), N(y)) \leq \gamma d(x,y), \quad \text{for each } x,\ y \in X.$$

(b) A contraction is defined if and only if it satisfies γ-Lipschitz with $\gamma < 1$.

Let $A : E \to E$ be a linear operator.

Definition 3.21. The resolvent set $\Lambda(1)$ of A consists of all complex numbers λ for which the linear operator $\lambda I - A$ is invertible, that is, $(\lambda I - A)^{-1}$ is a bounded linear operator in E: The family

$$R(\lambda, A) = (\lambda I - A)^{-1}, \quad \lambda \in \Lambda(1)$$

is called the resolvent of A. All complex numbers λ not in $\Lambda(1)$ form a set called the spectrum of A.

3.6 Iterative results

Theorem 3.11.[9] *Let (X, d) be a complete generalized metric space with $d : X \times X \longrightarrow \mathbb{R}^n$ and let $N : X \longrightarrow X$ be such that*

9 A. I. Perov, On the Cauchy problem for a system of ordinary differential equations, *Pviblizhen. Met. Reshen. Differ. Uvavn.* **2** (1964), 115–134.

$$d(N(x), N(y)) \le Md(x, y),$$

for all $x, y \in X$ and some square matrix M of nonnegative numbers. If the matrix M is convergent to zero, that is, $M^k \longrightarrow 0$ as $k \longrightarrow \infty$, then N has a unique fixed point $x_* \in X$,

$$d(N^k(x_0), x_*) \le M^k (I - M)^{-1} d(N(x_0), x_0),$$

for every $x_0 \in X$ and $k \ge 1$.

Theorem 3.12 ([5]). *Let E be a generalized Banach space, $C \subset E$ be a nonempty closed convex subset of E, and $N : C \to C$ be a continuous operator with a relatively compact range. Then N has at least one fixed point in C.*

As a consequence of the Schauder fixed-point theorem, we present the version of Schaefer's fixed-point theorem and the nonlinear alternative Leary–Schauder-type theorem in a generalized Banach space.

Theorem 3.13. *Let $(E, \|\cdot\|)$ be a generalized Banach space and $N : E \to E$ be a continuous compact mapping. Moreover, assume that the set*

$$\mathcal{A} = \{x \in E : x = \lambda N(x) \text{ for some } \lambda \in (0, 1)\}$$

is bounded. Then N has a fixed point.

Proof. Let $K > 0$ such that

$$\sum_{i=1}^{n} \|x\|_i < nK \quad \text{for each } x \in \mathcal{A}.$$

Set

$$M_* = (nK, \ldots, nK),$$

and we define

$$N_* : \overline{B(0, M_*)} \to \overline{B(0, M_*)},$$

by

$$N_*(x) = \begin{cases} N(x) & \text{if } \sum_{i=1}^{n} \|N(x)\|_i \le nK, \\ \frac{KnN(x)}{\sum_{i=1}^{n} \|N(x)\|_i} & \text{if } \sum_{i=1}^{n} \|N(x)\|_i > nK. \end{cases}$$

We show that N_* is continuous.

Let

$$x \in \overline{B(0, M_*)},$$

such that

$$\sum_{i=1}^{n} \|N(x)\|_i < nK,$$

then $N_*(x) = N(x)$.

If

$$(x_m)_{m\in\mathbb{N}} \in \overline{B(0, M_*)},$$

and

$$\sum_{i=1}^{n} \|N(x_m)\|_i \le M_*,$$

then the continuity of N implies that

$$\|N(x_m) - N(x)\| \to 0 \quad \text{as } m \to \infty \quad \Rightarrow \quad \|N_*(x_n) - N_*(x)\| \to 0 \quad \text{as } m \to \infty.$$

Now let

$$x \in \overline{B(0, M_*)},$$

such that

$$\sum_{i=1}^{n} \|N(x)\|_i > nK.$$

Then

$$N_*(x) = \frac{KnN(x)}{\sum_{i=1}^{n} \|N(x)\|_i}.$$

If

$$(x_m)_{m\in\mathbb{N}} \in \overline{B(0, M_*)},$$

and

$$\sum_{i=1}^{n} \|N(x_m)\|_i > nK,$$

then

$$N_*(x_m) = \frac{KnN(x_m)}{\sum_{i=1}^{n} \|N(x_m)\|_i}.$$

By the continuity of N we have, for every $j = 1, \dots, n$,

$$\begin{aligned}\|N_*(x_m) - N_*(x)\|_j &= \left\| \frac{KnN(x_m)}{\sum_{i=1}^n \|N(x_m)\|_i} - \frac{KnN(x)}{\sum_{i=1}^n \|N(x)\|_i} \right\|_j \\ &= \left\| \frac{KnN(x_m) \sum_{i=1}^n \|N(x)\|_i - KnN(x) \sum_{i=1}^n \|N(x_m)\|_i}{\sum_{i=1}^n \|N(x_m)\|_i \sum_{i=1}^n \|N(x)\|_i} \right\|_j \\ &\le \frac{Kn\|N(x_m) - N(x)\|_j}{\sum_{i=1}^n \|N(x_m)\|_i} + \frac{Kn\|N(x)\|_j \sum_{i=1}^n \|N(x_m) - N(x)\|_i}{\sum_{i=1}^n \|N(x_m)\|_i \sum_{i=1}^n \|N(x)\|_i}.\end{aligned}$$

Since

$$\sum_{i=1}^n \|N(x_m)\|_i > nM,$$

thus

$$\lim_{m\to\infty} \sum_{i=1}^n \|N(x_m)\|_i \ge nK,$$

and hence

$$\|N_*(x_m) - N_*(x)\| \to 0 \quad \text{as } m \to \infty.$$

Let

$$x \in \overline{B(0, M_*)},$$

such that

$$\sum_{i=1}^n \|N(x)\|_i = nK,$$

then

$$N_*(x) = N(x).$$

If

$$(x_m)_{m\in\mathbb{N}} \in \overline{B(0, M_*)},$$

and

$$\sum_{i=1}^n \|N(x_m)\|_i \le nK,$$

then the continuity of N implies that

$$\|N(x_m) - N(x)\| \to 0 \quad \text{as } m \to \infty \quad \Rightarrow \quad \|N_*(x_m) - N_*(x)\| \to 0 \quad \text{as } m \to \infty.$$

If

$$(x_m)_{m\in\mathbb{N}} \in \overline{B(0, M_*)},$$

and

$$\sum_{i=1}^{n} \|N(x_m)\|_i > M_*,$$

then

$$N_*(x_m) = \frac{KnN(x_m)}{\sum_{i=1}^{n} \|N(x_m)\|_i};$$

hence

$$\begin{aligned}
\|N_*(x_m) - N_*(x)\|_j &= \left\| \frac{KnN(x_m)}{\sum_{i=1}^{n} \|N(x_m)\|_i} - N(x) \right\|_j \\
&= \left\| \frac{KnN(x_m) - N(x)\sum_{i=1}^{n} \|N(x_m)\|_i}{\sum_{i=1}^{n} \|N(x_m)\|_i} \right\|_j \\
&\leq \frac{Kn\|N(x_m) - N(x)\|_j}{\sum_{i=1}^{n} \|N(x_m)\|_i} + \frac{\|N(x)\|_j |Kn - \sum_{i=1}^{n} \|N(x_m)\|_i|}{\sum_{i=1}^{n} \|N(x_m)\|_i}.
\end{aligned}$$

It is clear that

$$\lim_{m\to\infty} \sum_{i=1}^{n} \|N(x_m)\|_i = \sum_{i=1}^{n} \|N(x)\|_i = Kn.$$

Therefore,

$$\|N_*(x_m) - N_*(x)\|_j \to 0 \quad \text{as } m \to \infty.$$

Thus, we conclude that N_* is continuous. Consider the following map:

$$\rho : \overline{B(0, M_*)} \to \overline{B(0, M_*)},$$

defined by

$$\rho(x) = \begin{cases} x & \text{if } \sum_{i=1}^{n} \|x\|_i \leq nK, \\ \frac{Knx}{\sum_{i=1}^{n} \|x\|_i} & \text{if } \sum_{i=1}^{n} \|x\|_i > nK. \end{cases}$$

It is evident that ρ is continuous and $N_* = \rho \circ N$. The compactness of N implies that N_* is compact. By Theorem 3.12, there exists $x \in \overline{B(0,M_*)}$ such that $x = N_*(x)$. Notice that

$$\sum_{i=1}^{n} \|N(x)\|_i \leq Kn,$$

for otherwise,

$$x = \lambda N(x), \quad \lambda = \frac{Kn}{\sum_{i=1}^{n} \|N(x)\|_i} \quad \text{with } 0 < \lambda < 1 \Rightarrow x \in \mathcal{A}.$$

This implies that

$$\sum_{i=1}^{n} \|x\|_i < Kn,$$

but

$$x = \frac{KnN(x)}{\sum_i^n \|N(x)\|_i} \quad \Rightarrow \quad \sum_{i=1}^{n} \|x\|_i = Kn.$$

This yields a contradiction with $x \in \mathcal{A}$. Hence, we get that

$$x = N_*(x) = N(x). \qquad \square$$

Next, we state the nonlinear alternative of the Leray–Schauder type.

Lemma 3.13. *Let X be a generalized Banach space, $U \subset E$ be a bounded, convex open neighborhood of zero, and let $G : \overline{U} \to E$ be a continuous compact map. If G satisfies the boundary condition*

$$x \neq \lambda G(x),$$

for all $x \in \partial U$ and $0 \leq \lambda \leq 1$, then the set

$$Fix(G) = \{x \in U : x = G(x)\},$$

is nonempty.

Proof. Let p be the Minkowski function of U, and since $\overline{U}$ is bounded, there exists $M > 0$ such that

$$G(\overline{U}) \subseteq \frac{1}{2} B(0, M_*), \quad M_* = (K, \dots, K).$$

Consider

$$G_* : \overline{B(0, M_*)} \to \overline{B(0, M_*)},$$

defined by

$$G_*(x) = \begin{cases} G(x) & \text{if } x \in \overline{U}, \\ \frac{1}{p(x)} G(\frac{x}{p(x)}) & \text{if } x \in E \backslash \overline{U}. \end{cases}$$

It is clear that $\overline{B(0, M_*)}$ is a closed, convex, bounded subset of E, and G_* is a continuous compact operator. Then from Theorem 3.12, there exists $x \in \overline{B(0, M_*)}$ such that $\overline{G}(x) = x$. If $x \in E \backslash \overline{U}$, then

$$x = \frac{G(\frac{x}{p(x)})}{p(x)} \quad \Rightarrow \quad \frac{x}{p(x)} = \frac{1}{p^2(x)} G\left(\frac{x}{p(x)}\right).$$

Since $x \in E \backslash \overline{U}$, then

$$p(x) = 1 \text{ or } p(x) > 1 \quad \Rightarrow \quad x \in \partial U, \ \frac{x}{p(x)} \in \partial U.$$

This is a contradiction with

$$z \neq \lambda G(z), \quad \text{for each, } \lambda \in [0, 1], \ z \in \partial U.$$

Consequently, there exists $x_* \in U$ such that $G(x_*) = x_*$. □

Theorem 3.14. *Let $(E, \|\cdot\|)$ be a Banach space, $C \subset E$ a closed convex subset, $U \subset C$ a bounded set, open (with respect to the topology C), and such that $0 \in U$. Let $G : \overline{U} \to C$ be a compact continuous mapping. If the following assumption is satisfied:*

$$x \neq \lambda G(x), \quad \textit{for all } x \in \partial_C U \textit{ and all } \lambda \in (0, 1),$$

then f has a fixed point in U.

Proof. Let

$$C_* = \{x \in \overline{U} : x = \lambda G(x) \text{ for some } \lambda \in [0, 1]\}.$$

Since $0 \in U$ then C_* is a nonempty set, and by the continuity of G, we conclude that C_* is closed. It is clear that

$$\partial_C U \cap C_* = \emptyset.$$

From Theorem 3.4, there exists

$$f : \overline{U} \to [0, 1],$$

such that

$$f(x) = \begin{cases} 0 & \text{if } x \in \partial_C U, \\ 1 & \text{if } x \in C_*. \end{cases}$$

Consider $G_* : C \to C$ defined by

$$G_*(x) = \begin{cases} f(x)G(x) & \text{if } x \in U, \\ 0 & \text{if } x \in C \backslash U. \end{cases}$$

Since $G_*(x) = 0$, for each $x \in \partial_C U$, and G_* are continuous on $U, E \backslash U$, then G_* is continuous.

The set

$$\Omega = \overline{co}(\{0\} \cup G(\overline{U})),$$

is convex and compact. We can easily prove that

$$G_*(\Omega) \subset \Omega.$$

Then from Theorem 3.12, there exists $x \in \Omega$ such that $G_*(x) = x$. From the definition of G_*, we have $G(x) = x$. □

From the above theorem, we obtain the following.

Theorem 3.15. *Let $C \subset E$ be a closed convex subset and $U \subset C$ a bounded open neighborhood of zero (with respect to the topology of C). If $G : \overline{U} \to E$ is compact and continuous, then:*
1. *Either G has a fixed point in $\overline{U}$, or*
2. *There exists $x \in \partial U$ such that $x = \lambda G(x)$ or some $\lambda \in (0, 1)$.*

By the above lemma, we can easily prove the following so-called nonlinear alternatives of Leray and Schauder.

Lemma 3.14. *Let $(X, \|\cdot\|)$ be a generalized Banach space with $C \subset X$ a closed and convex subset of X. Assume U is an open subset of C, with $0 \in U$, and let $G : \overline{U} \longrightarrow C$ be a compact map. Then either:*
1. *G has a fixed point in $\overline{U}$, or*
2. *There is a point $u \in \partial U$ and $\lambda \in (0, 1)$, with $u \in \lambda G(u)$.*

The single-valued version may be stated as follows.

Lemma 3.15 ([8]). *Let $(X, \|\cdot\|)$ be a generalized Banach space, and*

$$G : X \longrightarrow \mathcal{P}_{cl,cv}(X),$$

be an upper semicontinuous and compact map. Then either:

1. *F has at least one fixed point, or*
2. *The set*

$$\mathcal{M} = \{x \in X \text{ and } \lambda \in (0,1), \text{ with } x \in \lambda G(u)\},$$

is unbounded.

Theorem 3.16. *Let (X,d) be a generalized complete metric space, and let $F : X \to \mathcal{P}_{cl}(X)$ be a multivalued map. Assume that there exists*

$$A, b, C \in \mathcal{M}_{n\times n}(\mathbb{R}_+),$$

such that

$$H_d(F(x), F(y)) \le Ad(x,y) + Bd(y, F(x)) + Cd(x, F(x)), \tag{3.2}$$

where $A + C$ converges to zero. Then there exist $x \in X$ such that $x \in F(x)$.

3.7 Semigroups

Let X be a Banach space and let $\mathcal{L}(X)$ be the set of all linear bounded operators from X to X. Endowed with the operator norm $\|.\|_{\mathcal{L}(X)}$, defined by

$$\|U\|_{\mathcal{L}(X)} = \sup_{\|x\|\le 1} \|Ux\|,$$

for each

$$U \in \mathcal{L}(X), \mathcal{L}(X),$$

is a Banach space.

Definition 3.22. A semigroup is a one-parameter family

$$\{S(t) : t \ge 0\} \subset \mathcal{L}(X),$$

satisfying the conditions:
1. $S(t) \circ S(s) = S(t+s)$, for $t, s \ge 0$.
2. $S(0) = I$ where I denotes the identity operator in X.

Definition 3.23. A semigroup $S(t)$ is uniformly continuous if

$$\lim_{t\to 0^+} \|S(t) - I\|_{\mathcal{L}(X)},$$

that is, if

$$\lim_{|t-s|\to 0^+} \|S(t) - S(s)\|_{\mathcal{L}(X)} = 0.$$

Definition 3.24. We say that the semigroup $\{S(t)\}_{t\geq 0}$ is strongly continuous (or a C_0-semigroup) if the map $t \to S(t)(x)$ is strongly continuous for each $x \in E$, that is,

$$\lim_{t\to 0^+} S(t)x = S(0)x \quad \text{for } x \in E.$$

Definition 3.25. Let $S(t)$ be a C_0-semigroup defined on X. The infinitesimal generator A in the linear operator of $S(t)$ is the linear operator defined by

$$Ax = \lim_{t\to 0^+} \frac{S(t)x - S(0)x}{t}, \qquad \text{for } x \in D(I),$$

where

$$D(I) = \left\{ x \in X \ \lim_{t\to 0^+} \frac{S(t)x - x}{t} \text{ exists in } X \right\},$$

and

$$Ax = \lim_{t\downarrow 0} \frac{1}{t}(S(t)x - x).$$

Equivalently, we say that A generates

$$\{S(t); t > 0\}.$$

Remark 3.3. If

$$A : D(I) \subseteq X \to X,$$

is the infinitesimal generator of a semigroup of linear operators, then $D(I)$ is a vector subspace of X, and A is a possibly unbounded linear operator.

Example 3.1. A first significant example of uniformly continuous semigroup is given by $t \longmapsto e^{tA}$ where e^{tA} is the exponential of the matrix tA. Namely, let $A \in \mathcal{M}_n(\mathbb{R})$ and let $S(t) = e^{tA}$ for each $t > 0$, where

$$e^{tA} = \sum_{n=0}^{\infty} \frac{t^n}{n!} A^n.$$

We can easily see that $\{S(t); t > 0\}$ is a uniformly continuous semigroup of linear operators. More that this, straightforward computations show that that $t \longmapsto S(t)$ is of class C^1 from $[0, +\infty)$ to X, and satisfies the first-order differential equation

$$\frac{d}{dt} S(t) = AS(t) = S(t)A, \tag{3.3}$$

for each $t > 0$.

The next example shows that there exist semigroups that are not uniformly continuous.

Example 3.2. Let $X = C_{ub}(\mathbb{R}_+)$ be the space of all bounded and uniformly continuous functions from $\mathbb{R}_+$ to $\mathbb{R}$, endowed with the sup-norm $\|.\|_\infty$, and let

$$\{S(t); t \geq 0\} \subset \mathcal{L}(X),$$

be defined by

$$[S(t)f](s) = f(t+s),$$

for each $f \in X$ and each $t, s \in \mathbb{R}_+$. One may easily verify that

$$\{S(t); t \geq 0\},$$

satisfies (1) and (1) in Definition 3.22 and, therefore, it is a semigroup of linear operators. As in this specific case, the uniform continuity of the semigroup is equivalent to the equicontinuity of the unit ball in X property, which obviously is not satisfied; the semigroup is not uniformly continuous.

The generator of the semigroup is given by

$$D(I) = \left\{ f \in X;\ \exists \lim_{t \downarrow 0^+} \frac{f(t+.) - f(.)}{t} = f' \text{ exists strongly in } X \right\},$$

and

$$Af = f'.$$

Proposition 3.1. *If*

$$\{S(t); t > 0\},$$

is a uniformly continuous semigroup of linear operators then, for each $t > 0$, $S(t)$ is invertible.

Proof. As much as

$$\lim_{t \downarrow 0} S(t) - I = 0,$$

in the norm topology of $\mathcal{L}(X)$, there exists $\delta > 0$ such that

$$\|S(t) - I\|_{\mathcal{L}(X)} \leq 1,$$

for each $t \in (0, \delta]$. Thus, for each $t \in (0, \delta]$, $S(t)$ is invertible. Let $t > \delta$. Then there exist $n \in \mathbb{N}^*$ and $\eta \in [0, \delta)$ such that $t = n\delta + \eta$. Therefore,

$$S(t) = S(\delta)^n S(\eta),$$

and so $S(t)$ is invertible. The proof is complete. □

3.7.1 Generators of uniformly continuous semigroups

Theorem 3.17. *A linear operator*

$$A : D(I) \subseteq X \to X,$$

is the generator of a uniformly continuous semigroup if and only if $D(I) = X$ *and* $A \in \mathcal{L}(X)$.

Proof. Let

$$\{S(t); t > 0\},$$

be uniformly continuous. Since

$$\lim_{t \downarrow 0} S(t) = I,$$

in the norm topology of $\mathcal{L}(X)$, there exists $\rho > 0$ such that

$$\left\| \frac{1}{\rho} \int_0^\rho S(t)dt - I \right\|_{\mathcal{L}(X)} < 1.$$

We notice that the integral here is the Riemann integral of the continuous function

$$S : [0, \rho] \to \mathcal{L}(X),$$

which is defined by a simple analogy with 1, its scalar counterpart. Consequently, the operator

$$\frac{1}{\rho} \int_0^\rho S(t)dt,$$

is invertible and accordingly

$$\int_0^\rho S(t)dt$$

has the same property. Let $h > 0$. Let us remark that

$$\frac{1}{h}(S(h) - I) \int_0^\rho S(t)dt = \frac{1}{h} \int_0^\rho S(t + h)dt - \frac{1}{h} \int_0^\rho S(t)dt.$$

The change of variable $t + h = s$ in the first integral on the right-hand side yields

$$\frac{1}{h}(S(h)-I)\int_0^{\rho} S(t)dt = \frac{1}{h}\int_h^{\rho+h} S(s)ds - \frac{1}{h}\int_0^{\rho} S(s)ds$$
$$= \frac{1}{h}\int_{\rho}^{\rho+h} S(s)ds - \frac{1}{h}\int_0^{h} S(s)ds.$$

Then

$$\frac{1}{h}(S(h)-I) = \left(\frac{1}{h}\int_{\rho}^{\rho+h} S(s)ds - \frac{1}{h}\int_0^{h} S(s)ds\right)\left(\frac{1}{h}\int_0^{\rho} S(t)dt\right)^{-1}.$$

But, the right-hand side of the equality above converges for h tending to 0 by positive values, and thus the left-hand side enjoys the same property. As the convergence in the uniform operator topology of $\mathcal{L}(X)$ implies the pointwise convergence, letting h to tend to 0 by positive values, we deduce

$$A = (S(\rho)-I)\left(\int_0^{\rho} S(t)dt\right)^{-1}.$$

Hence, $A \in \mathcal{L}(X)$, which proves the necessity.

The "if" part. Let $A \in \mathcal{L}(X)$, $t > 0$ and let

$$S(t) = \sum_{n=0}^{\infty} \frac{t^n}{n!}A^n,$$

where $A^n = A.A\ldots A$ times and $A^0 = I$.

We can easily see that $\{S(t); t > 0\}$ is a semigroup of linear operators. In order to prove that this semigroup is uniformly continuous let us remark that

$$\|S(t)-I\|_{\mathcal{L}(X)} = \left\|\sum_{n=0}^{\infty} \frac{t^n}{n!}A^n - I\right\|_{\mathcal{L}(X)} = \left\|\sum_{n=1}^{\infty} \frac{t^n}{n!}A^n\right\|_{\mathcal{L}(X)} \le t\sum_{n=1}^{\infty} \frac{t^{n-1}}{n!}\|A^n\|_{\mathcal{L}(X)}.$$

Since

$$\sum_{n=1}^{\infty} \frac{t^{n-1}}{n!}\|A^n\|_{\mathcal{L}(X)} \le \|A\|e^{t\|A\|},$$

we conclude that

$$\lim_{t\downarrow 0} S(t) = I,$$

in the norm topology of $\mathcal{L}(X)$, and thus

$$\{S(t); t > 0\},$$

is a uniformly continuous semigroup. To achieve the proof, we have to show that A is the infinitesimal generator of this semigroup. To this aim, it suffices to verify that

$$\lim_{t\downarrow 0}\left\|\frac{1}{t}(S(t)-I)-A\right\|_{\mathcal{L}(X)}=0.$$

But this follows from the obvious inequality

$$\left\|\frac{1}{t}(S(t)-I)-A\right\|_{\mathcal{L}(X)}=t\left\|\sum_{n=1}^{\infty}\frac{t^{n-2}}{n!}A^n\right\|_{\mathcal{L}(X)}\leq t\|A^2\|e^{t\|A\|},$$

thereby completing the proof. □

3.7.2 C_0-Semigroups: General properties

We introduce a class of semigroups of linear operators that is strictly larger than that of uniformly continuous semigroups, a class that proves to be very useful in the study of many partial differential equations of parabolic or hyperbolic type. The following properties are classical.

Definition 3.26. A semigroup of linear operators

$$\{S(t); t > 0\},$$

is called a semigroup of class C_0, or C_0-semigroup if, for each $x \in X$, we have

$$\lim_{t\downarrow 0} S(t)x = x.$$

Remark 3.4. Each uniformly continuous semigroup is of class C_0, but not conversely, as we can state from the example below.

Example 3.3. Let $X = C_{ub}(\mathbb{R}^+)$ be the space of all functions that are uniformly continuous and bounded from $\mathbb{R}^+$ to $\mathbb{R}$, endowed with the sup-norm $\|.\|_\infty$, and let

$$\{S(t); t > 0\},$$

be defined by

$$[S(t)f](s) = f(t+s),$$

for each $f \in X$ and each $t, s \in$. We know from Example 3.1 that

$$\{S(t); t > 0\},$$

is a semigroup. In addition, this is of class C_0. On the other hand, as we mentioned in Example 3.1, it is not uniformly continuous because the unit ball in X is not equicontinuous.

Theorem 3.18. *If $\{S(t); t > 0\}$ is a C_0-semigroup, then there exist $M > 1$, and $w \in \mathbb{R}$ such that*

$$\|S(t)\|_{\mathcal{L}(X)} \leq Me^{wt}, \quad \text{for } t \geq 0,$$

for each $t > 0$.

Proof. First, we will show that there exist $\eta > 0$ and $M > 1$ such that

$$\|S(t)\|_{\mathcal{L}(X)} \leq M, \quad \text{for } t \geq 0,$$

for each $t \in [0, \eta]$. To this aim, let us assume by contradiction that this is not the case. Then there exists at least one C_0-semigroup

$$\{S(t); t > 0\},$$

with the property that, for each $\eta > 0$ and each $M > 1$, there exists $t_{\eta,M} \in [0, \eta]$, such that

$$\|S(t)\|_{\mathcal{L}(X)} > M. \tag{3.4}$$

Taking $\eta = 1/n$, $M = n$ and denoting $t_{\eta,M} = t_n$ for $n \in \mathbb{N}^*$, we deduce

$$\|S(t_n)\|_{\mathcal{L}(X)} > n, \tag{3.5}$$

where $t_n \in [0, 1/n]$ for each $n \in \mathbb{N}^+$. Recalling that, for each $x \in X$,

$$\lim_{n \to \infty} S(t_n)x = x,$$

it follows that the family

$$\{S(t_n); n \in \mathbb{N}^+\},$$

of linear bounded operators is pointwise bounded, that is, for each $x \in X$, the set

$$\{S(t_n)x; n \in \mathbb{N}^+\},$$

is bounded. By the uniform boundedness principle (see Dunford and Schwartz [9], Corollary 21, p. 66), it follows that this family is bounded in the uniform operator norm $\|.\|_{\mathcal{L}(X)}$, which contradicts (3.5). This contradiction can be eliminated only if (3.4) holds. Next, let $t > 0$. Then there exist $n \in \mathbb{N}^+$ and $\delta \in [0, \eta)$, such that $t = n\eta + \delta$. We have

$$\|S(t)\|_{\mathcal{L}(X)} = \|S^n(\eta)S(\delta)\|_{\mathcal{L}(X)} \le \|S(\eta)\|^n_{\mathcal{L}(X)}\|S(\delta)\|_{\mathcal{L}(X)} \le MM^n.$$

But

$$n = \frac{t-\delta}{\eta} < \frac{t}{\eta},$$

and thus

$$\|S(t)\|_{\mathcal{L}(X)} \le MM^{\frac{t}{\eta}} = Me^{tw},$$

where

$$w = \frac{1}{\eta}\ln M.$$

The proof is complete. □

Remark 3.5. If

$$\{S(t); t > 0\},$$

is a uniformly continuous semigroup whose generator is A, then (3.4) holds with $M = 1$ and $w = \|A\|_{\mathcal{L}(X)}$.

Definition 3.27. A C_0-semigroup,

$$\{S(t); t > 0\},$$

is called of type (M, w) with $M > 1$ and $w \in \mathbb{R}$ if, for each $t > 0$, we have

$$\|S(t)\|_{\mathcal{L}(X)} \le Me^{wt}, \quad \text{for } t \ge 0.$$

A C_0-semigroup

$$\{S(t); t > 0\},$$

is called a C_0-semigroup of contractions, or of nonexpansive operators, that is, if for each $t > 0$, we have

$$\|S(t)\|_{\mathcal{L}(X)} \le 1, \quad \text{for } t \ge 0.$$

We shall also use the term contraction semigroup.

Corollary 3.2. *If $\{S(t); t > 0\}$ is a C_0-semigroup, then the mapping $(t, x) = S(t)x$ is jointly continuous from $[0, +\infty) \times X$ to X.*

Proof. Let $x, y \in X$, $t > 0$ and $h \in \mathbb{R}^*$ with $t + h > 0$. We distinguish between two cases $h > 0$, or $h < 0$. If $h > 0$, we have

$$\begin{aligned}\|S(t+h)y - S(t)x\| &\le \|S(t+h)y - S(t+h)x\| + \|S(t+h)x - S(t)x\| \\ &\le \|S(t+h)\|_{\mathcal{L}(X)}\|y - x\| + \|S(t+h)x - S(t)x\| \\ &\le Me^{(t+h)w}\|y - x\| + \|S(t)\|_{\mathcal{L}(X)}\|S(h)x - x\|,\end{aligned}$$

which shows that

$$\lim_{(t,y)\to(t+0,x)} S(\tau)y = S(t)x.$$

If $h < 0$, by Theorem 3.18, we deduce between two cases $h > 0$, or $h < 0$. If $h > 0$, we have

$$\begin{aligned}\|S(t+h)y - S(t)x\| &= \|S(t+h)y - S(t+h)S(-h)x\| \\ &\le \|S(t+h)\|_{\mathcal{L}(X)}\|y - S(-h)x\| \\ &\le Me^{(t+h)w}(\|y - x\| + \|S(-h)x - x\|),\end{aligned}$$

which implies that

$$\lim_{(t,y)\to(t-0,x)} S(\tau)y = S(t)x.$$

The proof is complete. □

Some basic properties of C_0-semigroups are listed below.

Theorem 3.19. *Let*

$$A : D(I) \subset X \to X,$$

be the infinitesimal generator of a C_0 semigroup $\{S(t); t > 0\}$. Then:

1. *For each $x \in X$ and each $t > 0$, we have*

$$\lim_{h\downarrow 0} \frac{1}{h} \int_{\tau}^{\tau+h} S(\tau)x d\tau = S(\tau)x.$$

2. *For each $\in X$ and each $t > 0$, we have*

$$\int_0^{\tau} S(\tau)d\tau x \in D(I),$$

and

$$A\left(\int_0^\tau S(\tau)xd\tau\right) = S(\tau)x - x.$$

3. *For each $x \in D(I)$ and each $t > 0$, we have $S(t)x \in D(I)$. In addition, the mapping $t \longmapsto S(t)x$ is of class C^1 on $[0, +\infty)$, and satisfies*

$$\frac{d}{dt}(S(t)x) = AS(t)x = S(t)Ax.$$

4. *For each $x \in D(I)$ and each $0 \leq s \leq t < +\infty$, we have*

$$\int_s^t AS(\tau)xd\tau = \int_s^t S(\tau)Axd\tau = S(t)x - S(s)x.$$

Proof. In order to prove (i), let us observe that

$$\left\|\frac{1}{h}\int_\tau^{\tau+h} S(\tau)x - S(t)xd\tau\right\| \leq \frac{1}{h}\int_\tau^{\tau+h} \|S(\tau)x - S(t)x\|d\tau.$$

The conclusion follows from Corollary 3.2.

Let $x \in X, t > 0$ and $h > 0$. We remark that

$$\frac{1}{h}(S(h) - I)\int_0^h S(\tau)d\tau x = \frac{1}{h}\int_0^t S(\tau + h)xd\tau - \frac{1}{h}\int_0^t S(\tau)xd\tau.$$

The change of variable $\tau + h = s$ in the first integral on the right-hand side yields

$$\frac{1}{h}(S(h) - I)\int_0^h S(\tau)d\tau x = \frac{1}{h}\int_h^{t+h} S(s)xds - \frac{1}{h}\int_0^t S(s)xds - \frac{1}{h}\int_0^h S(s)xds + \frac{1}{h}\int_t^{t+h} S(s)xds.$$

From this equality and from (1), we deduce that there exists

$$\lim_{h\downarrow 0}\frac{1}{h}\int_\tau^{\tau+h} S(\tau)xd\tau = S(\tau)x,$$

which proves (2).

Next, let $x \in D(A)$, $t > 0$, and $h > 0$. We have

$$\left\|\frac{1}{h}(S(t + h)x - S(t)x) - S(t)Ax\right\| \leq \|S(t)\|\left\|\frac{1}{h}(S(h)x - x) - Ax\right\|,$$

inequality, which proves that

$$S(t)x \in D(I), \quad t \longmapsto S(t)x,$$

is differentiable from the right, and that

$$\frac{d^+}{dt}(S(t)x) = AS(t)x = S(t)Ax. \tag{3.6}$$

On the other hand, for each $t > 0$ and $h < 0$ with $t + h > 0$, we have

$$\begin{aligned}\left\|\frac{1}{h}(S(t+h)x - S(t)x) - S(t)Ax\right\| &\leq \|S(t+h)\|\left\|\frac{1}{h}(x - S(-h)x) - S(-h)Ax\right\| \\ &\leq \|S(t+h)\|\left(\left\|\frac{-1}{h}(S(-h)x - x) - Ax\right\| + \|S(-h)Ax - Ax\|\right).\end{aligned}$$

This inequality shows that $t \longmapsto S(t)x$ is differentiable from the left as well. From (3.6) and the continuity of the function $t \longmapsto S(t)Ax$ on $[0, +\infty)$, we deduce that $t \longmapsto S(t)x$ is of class C^1 on $[0, +\infty)$, which completes the proof of (3).

Since (4) follows from (3) by integrating both sides from s to t in (3.6), the proof is complete. □

3.7.3 The infinitesimal generator

In this section, we shall prove two basic properties of the generator of a C_0-semigroup: the density of the domain and the closeness of the graph. First, let us recall the following.

Definition 3.28. An operator

$$A : D(I) \subseteq X \longrightarrow X,$$

is called closed if its graph is closed in $X \times X$.

Theorem 3.20. *Let*

$$A : D(I) \subseteq X \longrightarrow X,$$

be the infinitesimal generator of a C_0-semigroup

$$\{S(t); t > 0\}.$$

Then $D(I)$ is dense in X, and A is a closed operator.

Proof. Let $x \in X$ and $\epsilon > 0$. Then, by virtue of (1) in Theorem 3.19, we have

$$\frac{1}{\epsilon}\int_0^\epsilon S(\tau)x d\tau \in D(I),$$

and

$$\lim_{\epsilon\downarrow 0}\frac{1}{\epsilon}\int_0^{\epsilon} S(\tau)x d\tau = x.$$

Consequently, $D(I)$ is dense in X.

Next, let $(x_n)_{n\in\mathbb{N}^*}$ be a sequence in $D(I)$ such that

$$\lim_{n\to} x_n = x \quad \text{and} \quad \lim_{n\to} Ax_n = y.$$

From (4) in Theorem 3.19, it follows that

$$S(h)x - x = \int_0^h S(\tau)y d\tau.$$

By virtue of (1) in Theorem 3.19, it follows that there exists

$$\lim_{h\downarrow 0}\frac{1}{h}\int_0^{\epsilon} S(\tau)y d\tau = y.$$

From this relation and the preceding one, we deduce that $x \in D(I)$ and $Ax = y$. The proof is complete. □

3.7.4 Hille–Yosida theorem

This section begins with the presentation of the most fundamental result within the theory of C_0-semigroups, known as the Hille–Yosida generation theorem. This gives a very precise delimitation of the class of linear operators A, acting in a Banach space X, that generate C_0-semigroups containing only operators whose norms do not exceed.

3.7.4.1 Necessity

The goal of the next two sections is to prove the most important result in the theory of C_0-semigroups: the Hille–Yosida theorem. More precisely, we shall present a necessary and sufficient condition such that a linear operator A generates a C_0-semigroup of contractions. We recall that if

$$A : D(I) \subseteq X \longrightarrow X,$$

is a linear operator, the resolvent set $\rho(1)$ is the set of all those complex numbers A, called regular values, for which $R(\lambda I - A)$ is dense in X and

$$R(\lambda; A) = (\lambda I - A)^{-1},$$

is continuous from $R(\lambda I - A)$ to X.

Theorem 3.21 (Hille–Yosida). *A linear operator*

$$A : D(I) \subseteq X \longrightarrow X,$$

is the infinitesimal generator of a C_0-semigroup of contractions if and only if:
1. *A is densely defined and closed, and*
2. *$(0, +\infty) \subseteq \rho(1)$, and for each $\lambda > 0$*

$$\|R(\lambda; A)\|_{\mathcal{L}(X)} \le \frac{1}{\lambda}.$$

Proof. We begin with the necessity. Let

$$A : D(I) \subseteq X \longrightarrow X,$$

be the infinitesimal generator of a C_0-semigroup of contractions 0,

$$\{S(t); t > 0\}.$$

In view of Theorem 3.20, A is densely defined and closed. Thus, (1) holds. In order to prove (ii), let $\lambda > 0$, $x \in X$, and let us define

$$R(\lambda)x = \int_0^\infty e^{-\lambda x} S(t) x dt.$$

We notice that the integral on the right-hand side of the equality above is convergent. Indeed, for each $a, b > 0$, $a < b$, we have

$$\left\| \int_a^b e^{-\lambda x} S(t) x dt \right\| \le \int_a^b e^{-\lambda t} \|S(t)\|_{\mathcal{L}(X)} \|x\| dt,$$

and

$$\int_a^b e^{-\lambda t} \|x\| dt = \frac{e^{-\lambda a} - e^{-\lambda b}}{\lambda} \|x\|.$$

Accordingly, we are in the hypotheses of the Catchy test, and thus the integral is convergent. Clearly, $R(\lambda) \in \mathcal{L}(X)$ and

$$
\begin{aligned}
\|R(\lambda)x\| &= \left\| \int_0^\infty e^{-\lambda t} S(t)x dt \right\| \\
&\le \int_0^\infty e^{-\lambda t} \|S(t)\|_{\mathcal{L}(X)} \|x\| dt \\
&\le \frac{1}{\lambda} \|x\|.
\end{aligned}
$$

Hence,

$$
\|R(\lambda)\| \le \frac{1}{\lambda}.
$$

We prove next that $R(\lambda)$ coincides with $R(\lambda; A)$. To this aim, we show that $R(\lambda)$ is both the right and the left inverse of the operator $\lambda I - A$. Let $x \in X$, $\lambda > 0$, and $h > 0$. We have

$$
\begin{aligned}
\frac{1}{h}(S(h) - I)R(\lambda)x &= \frac{1}{h} \int_0^\infty e^{-\lambda t} S(t + h)x dt - \frac{1}{h} \int_0^\infty e^{-\lambda t} S(t)x dt \\
&= \frac{e^{\lambda h} - 1}{h} \int_0^\infty e^{-\lambda t} S(t)x dt - \frac{e^{\lambda h}}{h} \int_0^\infty e^{-\lambda t} S(t)x dt.
\end{aligned}
$$

As the right-hand side of the above equality converges to $\lambda R(\lambda)x - x$, it follows that

$$
R(\lambda)x \in D(I),
$$

and

$$
AR(\lambda) = \lambda R(\lambda) - I,
$$

which proves that

$$
(\lambda I - A)R(\lambda) = I.
$$

So, $R(\lambda)$ is the right inverse of $I\lambda - A$. Next, let $x \in D(I)$. Let us remark that

$$
\begin{aligned}
R(\lambda)Ax &= \int_0^\infty e^{-\lambda t} S(t)Ax dt \\
&= \int_0^\infty e^{-\lambda t} \frac{d}{dt}(S(t)x) dt \\
&= \lim_{t \to \infty} e^{-\lambda t} S(t)x - x + \lambda \int_0^\infty e^{-\lambda t} S(t)x dt \\
&= \lambda R(\lambda)x - x.
\end{aligned}
$$

This equality may be equivalently rewritten as

$$R(\lambda)(\lambda I - A) = I,$$

which shows that $R(\lambda)$ is the left inverse of $\lambda I - A$, and this completes the proof of necessity. □

Remark 3.6. Using similar arguments, one may prove that whenever A generates a C_0-semigroup of contractions, then

$$\{\lambda \in \mathbb{C}; \operatorname{Re}(\lambda) > 0\} \subseteq \rho(1),$$

and for each $\lambda \in \mathbb{C}$ with $\operatorname{Re}(\lambda) > 0$, we have

$$\|R(\lambda; A)\|_{\mathcal{L}(X)} \leq \frac{1}{\operatorname{Re}(\lambda)}.$$

3.7.4.2 Sufficiency

In order to prove the sufficiency, some preliminary lemmas are needed. First, let us observe that by (1), it follows that for each

$$\lambda > 0, \quad R(\lambda; A) \in \mathcal{L}(X).$$

Definition 3.29. Let

$$A : D(I) \subseteq X \longrightarrow X,$$

be a linear operator satisfying (1) and (2) in Theorem 3.21, and let $\lambda > 0$. The operator $A_\lambda : X \to X$, defined by

$$A_\lambda = \lambda A R(\lambda, A),$$

is called the Yosida approximation of A.

Lemma 3.16. *Let*

$$A : D(I) \subseteq X \longrightarrow X$$

be a linear operator, which satisfies (1) and (2) in Theorem 3.21. Then

$$\lim_{\lambda \to \infty} \lambda R(\lambda; A)x = x, \tag{3.7}$$

for each $x \in X$,

$$A_\lambda x = \lambda^2 R(\lambda; A)x - x, \tag{3.8}$$

for each $x \in X$, and

$$\lim_{\lambda\to\infty} A_\lambda x = Ax, \tag{3.9}$$

for each $x \in D(I)$.

Proof. Let $x \in D(I)$ and $\lambda > 0$. We have

$$\|\lambda R(\lambda; A)x - x\| = \|AR(\lambda; A)x\| = \|R(\lambda; A)Ax\| \leq \frac{1}{\lambda}\|Ax\|,$$

and consequently

$$\lim_{\lambda\to\infty} \lambda R(\lambda; A)x = x,$$

for each $x \in D(I)$. Since $D(I)$ is dense in X and

$$\|R(\lambda; A)x\| \leq 1,$$

from the last relation, we deduce (3.7). To check (3.8), let us remark that we have successively

$$\lambda^2 R(\lambda; A) - \lambda I = \lambda^2 R(\lambda; A) - \lambda(\lambda I - A)R(\lambda; A) = \lambda AR(\lambda; A) = A_\lambda.$$

Finally, if $x \in D(I)$, by (3.7), we have

$$\lim_{\lambda\to\infty} A_\lambda x = \lim_{\lambda\to\infty} \lambda AR(\lambda; Ax) = \lambda R(\lambda; A)Ax = Ax,$$

which concludes the proof of Lemma 3.16. □

Lemma 3.17. *Let*

$$A : D(I) \subseteq X \longrightarrow X,$$

be a linear operator that satisfies (1) and (2) in Theorem 3.21. Then, for each $\lambda > 0$, A_λ is the infinitesimal generator of a uniformly continuous semigroup

$$\{e^{tA_\lambda}; t \geq 0\},$$

satisfying

$$\|e^{tA_\lambda}\|_{\mathcal{L}(X)} \leq 1, \tag{3.10}$$

for each $t \geq O$. In addition, for each $x \in X$ and each $\lambda, \mu > O$, we have

$$\|e^{tA_\lambda}x - e^{tA_\mu}x\|_{\mathcal{L}(X)} \leq t\|A_\lambda x - A_\mu x\|. \tag{3.11}$$

Proof. As $A_\lambda \in \mathcal{L}(X)$, by Theorem 3.17, it follows that it generates a uniformly continuous semigroup

$$\{e^{tA_\lambda}; t \geq 0\}.$$

In order to check (3.10), let us remark that, by virtue of (3.8) and (ii), we have

$$\begin{aligned}\|e^{tA_\lambda}\|_{\mathcal{L}(X)} = \|e^{t\lambda^2 R(\lambda;A)-t\lambda I}\|_{\mathcal{L}(X)} &\leq \|e^{t\lambda^2 R(\lambda;A)}\|_{\mathcal{L}(X)}\|e^{-t\lambda I}\|_{\mathcal{L}(X)}\\ e^{t\lambda^2\|R(\lambda;A)\|_{\mathcal{L}(X)}}e^{-t\lambda} &\leq e^{-t\lambda}e^{t\lambda} = 1.\end{aligned}$$

Since A_λ, a_μ, e^{tA_λ}, and e^{tA_μ} commute with one another, we have

$$\begin{aligned}\|e^{tA_\lambda}x - e^{tA_\mu}x\|_{\mathcal{L}(X)} &= \left\|\int_0^1 \frac{d}{ds}(e^{stA_\lambda})e^{(1-s)tA_\mu}xds\right\|\\ &\leq \int_0^1 t\|(e^{stA_\lambda})e^{(1-s)tA_\mu}(A_\lambda x - A_\mu x)\|ds\\ &\leq t\|A_\lambda x - A_\mu x\|,\end{aligned}$$

which completes the proof. □

We proceed to the proof of the sufficiency of Theorem 3.21.

Proof of Theorem 3.21 *(continued).* From (3.9) and (3.11), it follows that, for each $t \geq 0$, there exists a linear operator

$$S(t) : D(I) \subseteq X \longrightarrow X,$$

such that, for each $x \in D(I)$,

$$\lim_{\lambda\to\infty} e^{tA_\lambda}x = S(t)x,$$

uniformly on compact subsets in $\mathbb{R}^+$. By (3.10) we deduce that

$$\|S(t)x\| \leq \|x\|,$$

for each $t \geq 0$ and $x \in D(I)$. Since $D(I)$ is dense in X, it follows that $S(t)$ can be extended by continuity to the whole space X. It is easy to see that the family of linear bounded operators thus obtained is a semigroup, denoted for simplicity again by

$$\{S(t); t > 0\}.$$

Clearly, it satisfies

$$\|S(t)\| \leq 1.$$

In addition, for each $t > 0$ and $x, y \in X$, we have

$$\begin{aligned}\|S(t)x - x\| &\leq \|S(t)x - S(t)y\| + \|S(t)y - e^{tA_\lambda}y\| + \|e^{tA_\lambda}y - y\| + \|y - x\| \\ &\leq \|S(t)y - e^{tA_\lambda}y\| + \|e^{tA_\lambda}y - y\| + 2\|y - x\|.\end{aligned}$$

Let $T > 0$ and $\epsilon > 0$. Fix $y = x_\epsilon \in D(I)$, with $\|x - x_\epsilon\| < \epsilon$, and a sufficiently large λ, such that

$$\|S(t)x_\epsilon - e^{tA_\lambda}x_\epsilon\| \leq \epsilon,$$

for each $t \in [0, T]$. By this inequality, we deduce

$$\|S(t)x - x\| \leq 3\epsilon + \|e^{tA_\lambda}x_\epsilon - x_\epsilon\|. \tag{3.12}$$

Inasmuch as

$$\{e^{tA_\lambda}x_\epsilon; t > 0\},$$

is a uniformly continuous semigroup, for the very same $\epsilon > 0$, there exists $\delta(\epsilon) > 0$, such that

$$\|e^{tA_\lambda} - I\|_{\mathcal{L}(x)} \leq \epsilon,$$

for each $t \in (0, \delta(\epsilon))$. Consequently,

$$\begin{aligned}\|e^{tA_\lambda}x_\epsilon - x_\epsilon\| &\leq \|e^{tA_\lambda} - I\|_{\mathcal{L}(x)}\|x_\epsilon\| \\ &\leq \epsilon\|x_\epsilon\|,\end{aligned}$$

for each $t \in (0, \delta(\epsilon))$. Since $\{x_\epsilon; \epsilon > 0\}$ is bounded, this inequality, along with (3.12), shows that $\{S(t); t > 0\}$ is a semigroup of class C_0. To conclude the proof, we have merely to show that the infinitesimal generator,

$$B : D(2) \subseteq X \to X,$$

of this semigroup coincides with

$$A : D(I) \subseteq X \to X.$$

To this aim, let $x \in D(I)$ and $h > 0$. We have

$$\lim_{\lambda\to\infty} e^{tA_\lambda}A_\lambda x = S(t)Ax,$$

uniformly on compact subsets in $\mathbb{R}_+$. Indeed,

$$\begin{aligned}\|e^{tA_\lambda}A_\lambda x - S(t)Ax\| &\le \|e^{tA_\lambda}A_\lambda x - e^{tA_\lambda}Ax\| + \|e^{tA_\lambda}Ax - S(t)Ax\| \\ &\le \|e^{tA_\lambda}\|_{\mathcal{L}(X)}\|A_\lambda x - Ax\| + \|e^{tA_\lambda}Ax - S(t)Ax\|.\end{aligned}$$

But this relation, along with (3.9) and with the partial conclusions above, proves that

$$S(h)x - x = \lim_{\lambda\to\infty}(e^{hA_\lambda}x - x) = \lim_{\lambda\to\infty}\int_0^h e^{tA_\lambda}A_\lambda x dt = \int_0^h S(t)Axdt.$$

Dividing both sides of this equality by h and letting h tend to 0 with positive values, we deduce that $x \in D(2)$ and $Bx = Ax$. Finally, we show that $D(I) = D(2)$. Since B is the infinitesimal generator of a C_0-semigroup of contractions, it follows from necessity that $1 \in \rho(2)$. Accordingly, $I - B$ is invertible and

$$(I - B)^{-1}X = D(2).$$

As

$$(I - B)D(I) = (I - A)D(I),$$

and by (2),

$$(I - A)D(I) = X,$$

it follows that

$$(I - B)D(I) = X,$$

or equivalently

$$(I - B)^{-1}X = D(I).$$

Hence, $D(I) = D(2)$, which completes the proof of Theorem 3.21. □

4 Impulses stochastic differential equations

4.1 Basic tools on stochastic calculus

For a stochastic process

$$x(\cdot,\cdot) : [0,T] \times \Omega \to X,$$

we will write $x(t)$ (or simply x when no confusion is possible) instead of $x(t,\omega)$.

Definition 4.1. Let

$$\phi^j(s), \quad s \in [0,T],$$

be a function with values in $L^0_{Q_{H^j}}(Y,X)$. The Wiener integral of ϕ^j with respect to *fBm* given by (2.48) is defined by

$$\begin{aligned}\int_0^T \phi^j(s)dB^{H_j}(s) &= \sum_{n=1}^{\infty}\int_0^t \sqrt{\lambda_n}\phi^j(s)e_n d\beta_n^{H_j}\\ &= \sum_{n=1}^{\infty}\int_0^T \sqrt{\lambda_n}K^*_{H_j}(\phi^j e_n)(s)d\beta_n.\end{aligned}$$

Notice that if

$$\sum_{n=1}^{\infty}\|\phi Q^{1/2}e_n\|_{L^{1/H_j}([0,T];X)} < \infty, \tag{4.1}$$

the next result ensures the convergence of the series.

Lemma 4.1. *For any*

$$\phi^j : [0,T] \to L^0_{Q_{H^j}}(Y,X),$$

such that (4.1) *holds, and for any* $\alpha,\beta \in [0,T]$ *with* $\alpha > \beta$, *for each* $j = 1,2$

$$E\left|\int_\alpha^\beta \phi^j(s)dB^{H_j}(s)\right|_X^2 \leq c_2(H_j)H_j(2H_j-1)(\alpha-\beta)^{2H_j-1}\sum_{n=1}^{\infty}\int_\alpha^\beta |\phi^j(s)Q^{1/2}e_n|_X^2 ds,$$

where $c_2(H_j)$ *is a constant depending on* H_j. *If, in addition,*

$$\sum_{n=1}^{\infty}|\phi^j Q^{1/2}e_n|_X \quad \textit{is uniformly convergent for } t \in [0,T],$$

then

https://doi.org/10.1515/9783112251898-004

$$E\left|\int_{\alpha}^{\beta} \phi^j(s)dB^{H_j}(s)\right|_X^2 \le c_2(H_j)H_j(2H_j-1)(\alpha-\beta)^{2H_j-1}\int_{\alpha}^{\beta} \|\phi^j(s)\|^2_{L^0_{Q_{H^j}}} ds.$$

4.2 Nonsmooth analysis

Let $x, y \in X$; the projection of x, y into $C_j \subset X$ is the set

$$\mathrm{Proj}(y, c_j) = \{z \in C_j : d(z, c_j) = \|z - y\|\}.$$

This set is nonempty if, for example, C_j is weakly closed. Let C_j be a closed subset of space X, and let $x, y \in C_i$: We say that a vector $v \in X$ is a proximal normal to C_j at z if $v = y - z$ for some $y \in X$ with $z \in \mathrm{Proj}(y, c_j)$. We denote by $N^p(z, c_j)$ the normal cone. One can show that $\eta \in N^p(y, c_j)$ if and only if there exists M such that the following proximal normal inequality holds:

$$\langle \eta, z - y\rangle \le M\|z - y\|,$$

for all $z \in C_j$ (in general, M will depend on x). On the other hand,

$$N^p(z, c_j) = \bigcup_{n=1}^{\infty}\left\{v \in X : d\left(y + \frac{v}{n}\right) = \frac{\|v\|}{n}\right\}.$$

This cone is convex, but in general not closed. A useful characterization of the proximal normal cone is the following:

$$N^p(z, c_j) = \bigcup_{\mu>0}\{v \in X : \langle v, a - z\rangle \le \mu\|z - y\|^2,\ a \in C_j\}.$$

If C_j is closed and convex, then we have

$$z \in N^p(z, c_j) \Longleftrightarrow y \in C_j \quad \text{and} \quad \langle z, y\rangle = \sigma(z, c_i) \Longleftrightarrow y \in C_j,\ x \in \partial\varphi_{C_j}(y)$$

where σ is the support function of a subset C_j of X, $\partial\varphi_{C_j}$ is the subdifferential in the sense of convex analysis, and C_i is the indicator function of a subset C_j of X,

$$\partial\varphi_{C_j}(y) = \begin{cases} 0, & \text{if } y \in C_j, \\ \emptyset, & \text{if } y \in C_j. \end{cases}$$

We define the Bouligand cone by

$$T_{C_j}(x) = \left\{v \in X : \liminf_{h\to 0}\frac{d(z + hv, c_j)}{h}\right\} = \bigcap_{\epsilon>0}\bigcap_{\delta>0}\bigcup_{0<h<\delta}\left(\frac{C_j - z}{h} + \epsilon\overline{B}(0,1)\right).$$

If the multivalued map F is completely continuous with nonempty compact values, then F is u. s. c. if and only if F has a closed graph, that is, $x_n \longrightarrow x_*, y_n \longrightarrow y_*, y_n \in F(x_n)$ imply $y_* \in F(x_*)$.

A multivalued map,

$$F : J \longrightarrow \mathcal{P}_{cp,c},$$

is said to be measurable if, for each $y \in X$, the mean-square distance between y and $F(t)$ is measurable.

Definition 4.2. A set-valued operator

$$G : J \longrightarrow \mathcal{P}_{cl}(X)$$

is said to be a contraction if there exists $0 \leq \gamma < 1$, such that

$$H_d(G(x), G(y)) \leq \gamma d(x, y), \quad \text{for all } x, y \in X,$$

Lemma 4.2. *Let the function $C : [0, T] \to \mathcal{P}_{cl}(X)$ be such that:*
1. *C is Hausdorff lower semicontinuous at $t = 0$;*
2. *∂C is Hausdorff upper semicontinuous at $t = 0$;*
3. *There exist $x \in X$ and $r_0 > 0$ such that $B(x, r_0) \subseteq C(0)$.*

Then, for every $r \in (0, r_0)$, there exists $\delta > 0$ such that $B(x, r) \subset C(r)$ for all $t \in [0, \delta]$.

We are going to establish sufficient conditions for the existence of solutions for the system of stochastic impulsive functional equations with infinite Brownian motions. The approach is based on Perov's fixed-point theorem and a new version of Schaefer's fixed point in generalized Banach spaces.

Consider the following problem:

$$\begin{cases} dx(t) = \sum_{l=1}^{\infty} f_l^1(t, x(t), y(t))dW^l(t) + g^1(t, x(t), y(t))dt, & t \in J, \ t \neq t_k \\ dy(t) = \sum_{l=1}^{\infty} f_l^2(t, x(t), y(t))dW^l(t) + g^2(t, x(t), y(t))dt, & t \in J, \ t \neq t_k \\ x(t_k^+) - x(t_k) = I_k(x(t_k)), & t = t_k, \ k = 1, 2, \ldots, m \\ y(t_k^+) - y(t_k) = \overline{I}_k(y(t_k)), \\ x(0) = x_0, \\ y(0) = y_0, \end{cases} \tag{4.2}$$

where

$$0 = t_0 < t_1 < \cdots < t_m < t_{m+1} = T, \quad J = [0, T],$$

and

$$f_l^1, f_l^2 : J \times \mathbb{R}^2 \to \mathbb{R},$$
$$g^1, g^2 : J \times \mathbb{R}^2 \to \mathbb{R}$$

are Carathéodory functions, and W^l are an infinite sequence of independent standard Brownian motions, $l = 1, 2, \ldots$

$$I_k, \bar{I}_k \in C(\mathbb{R}, \mathbb{R}) \quad (k = 1, \ldots, m),$$

and

$$\Delta x|_{t=t_k} = x(t_k^+) - x(t_k^-),$$
$$\Delta y|_{t=t_k} = y(t_k^+) - y(t_k^-).$$

The notation

$$y(t_k^+) = \lim_{h \to 0^+} y(t_k + h)$$

and

$$y(t_k^-) = \lim_{h \to 0^+} y(t_k - h)$$

stand for the right and left limits of the function y at $t = t_k$, respectively.

Set

$$\begin{cases} f_i(., x, y) = (f_1^i(., x, y), f_2^i(., x, y), \ldots), \\ \|f_i(., x, y)\| = (\sum_{l=1}^{\infty} (f_l^i)^2(., x, y))^{\frac{1}{2}}, \end{cases}$$

where

$$i = 1, 2, \quad f_i(., x, y) \in l^2, \quad x \in \mathbb{R}.$$

4.3 Unique solution via iterative methods

Let

$$J_k = (t_k, t_{k+1}], \quad k = 1, 2, \ldots, m.$$

In order to define a solution for Problem (4.2), consider the space of piecewise continuous functions

$$PC = \{x : \Omega \times J \longrightarrow \mathbb{R},\ x(w, .) \in C(J_k, \mathbb{R}),\ k = 1, \ldots, m \text{ such that } x(t_k^+, .) \text{ and } x(t_k^-, .) \text{ exist with } x(t_k^-, .) = x(t_k, .)\},$$

endowed with the norm

$$\|x\|_{PC}^2 = \sup_{t\in J} \mathbb{E}\big|x(t,.)\big|^2,$$

PC is a Banach space with norm $\|\cdot\|_{PC}$.

Definition 4.3. $\mathbb{R}$-valued stochastic process

$$u = (x,y) \in PC \times PC,$$

is said to be a solution of (4.2) with respect to the probability space $(\Omega, \mathcal{F}, \mathbb{P})$, if:
1. $u(t)$ is $\mathcal{F}_t$-adapted for all $t \in J_k = (t_k, t_{k+1}]$ $\quad k = 1, 2, \dots, m$;
2. $u(t)$ is right continuous and has a limit on the left;
3. $u(t)$ satisfies that

$$\begin{cases} x(t) = x_0 + \sum_{l=1}^{\infty}\int_0^t f_l^1(s, x(s), y(s))dW^l(s) + \int_0^t g^1(s, x(s), y(s))ds + \sum_{0\le t_k\le t} I_k(x(t_k)), \\ \quad t \in J, \\ y(t) = y_0 + \sum_{l=1}^{\infty}\int_0^t f_l^2(s, x(s), y(s))dW^l(s) + \int_0^t g^2(s, x(s), y(s))ds + \sum_{0\le t_k\le t} \overline{I}_k(y(t_k)), \\ \quad t \in J. \end{cases}$$

We are now in a position to state and prove our existence result. First, we will list the following hypotheses that will be imposed in our main theorem:

(H_1) There exist nonnegative numbers a_i and b_i for each $i \in \{1, 2\}$,

$$\begin{cases} \mathbb{E}(\|f^1(t,x,y) - f^1(t,\overline{x},\overline{y})\|)^2 \le a_1\mathbb{E}(|x-\overline{x}|)^2 + b_1\mathbb{E}(|y-\overline{y}|)^2, \\ \mathbb{E}(\|f^2(t,x,y) - f^2(t,\overline{x},\overline{y})\|)^2 \le a_2\mathbb{E}(|x-\overline{x}|)^2 + b_2\mathbb{E}(|y-\overline{y}|)^2, \end{cases}$$

for all $x, y, \overline{x}, \overline{y} \in \mathbb{R}$.

(H_2) There exist positive constants α_i and β_i for each $i = 1, 2$,

$$\begin{cases} \mathbb{E}(|g^1(t,x,y) - g^1(t,\overline{x},\overline{y})|)^2 \le \alpha_1\mathbb{E}(|x-\overline{x}|)^2 + \beta_1\mathbb{E}(|y-\overline{y}|)^2, \\ \mathbb{E}(|g^2(t,x,y) - g^2(t,\overline{x},\overline{y})|)^2 \le \alpha_2\mathbb{E}(|x-\overline{x}|)^2 + \beta_2\mathbb{E}(|y-\overline{y}|)^2, \end{cases}$$

for all $x, y, \overline{x}, \overline{y} \in \mathbb{R}$ and $t \in J$.

(H_3) There exist constants $d_k \ge 0$ and $\overline{d}_k \ge 0$, $k = 1, \dots, m$ such that

$$\mathbb{E}\big(\big|I_k(x) - I_k(\overline{x})\big|\big)^2 \le d_k\mathbb{E}(|x-\overline{x}|)^2,$$
$$\mathbb{E}\big(\big|\overline{I}_k(y) - \overline{I}_k(\overline{y})\big|\big)^2 \le \overline{d}_k\mathbb{E}(|y-\overline{y}|)^2,$$

for all $x, y, \overline{x}, \overline{y} \in \mathbb{R}$.

For our main consideration of Problem (4.2), a Perov fixed point is used to investigate the existence and uniqueness of solutions for a system of impulsive stochastic differential equations.

Theorem 4.1. *Assume that* (H_1)–(H_3) *are satisfied and the matrix*

$$M = \sqrt{3}\begin{pmatrix} \sqrt{C_2 a_1 + \alpha_1 T + l_1} & \sqrt{C_2 b_1 + \beta_1 T} \\ \sqrt{C_2 a_2 + \alpha_2 T} & \sqrt{C_2 b_2 + \beta_2 T + l_2} \end{pmatrix},$$

$$l_1 = \sum_{k=1}^{m} d_k,$$

$$l_2 = \sum_{k=1}^{m} \overline{d}_k,$$

where $C_2 \geq 0$ *is defined in Theorem 2.10. If* M *converges to zero. Then the problem* (4.2) *has a unique solution.*

Proof. Consider the operator

$$N : PC \times PC \to PC \times PC,$$

defined by

$$N(x,y) = (N_1(x,y), N_2(x,y)), \quad (x,y) \in PC \times PC,$$

where

$$N_1(x,y) = x_0 + \sum_{l=1}^{\infty} \int_0^t f_l^1(s, x(s), y(s))dW^l(s) + \int_0^t g^1(s, x(s), y(s))ds + \sum_{0<t_k\leq t} I_k(x(t_k)),$$

and

$$N_2(x,y) = y_0 + \sum_{l=1}^{\infty} \int_0^t f_l^2(s, x(s), y(s))dW^l(s) + \int_0^t g^2(s, x(s), y(s))ds + \sum_{0<t_k\leq t} \overline{I}_k(y(t_k)).$$

We use Theorem 3.11 to prove that N has a fixed point.

Indeed, let

$$(x,y), (\overline{x}, \overline{y}) \in PC \times PC.$$

Then we have for each $t \in [0, T]$,

$$
\begin{aligned}
&|N_1(x(t),y(t)) - N_1(\overline{x}(t),\overline{y}(t))|^2 \\
&= \Bigg| \sum_{l=1}^{\infty} \int_0^t f_l^1(s,x(s),y(s))dW^l(s) + \int_0^t g^1(s,x(s),y(s))ds + \sum_{k=1}^{m} I_k(x(t_k)) \\
&\quad - \sum_{l=1}^{\infty} \int_0^t f_l^1(s,\overline{x}(s),\overline{y}(s))dW^l(s) - \int_0^t g^1(s,\overline{x}(s),\overline{y}(s))ds - \sum_{k=1}^{m} I_k(\overline{x}(t_k)) \Bigg|^2 .
\end{aligned}
$$

Then

$$
\begin{aligned}
&|N_1(x(t),y(t)) - N_1(\overline{x}(t),\overline{y}(t))|^2 \\
&\le 3\left| \sum_{l=1}^{\infty} \int_0^t (f_l^1(s,x(s),y(s)) - f_l^1(s,\overline{x}(s),\overline{y}(s)))dW^l(s) \right|^2 \\
&\quad + 3\left| \int_0^t (g^1(s,x(s),y(s) - g^1(s,\overline{x}(s),\overline{y}(s))ds \right|^2 + 3\sum_{k=1}^{m} |I_k(x(t_k)) - I_k(\overline{x}(t_k))|^2 .
\end{aligned}
$$

By the generalized Burkholder–Davis–Gundy (BDG) inequality,[1] we get

$$
\begin{aligned}
\mathbb{E}|N_1(x(t),y(t)) - N_1(\overline{x}(t),\overline{y}(t))|^2 &\le 3C_2 \int_0^t \mathbb{E}|f^1(s,x(s),y(s)) - f^1(s,\overline{x}(s),\overline{y}(s))|^2 ds \\
&\quad + 3t \int_0^t \mathbb{E}|g^1(s,x(s),y(s)) - g^1(s,\overline{x}(s),\overline{y}(s))|^2 ds \\
&\quad + \sum_{k=1}^{m} \mathbb{E}|I_k(x(t_k)) - I_k(\overline{x}(t_k))|^2 .
\end{aligned}
$$

Thus,

$$
\begin{aligned}
\sup_{t\in J} \mathbb{E}|N_1(x(t),y(t)) - N_1(\overline{x}(t),\overline{y}(t))|^2 &\le C_2 \int_0^t a_1\mathbb{E}(|x(s) - \overline{x}(s)|)^2 + b_1\mathbb{E}(|y(s) - \overline{y}(s)|)^2 ds \\
&\quad + 3t \int_0^t \alpha_1 E(|x(s) - \overline{x}(s)|)^2 + \beta_1\mathbb{E}(|y(s) - \overline{y}(s)|)^2 ds
\end{aligned}
$$

1

$$
\mathbb{E}\left(\sup_{0\le t\le T} \left\| \int_0^t \Phi(s)\, dW(s) \right\|_X^p \right) \le C_p\, \mathbb{E}\left(\int_0^T \|\Phi(s)\|^2_{\mathcal{L}_2(U,X)}\, ds \right)^{\frac{p}{2}},
$$

where $p \ge 2$, $C_p > 0$ is a constant depending only on p, $W(t)$ is a cylindrical Wiener process on a Hilbert space U, and $\mathcal{L}_2(U,X)$ denotes the space of Hilbert–Schmidt operators from U into X.

$$+3\sum_{k=1}^{m} d_k \mathbb{E}(|x(t_k)-\overline{x}(t_k)|)^2.$$

Therefore,

$$\begin{aligned}&\sup_{t\in J}\mathbb{E}|N_1(x(t),y(t))-N_1(\overline{x}(t),\overline{y}(t))|^2\\ &\quad\le 3(C_2a_1+\alpha_1T+l_1)\|x-\overline{x}\|_{PC}^2+3(C_2b_1+\beta_1T)\|y-\overline{y}\|_{PC}^2.\end{aligned}$$

Similarly, we have

$$\|N_2(x,y)-N_2(\overline{x},\overline{y})\|_{PC}\le 3(C_2a_2+\alpha_2T)\|x-\overline{x}\|_{PC}+(C_2b_2+\beta_2T+l_2)\|y-\overline{y}\|_{PC}.$$

Hence,

$$\begin{aligned}\|N(x,y)-N(\overline{x},\overline{y})\|_{PC}&=\begin{pmatrix}\|N_1(x,y)-N_1(\overline{x},\overline{y})\|_{PC}\\ \|N_2(x,y)-N_2(\overline{x},\overline{y})\|_{PC}\end{pmatrix}\\ &\le\sqrt{3}\begin{pmatrix}\sqrt{C_2a_1+\alpha_1T+l_1} & \sqrt{C_2b_1+\beta_1T}\\ \sqrt{C_2a_2+\alpha_2T} & \sqrt{C_2b_2+\beta_2T+l_2}\end{pmatrix}\begin{pmatrix}\|x-\overline{x}\|_{PC}\\ \|y-\overline{y}\|_{PC}\end{pmatrix}.\end{aligned}$$

Therefore,

$$\|N(x,y)-N(\overline{x},\overline{y})\|_{PC}\le M\begin{pmatrix}\|x-\overline{x}\|_{PC}\\ \|y-\overline{y}\|_{PC}\end{pmatrix},\quad \text{for all}, (x,y),(\overline{x},\overline{y})\in PC\times PC.$$

From the Perov fixed-point theorem, the mapping N has a unique fixed $(x,y)\in PC\times PC$, which is the unique solution to problem (4.2). □

4.3.1 Nagumo type growth conditions

In this section, we present the existence result under a nonlinearity f^i and g^i, $i=1,2$ satisfying Nagumo-type growth conditions:

(H_4) There exists a function $p_i\in L^1(J,\mathbb{R}^+)$ and a continuous nondecreasing function $\psi_i:[0,\infty)\to(0,\infty)$ for each $i=1,2$, such that

$$\begin{cases}\mathbb{E}(\|f^1(t,x,y)\|)^2\le p_1(t)\psi_1(\mathbb{E}(|x|^2+|y|^2)),\\ \mathbb{E}(\|f^2(t,x,y)\|)^2\le p_2(t)\psi_2(\mathbb{E}(|x|^2+|y|^2)),\end{cases}$$

with

$$\int_0^T m_1(s)ds<\int_{v_1}^{\infty}\frac{ds}{\psi_1(s)+\psi_2(s)},$$

where

$$m_1(t) = \max\{4C_2p_1(t),\ 4Tp_2(t)\},$$

$$v_1 = 4E|x_0|^2 + 4\sum_{k=1}^{m} c_k,$$

(H_5) There exists a function $p_i \in L^1(J, \mathbb{R}^+)$ and a continuous nondecreasing function $\psi_i : [0,\infty) \to (0,\infty)$ for each $i = 3, 4$, such that

$$\begin{cases} \mathbb{E}(|g^1(t,x,y)|)^2 \le p_3(t)\psi_3(\mathbb{E}(|x|^2 + |y|^2)), \\ \mathbb{E}(|g^2(t,x,y)|)^2 \le p_4(t)\psi_4(\mathbb{E}(|x|^2 + |y|^2)), \end{cases}$$

with

$$\int_0^T m_2(s)ds < \int_{v_2}^{\infty} \frac{ds}{\psi_3(s) + \psi_4(s)},$$

where

$$m_2(t) = \max\{4C_2p_3(t), 4Tp_4(t)\},$$

$$v_2 = 4\mathbb{E}|y_0|^2 + \sum_{k=1}^{m} \widetilde{c}_k.$$

(H_6) There exist positive constants $c_k, \widetilde{c}_k, k = 1, \dots, m$, such that

$$\mathbb{E}(|I_k(x_k)|)^2 \le c_k \quad \text{for all } x \in \mathbb{R},$$

$$\mathbb{E}(|\bar{I}_k(y_k)|)^2 \le \widetilde{c}_k \quad \text{for all } y \in \mathbb{R}.$$

Theorem 4.2. *Assume that* (H_4)–(H_6) *holds. Then* (4.2) *has at least one solution on* J.

Proof. Clearly, the fixed point of N are solutions to (4.2), where N is defined in Theorem 4.1. In order to apply Theorem 3.13, we first show that N is completely continuous. The proof will be given in several steps.

Step 1. $N = (N_1, N_2)$ is continuous.

Let (x_n, y_n) be a sequence such that

$$(x_n, y_n) \to (x, y) \in PC \times PC,$$

as $n \to \infty$. Then

$$|N_1(x_n(t), y_n(t)) - N_1(x(t), y(t))|^2$$
$$= \left|\sum_{l=1}^{\infty} \int_0^t f_l^1(s, x_n(s), y_n(s))dW^l(s) + \int_0^t g^1(s, x_n(s), y_n(s))ds + \sum_{0<t_k\le t} I_k(x_n(t_k))\right.$$

$$-\sum_{l=1}^{\infty}\int_0^t f_l^1(s,x(s),y(s))dW^l(s)-\int_0^t g^1(s,x(s),y(s))ds-\sum_{0<t_k\le t} I_k(x(t_k))\Bigg|^2.$$

Since f^1, g^1 is a Carathéodory function and $I_k, \bar{I}_k$ are continuous functions, by Lebesgue dominated convergence theorem, we get

$$\begin{aligned}
&\sup_{t\in J}\mathbb{E}|N_1(x_n(t),y_n(t))-N_1(x(t),y(t))|^2\\
&\quad\le 3C_2\mathbb{E}\|f^1(.,x_n,y_n)-f^1(.,x,y)\|_{L^2}^2+3TE\|g^1(.,x_n,y_n)-g^1(.,x,y)\|_{L^2}^2\\
&\qquad+3\sum_{k=1}^{m}\mathbb{E}|I_k(x_n(t_k))-I_k(x(t_k))|^2\to 0\quad\text{as } n\to\infty.
\end{aligned}$$

Similarly,

$$\begin{aligned}
&\sup_{t\in J}\mathbb{E}|N_2(x_n(t),y_n(t))-N_1(x(t),y(t))|^2\\
&\quad\le 3C_2E\|f^2(.,x_n,y_n)-f^2(.,x,y)\|_{L^2}^2+3TE\|g^2(.,x_n,y_n)-g^2(.,x,y)\|_{L^2}\\
&\qquad+3\sum_{k=1}^{m}\mathbb{E}|\bar{I}_k(y_n(t_k))-\bar{I}_k(y(t_k))|^2\to 0\quad\text{as } n\to\infty.
\end{aligned}$$

Thus N is continuous.

Step 2. N maps bounded sets into bounded sets in $PC\times PC$. Indeed, it is enough to show that for any $q>0$ there exists a positive constant l such that for each

$$(x,y)\in B_q=\{(x,y)\in PC\times PC:\|x\|_{PC}\le q,\ \|y\|\le q\},$$

we have

$$\|N(x,y)\|_{PC}\le l=(l_1,l_2).$$

Then for each $t\in J$, we get

$$\begin{aligned}
&|N_1(x(t),y(t))|^2\\
&=\left|x_0+\sum_{l=1}^{\infty}\int_0^t f_l^1(s,x(s),y(s))dW^l(s)+\int_0^t g^1(s,x(s),y(s))ds+\sum_{k=1}^{m}I_k(x(t_k))\right|^2\\
&\le 4|x_0|^2+4\left|\sum_{l=1}^{\infty}\int_0^t f_l^1(s,x(s),y(s))dW^l(s)\right|^2+4\left|\int_0^t g^1(s,x(s),y(s))ds\right|^2\\
&\quad+4\left|\sum_{k=1}^{m}I_k(x(t_k))\right|^2.
\end{aligned}$$

Using the BDG inequality, we get

$$\begin{aligned}\mathbb{E}|N_1(x(t),y(t))|^2 &\le 4\mathbb{E}|x_0|^2 + 4C_2\int_0^t E\|f^1(s,x(s),y(s))\|^2 ds\\ &\quad + 4T\int_0^t E|g^1(s,x(s),y(s))|^2 d(s)\\ &\quad + 4E\sum_{k=1}^m |I_k(x(t_k))|^2\\ &\le 4\mathbb{E}|x_0|^2 + 4C_2\|p_1\|_{L^1}\psi_1(2q) + 4T\|p_3\|_{L^1}\psi_2(2q)ds + 4\sum_{k=1}^m c_k.\end{aligned}$$

Therefore,

$$\|N_1(x,y)\|_{PC} \le 4\mathbb{E}|x_0|^2 + 4C_2\|p_1\|_{L^1}\psi_1(2q) + 4\|p_2\|_{L^1}\psi_2(2q)ds + 4\sum_{k=1}^m c_k = l_1.$$

Similarly, we have

$$\|N_2(x,y)\|_{PC} \le 4\mathbb{E}|x_0|^2 + 4C_2\|p_3\|_{L^1}\psi_2(q) + 4\|p_4\|_{L^1}\psi_4(q)ds + 4\sum_{k=1}^m \tilde{c}_k = l_2.$$

Step 3. N maps bounded sets into equicontinuous sets of $PC \times PC$.
Let B_q be a bounded set in $PC \times PC$ as in Step 2. Let $r_1, r_2 \in J$, $r_1 < r_2$, and $u \in B_q$. Thus, we have

$$\begin{aligned}&|N_1(x(r_2),y(r_2)) - N_1(x(r_1),y(r_1))|^2\\ &\le 3\left|\sum_{l=1}^{\infty}\int_{r_1}^{r_2} f_l^1(s,x(s),y(s))dW^l(s)\right|^2 + 3\left|\int_{r_1}^{r_2} g^1(s,x(s),y(s))ds\right|^2 + 3\sum_{r_1\le t_k\le r_2}|I_k(x(t_k))|^2.\end{aligned}$$

Hence,

$$\begin{aligned}&\mathbb{E}|N_1(x(r_2),y(r_2)) - N_1(x(r_1),y(r_1))|^2\\ &\le 3C_2\int_{r_1}^{r_2}\mathbb{E}\|f^1(s,x(s),y(s))\|^2 ds + 3T\int_{r_1}^{r_2}\mathbb{E}|g^1(s,x(s),y(s))|^2 ds\\ &\quad + 3\sum_{r_1\le t_k\le r_2} E|I_k(x(t_k))|^2\\ &\le 3C_2\psi_1(q)\int_{r_1}^{r_2} p_1(s)ds + T\psi_2(q)\int_{r_1}^{r_2} p_2(s)ds + 3\sum_{r_1\le t_k\le t_2} c_k.\end{aligned}$$

The right-hand term tends to zero as

$$|r_2 - r_1| \to 0.$$

As a consequence of Steps 1 to 3 together with the Arzelà–Ascoli theorem, we conclude that N maps B_q into a precompact set in $PC \times PC$.

Step 4. It remains to show that

$$\mathcal{A} = \{(x,y) \in PC \times PC : (x,y) = \lambda N(x,y),\ \lambda \in (0,1)\},$$

is bounded.

Let $(x,y) \in \mathcal{A}$. Then $x = \lambda N_1(x,y)$ and $y = \lambda N_2(x,y)$ for some $0 < \lambda < 1$. Thus, for $t \in J$, we have

$$\begin{aligned}
\mathbb{E}|x(t)|^2 &\le 4\mathbb{E}|x_0|^2 + 4\mathbb{E}\left|\sum_{l=1}^{\infty}\int_0^t f_l^1(s,u(s))dW^l(s)\right|^2 \\
&\quad + 4\mathbb{E}\left|\int_0^t g^1(s,x(s),y(s))ds\right|^2 + 4\mathbb{E}\left|\sum_{k=1}^{m} I_k(x(t_k))\right|^2 \\
&\le 4\mathbb{E}|x_0|^2 + 4C_2\int_0^t E\|f^1(s,x(s),y(s))\|^2 ds \\
&\quad + 4T\int_0^t \mathbb{E}|g^1(s,x(s),y(s))|^2 ds + E\sum_{k=1}^{m}|I_1(x(t_k))|^2 \\
&\le 4\mathbb{E}|x_0|^2 + 4C_2\int_0^t p_1(s)\psi_1(\mathbb{E}|x(s)|^2 + \mathbb{E}|y(s)|^2)ds \\
&\quad + 4T\int_0^t p_2(s)\psi_2(\mathbb{E}|x(s)|^2 + \mathbb{E}|y(s)|^2)ds + 4\sum_{k=1}^{m} c_k.
\end{aligned}$$

Hence,

$$\begin{aligned}
\mathbb{E}|x(t)|^2 &\le 4\mathbb{E}|x_0|^2 + 4C_2\int_0^t p_1(s)\psi_1(\mathbb{E}|x(s)|^2 + \mathbb{E}|y(s)|^2)ds \\
&\quad + 4T\int_0^t p_2(s)\psi_2(\mathbb{E}|x(s)|^2 + \mathbb{E}|y(s)|^2)ds + 4\sum_{k=1}^{m} c_k,
\end{aligned}$$

and

$$\mathbb{E}|y(t)|^2 \le 4\mathbb{E}|x_0|^2 + 4C_2\int_0^t p_3(s)\psi_2(\mathbb{E}|x(s)|^2 + \mathbb{E}|y(s)|^2)ds$$

$$+4T\int_0^t p_4(s)\psi_3(\mathbb{E}|x(s)|^2+\mathbb{E}|y(s)|^2)ds+4\sum_{k=1}^m \widetilde{c}_k.$$

Therefore,

$$\mathbb{E}|x(t)|^2+\mathbb{E}|y(t)|^2\le\gamma+\int_0^t p(s)\phi(\mathbb{E}|x(s)|^2+\mathbb{E}|y(s)|^2)ds,$$

where

$$\gamma=8\mathbb{E}|x_0|^2+4\sum_{k=1}^m(c_k+\widetilde{c}_k),$$
$$p(t)=m_1(t)+m(t),$$

and

$$\phi(t)=\sum_{i=1}^m\psi_i(t).$$

By Lemma 3.1, we have

$$\mathbb{E}|x(t)|^2+\mathbb{E}|y(t)|^2\le\Gamma^{-1}\left(\int_\gamma^T p(s)ds\right)=K,\quad\text{for each } t\in J,$$

where

$$\Gamma(z)=\int_\gamma^z\frac{du}{\phi(u)}.$$

Consequently,

$$\|x\|_{PC}\le K\quad\text{and}\quad\|y\|_{PC}\le K.$$

This shows that $\mathcal{E}$ is bounded. As a consequence of Theorem 3.13, we deduce that N has a fixed point (x,y), which is a solution to the problem (4.2). □

The goal of the second result of this section is to apply the Schauder fixed-point theorem. For the study of this problem, we first introduce the following hypotheses:
(H_7) There exist nonnegative numbers $\overline{a}_i$ and $\overline{b}_i$, c_i for each $i=1,2$,

$$\begin{cases}\mathbb{E}(|f_1(t,x,y)|^2)\le\overline{a}_1\mathbb{E}(|x|)^2+\overline{b}_1E(|y|)^2+c_1,\\ \mathbb{E}(|f_2(t,x,y)|^2)\le\overline{a}_2\mathbb{E}(|x|)^2+\overline{b}_2\mathbb{E}(|y|)^2+c_2.\end{cases}$$

(H_8) There exist positive constants $\overline{\alpha}_i$ and $\overline{\beta}_i$, λ_i for each $i = 1,2$,

$$\begin{cases} \mathbb{E}(|g_1(t,x,y)|^2) \le \overline{\alpha}_1 \mathbb{E}(|x|)^2 + \overline{\beta}_1 \mathbb{E}(|y|)^2 + \lambda_1, \\ E(|g_2(t,x,y)|^2) \le \overline{\alpha}_2 \mathbb{E}(|x|)^2 + \overline{\beta}_2 \mathbb{E}(|y|)^2 + \lambda_2. \end{cases}$$

(H_9) There exist constants $d \ge 0, \overline{d} \ge 0$ and $e_i \ge 0$ for each $i \in \{1,2\}$ and $k = 1,\dots,m$, such that

$$\begin{cases} \sum_{k=1}^m \mathbb{E}|I_k(x)|^2 \le d\mathbb{E}|x|^2 + e_1, \\ \sum_{k=1}^m \mathbb{E}|\overline{I}_k(2)|^2 \le \overline{d}\mathbb{E}|x|^2 + e_2, \end{cases}$$

for all $x, y, \in \mathbb{R}$.

Theorem 4.3. *Assume* (H_7)–(H_9) *holds and*

$$M_{a,b} = \begin{pmatrix} C_2\overline{a}_1 + \overline{a}_1 T + d & C_2\overline{b}_1 + \overline{\beta}_1 T \\ C_2\overline{a}_2 + \overline{a}_2 T & C_2\overline{b}_2 + \overline{\beta}_2 T + \overline{d} \end{pmatrix},$$

converges to zero; then problem (4.2) *has at least one solution.*

Proof. Consider the operator

$$N = (N_1, N_2) : PC \times\times PC \longrightarrow PC \times\times PC,$$

defined for $x, y \in PC$ by

$$N_1(x,y) = x_0 + \sum_{l=1}^{\infty} \int_0^t f_l^1(s,x(s),y(s))dW^l(s) + \int_0^t g^1(s,x(s),y(s))ds + \sum_{0<t_k<t} I_k(x(t_k)),$$

and

$$N_2(x,y) = y_0 + \sum_{l=1}^{\infty} \int_0^t f_l^2(s,x(s),y(s))dW^l(s) + \int_0^t g^2(s,x(s),y(s))ds + \sum_{0<t_k<t} \overline{I}_k(y(t_k)),$$

$$D = \{(x,y) \in PC \times PC : \|x\|_{PC} \le R_1, \|y\|_{PC} \le R_2\}.$$

Obviously, the set D is a bounded, closed, convex set in space $PC \times PC$.

Using the BDG inequality, we have

$$\begin{aligned} |N_1(x,y)|^2 &= \left| x_0 + \sum_{l=1}^{\infty} \int_0^t f_l^1(s,x(s),y(s))dW^l(s) + \int_0^t g^1(s,x(s),y(s))ds + \sum_{0<t_k<t} I_k(x(t_k)) \right|^2 \\ &\le 4|x_0|^2 + 4\left| \sum_{l=1}^{\infty} \int_0^t f_l^1(s,x(s),y(s))dW^l(s) \right|^2 + 4\left| \int_0^t g^1(s,x(s),y(s))ds \right|^2 \end{aligned}$$

$$+\left|4\sum_{k=1}^{m} I_k(x(t_k))\right|^2.$$

Using the BDG inequality, we get

$$\begin{aligned}
\mathbb{E}|N_1(x,y)|^2 &\le 4\mathbb{E}|x_0|^2 + 4E\left|\sum_{l=1}^{\infty}\int_0^t f_l^1(s,x(s),y(s))dW^l(s)\right|^2 \\
&\quad + 4E\left|\int_0^t g^1(s,x(s),y(s))ds\right|^2 + 4\mathbb{E}\left|\sum_{k=1}^{m} I_k(x(t_k))\right|^2 \\
&\le 4\mathbb{E}|x_0|^2 + 4C_2\int_0^t \mathbb{E}|f^1(s,x(s),y(s))|^2 ds \\
&\quad + 4T\int_0^t E|g^1(s,x(s),y(s))|^2 ds + 4E\sum_{k=1}^{m}|I_k(x(t_k)|^2 \\
&\le 4E|x_0|^2 + 4C_2\int_0^t (\overline{a}_1\mathbb{E}|x(s)|^2 + 4\overline{b}_1\mathbb{E}|y(s)|^2 + c_1)ds \\
&\quad + 4T\int_0^t (\overline{a}_1\mathbb{E}|x(s)|^2 + \overline{\beta}_1 E|y(s)|^2 + \lambda_1)ds + 4dE|x|^2 + 4e_1 \\
&\le 4E|x_0|^2 + 4C_2\overline{a}_1\int_0^t \mathbb{E}|x(s)|^2 ds + 4\overline{b}_1 C_2\int_0^t \mathbb{E}|y(s)|^2 ds + 4c_1 T \\
&\quad + 4T\overline{a}_1\int_0^t \mathbb{E}|x(s)|^2 d(s) + 4\overline{\beta}_1 T\int_0^t \mathbb{E}|y(s)|^2 ds + 4\lambda_1 T + 4d\mathbb{E}|x|^2 + 4e_1.
\end{aligned}$$

Thus,

$$\begin{aligned}
\sup_{t\in J} E|N_1(x,y)|^2 &\le 4(C_2\overline{a}_1 + \overline{a}_1 T + d)\|x\|_{PC}^2 + 4(C_2\overline{b}_1 + \overline{\beta}_1 T)\|y\|_{PC}^2 \\
&\quad + 4E|x_0|^2 + 4e_1 + 4Tc_1 + 4T\lambda_1.
\end{aligned} \tag{4.3}$$

From (4.3), we obtain that

$$\|N_1(x,y))\|_{PC}^2 \le \widetilde{a}_1\|x\|_{PC}^2 + \widetilde{b}_1\|y\|_{PC}^2 + \widetilde{c}_1, \tag{4.4}$$

where

$$\begin{aligned}
\widetilde{a}_1 &= 4C_2\overline{a}_1 + 4\overline{a}_1 T + 4d, \\
\widetilde{b}_1 &= 4C_2\overline{b}_1 + 4\overline{\beta}_1 T,
\end{aligned}$$

$$\widetilde{c}_1 = 4E|x_0 \text{ and } |^2 + 4e_1 + 4Tc_1 + 4T\lambda_1.$$

Similarly, we have

$$\|N_2(x,y)\|_{PC}^2 \leq \widetilde{a}_2\|x\|_{PC}^2 + \widetilde{b}_2\|y\|_{PC}^2 + \widetilde{c}_2, \tag{4.5}$$

where

$$\widetilde{a}_2 = 4C_2\overline{a}_2 + 4\overline{a}_2T,$$
$$\widetilde{b}_2 = 4C_2\overline{b}_2 + 4\overline{\beta}_2T + 4\overline{d},$$

and

$$\widetilde{c}_2 = 4E|y_0|^2 + 4e_2 + 4Tc_2 + 4T\lambda_2.$$

Now, (4.4), (4.5) can be

$$\|N(x,y)\|_{PC} = \begin{pmatrix} \|N_1(x,y)\|_{PC} \\ \|N_2(x,y)\|_{PC} \end{pmatrix} \leq 2\begin{pmatrix} \sqrt{C_2\overline{a}_1 + \overline{a}_1T + d} & \sqrt{C_2\overline{b}_1 + \overline{\beta}_1T} \\ \sqrt{C_2\overline{a}_2 + \overline{a}_2T} & \sqrt{C_2\overline{b}_2 + \overline{\beta}_2T + \overline{d}} \end{pmatrix}\begin{pmatrix} \|x\|_{PC} \\ \|y\|_{PC} \end{pmatrix} + \begin{pmatrix} \sqrt{\widetilde{c}_1} \\ \sqrt{\widetilde{c}_2} \end{pmatrix}.$$

Therefore,

$$\|N(x,y)\|_{PC} \leq M_{a,b}\begin{pmatrix} \|x\|_{PC} \\ \|y\|_{PC} \end{pmatrix} + \begin{pmatrix} \sqrt{\widetilde{c}_1} \\ \sqrt{\widetilde{c}_2} \end{pmatrix}.$$

Since $M_{a,b} \in \mathcal{M}_{n\times n}(\mathbb{R}_+)$ converges to zero. Next, we look for two positive numbers R_1, R_2 such that if

$$\|x\|_{PC}^2 \leq R_1,$$
$$\|y\|_{PC}^2 \leq R_2,$$

then

$$\|N_1(x,y)\|_{PC}^2 \leq R_1,$$
$$\|N_2(x,y)\|_{PC}^2 \leq R_1.$$

To this end, it is sufficient that

$$\begin{pmatrix} \sqrt{R_1} \\ \sqrt{R_2} \end{pmatrix} \leq M_{a,b}\begin{pmatrix} \sqrt{R_1} \\ \sqrt{R_2} \end{pmatrix} + \begin{pmatrix} \sqrt{\widetilde{c}_1} \\ \sqrt{\widetilde{c}_2} \end{pmatrix},$$

whence

$$(I - M_{a,b}) \begin{pmatrix} \sqrt{R_1} \\ \sqrt{R_2} \end{pmatrix} \leq \begin{pmatrix} \sqrt{\tilde{c}_1} \\ \sqrt{\tilde{c}_2} \end{pmatrix},$$

that is,

$$\begin{pmatrix} \sqrt{R_1} \\ \sqrt{R_2} \end{pmatrix} \leq (I - M_{a,b})^{-1} \begin{pmatrix} \sqrt{\tilde{c}_1} \\ \sqrt{\tilde{c}_2} \end{pmatrix}.$$

Set

$$D = \{(x,y) \in PC \times PC : \|x\|_{PC}^2 \leq R_1, \|y\|_{PC}^2 \leq R_2\}.$$

It is clear that $N(D) \subset D$.. Hence, by Theorem 3.5, the operator N has at least one fixed point, which is the solution of (4.2). □

4.3.2 Example

We consider the following problem:

$$\begin{cases} dx(t) = \sum_{l=1}^{\infty}(a_{2l+1}\sin k^2 x + a_{2l}\cos l^2 y)dW^l(t) + d_1(t + x(t) + y(t))dt, & t \in [0,1],\ t \neq \frac{1}{2}, \\ dy(t) = \sum_{l=1}^{\infty}(b_{2l+1}\sin k^2 x + b_{2l}\cos l^2 y)dW^l(t) + d_2(t + x(t) + y(t))dt, & t \in [0,1],\ t \neq \frac{1}{2}, \\ \Delta x(t) = c_1 \frac{x(t)}{1+|x(t)|}, & t = \frac{1}{2}, \\ \Delta y(t) = c_1 \frac{y(t)}{1+|y(t)|}, & t = \frac{1}{2}, \\ x(0) = x_0, \\ y(0) = y_0, \end{cases}$$

where

$$c_1, c_2 \in \mathbb{R}, \quad (a_l)_{l\in\mathbb{N}}, (b_l)_{l\in\mathbb{N}} \in l^2,$$

and

$$f_1, f_2 : [0,1] \times \mathbb{R} \times \mathbb{R} \to \mathbb{R}$$

are defined by

$$f_1(t,x,y) = \sum_{k=1}^{\infty}(a_{2k+1}\sin k^2 x + a_{2k}\cos k^2 y),$$
$$f_2(t,x,y) = \sum_{k=1}^{\infty}(b_{2k+1}\sin k^2 x + b_{2k}\cos k^2 y).$$

We deduce

$$\begin{aligned}\|f_1(t,x,y)\|^2 &= \sum_{k=1}^{\infty}(a_{2k+1}\sin k^2x + a_{2k}\cos k^2y)^2\\ &\le 2\sum_{k=1}^{\infty}(a_{2k+1}^2 + a_{2k}^2)\\ &\le 4\sum_{k=1}^{\infty} a_k^2 < \infty,\end{aligned}$$

and

$$\begin{aligned}\|f_1(t,x,y)\|^2 &= \sum_{k=1}^{\infty}(b_{2k+1}\sin k^2x + b_{2k}\cos k^2y)^2\\ &\le 2\sum_{k=1}^{\infty}(b_{2k+1}^2 + b_{2k}^2)\\ &\le 4\sum_{k=1}^{\infty} b_k^2 < \infty.\end{aligned}$$

Hence,

$$\mathbb{E}\left|f_1(t,x,y)\right|^2 \le 4\sum_{k=1}^{\infty} a_k^2 + \mathbb{E}(|x|^2 + |y|^2) \quad \text{for all } x,y \in \mathbb{R},$$

and

$$\mathbb{E}\left|f_2(t,x,y)\right|^2 \le 4\sum_{k=1}^{\infty} b_k^2 + \mathbb{E}(|x|^2 + |y|^2) \quad \text{for all } x,y \in \mathbb{R}.$$

Also, we have

$$I_1(x) = c_1\frac{x(t)}{1+|x(t)|}, \quad I_2(y) = c_2\frac{y}{1+|y|} \quad \Rightarrow \quad E\left|I_1(x)\right|^2 \le c_1, \quad \mathbb{E}\left|I_2(x)\right|^2 \le c_2,$$

and

$$g^1(t,x,y) = d_1(t+x+y), \quad g^2(t,x,y) = d_2(t+x+y), \quad x,y \in \mathbb{R}, \quad t \in [0,1].$$

Hence,

$$\mathbb{E}\left|g^1(t,x,y)\right|^2 \le 3d_1^2(1+\mathbb{E}|x|^2+\mathbb{E}|y|^2), \quad \mathbb{E}\left|g^2(t,x,y)\right|^2 \le 3d_2^2(1+E|x|^2+E|y|^2).$$

Thus, all the conditions of Theorem 4.2 hold; therefore, problem (4.2) has at least one solution.

Remark 4.1. In the case where $f_2 = 0$, $g^1 = g^2 = 0$, and $I_1 = I_2 = 0$ are in our example, we include the equation for Brownian motion on the group of diffeomorphisms of the circle.

5 Impulsive functional stochastic differential with multiple delay

In this chapter, we establish sufficient conditions for membership in some classes concerned with the existence of solutions for a system of stochastic impulsive differential equations with fixed moments and multiple delays of the following type:

$$\begin{cases} dx(t) = \sum_{i=1}^{n_*} x(t-T_i) + \sum_{l=1}^{\infty} f_l^1(t,x_t,y_t)dW^l(t) + g^1(t,x_t,y_t)dt, & t \in J,\ t \neq t_k, \\ dy(t) = \sum_{i=1}^{n_*} y(t-T_i) + \sum_{l=1}^{\infty} f_l^2(t,x_t,y_t)dW^l(t) + g^2(t,x_t,y_t)dt, & t \in J,\ t \neq t_k, \\ \Delta x(t) = I_k(x(t_k)), & t = t_k,\ k = 1,2,\ldots,m, \\ \Delta y(t) = \bar{I}_k(y(t_k)), & \\ x(t) = \phi(t) \in \mathcal{D}_{\mathcal{F}_0}, & t \in [-r,0], \\ y(t) = \bar{\phi}(t) \in \mathcal{D}_{\mathcal{F}_0}, & t \in [-r,0], \end{cases} \tag{5.1}$$

where

$$n_* \in \{1,2,\ldots\}, \quad r = \max_{1 \le i \le n_*} T_i,$$

and with respect to a complete probability space $(\Omega, \mathcal{F}, \mathcal{F}_t, P)$ with a family of right continuous and increasing σ-algebras

$$\{\mathcal{F}_t,\ t \in J\},$$

satisfying $\mathcal{F}_t \subset \mathcal{F}$. The impulse times t_k satisfy

$$0 = t_0 < t_1 < t_2 < \cdots t_m < T.$$

As for y_t, we mean the segment solution, which is defined in the usual way, that is, if

$$y(\cdot,\cdot) : [-r,b] \times \Omega \to \mathbb{R}^n,$$

then for any $t \ge 0$,

$$y_t(\cdot,\cdot) : [-r,0] \times \Omega \to \mathbb{R}^n,$$

is given by

$$y_t(\theta,\omega) = y(t+\theta,\omega), \quad \text{for } \theta \in [-r,0],\ \omega \in \Omega.$$

Here, $y_t(\cdot)$ represents the history of the state from time $t-r$ up to the present time t. Before describing the properties fulfilled by the operators f^i, g^i and I_k, $\bar{I}_k$, we need to introduce some notation and describe some spaces.

https://doi.org/10.1515/9783112251898-005

Let $\mathcal{D}_{\mathcal{F}_0}$ be the following space defined by

$$\mathcal{D}_{\mathcal{F}_0} = \{\phi\colon [-r,0]\times\Omega \to \mathbb{R}^n\}$$

is continuous everywhere except for a finite number of points

$$\phi(t_k^-) \quad \text{and} \quad \phi(t_k^+) \quad \text{with } \phi(t_k) = \phi(t_k^-),$$

endowed with the norm

$$\|\phi(t)\|_{\mathcal{D}_{\mathcal{F}_0}} = \int_{-r}^{0} |\phi(t)|^2 dt.$$

Now, for a given $b > 0$, we define

$$\mathcal{D}_{\mathcal{F}_b} = \{y\colon [-r,b]\times\Omega \to \mathbb{R}^n,\ y_k \in C(J_k,\mathbb{R}^n)\},$$

for $k = 1,\ldots m$, $y_0 \in \mathcal{D}_{\mathcal{F}_0}$, and there exists $y(t_k^-)$, and $y(t_k^+)$, with

$$y(t_k) = y(t_k^-), \quad k = 1,\ldots,m,$$

and

$$\sup_{t\in[0,b]} E(|y(t)|^2) < \infty,$$

$$\int_{-r}^{0} |\phi(t)|^2 dt < \infty,$$

endowed with the norm

$$\|y\|_{\mathcal{D}_{\mathcal{F}_b}} = \|\phi\|_{\mathcal{D}_{\mathcal{F}_0}} + \sup_{0\le s\le b} (E\|y(s)\|^2)^{\frac{1}{2}},$$

where y_k denotes the restriction of y to

$$J_k = (t_{k-1},t_k], \quad k = 1,2,\ldots,m,$$

and $J_0 = [-r,0]$.

Then we will consider our initial data $\phi \in \mathcal{D}_{\mathcal{F}_0}$ where $J = [0,b]$,

$$f_l^1, f_l^2 : J\times\mathcal{D}_{\mathcal{F}_0}\times\mathcal{D}_{\mathcal{F}_0} \to \mathbb{R}^n,$$
$$g^1, g^2 : J\times\mathcal{D}_{\mathcal{F}_0}\times\mathcal{D}_{\mathcal{F}_0} \to \mathbb{R}^n,$$

are Carathéodory functions, and W^l is an infinite sequence of independent standard Brownian motions, $l = 1, 2, \ldots$ and

$$I_k, \bar{I}_k \in C(\mathbb{R}^n, \mathbb{R}^n) \quad (k = 1, \ldots, m),$$

and

$$\begin{aligned} \Delta x|_{t=t_k} &= x(t_k^+) - x(t_k^-), \\ \Delta y|_{t=t_k} &= y(t_k^+) - y(t_k^-). \end{aligned}$$

The notation

$$y(t_k^+) = \lim_{h \to 0^+} y(t_k + h),$$

and

$$y(t_k^-) = \lim_{h \to 0^+} y(t_k - h),$$

stand for the right and left limits of the function y at $t = t_k$, respectively. Set

$$\begin{cases} f(t,x,y) = (f_1(t,x,y), f_2(t,x,y), \ldots), \\ \|\sigma(t,x,y)\|^2 = \sum_{j=1}^{\infty} |f_j(t,x,y)|^2 < \infty, \end{cases}$$

where $\sigma(\cdot,\cdot,\cdot) \in \ell^2$, and ℓ^2 are given by

$$\ell^2 = \left\{ \phi = (\phi_j)_{j \geq 1} : [0,b] \times \mathbb{R}^n \times \mathbb{R}^n \to \mathbb{R}^n : \|\phi(t,x,y)\|^2 = \sum_{j=1}^{\infty} |\phi_j(t,x,y)|^2 < \infty \right\}.$$

We denote

$$a \wedge b = \min(a,b),$$

and

$$a \vee b = \max(a,b).$$

It is obvious that the system (5.1) can be seen as a fixed-point problem

$$\begin{cases} dz(t) = \sum_{i=1}^{n_*} z(t - T_i) + g(t, z_t)dt + \sum_{l=1}^{\infty} f_l(t, z_t))dW^l(t), & t \in [0,b],\ t \neq t_k, \\ \Delta z(t) = I_k^*(z(t_k)), & t = t_k,\ k = 1, 2, \ldots, m \\ z(t) = \tilde{\phi}(t) \in \mathcal{D}_{\mathcal{F}_0}, & t \in [-r, 0], \end{cases}$$

where

$$z(t) = \begin{bmatrix} x(t) \\ y(t) \end{bmatrix},$$

$$g(t,z) = \begin{bmatrix} g^1(t,x_t,y_t) \\ g^2(t,x_t,y_t) \end{bmatrix},$$

$$f_l(t,z) = \begin{bmatrix} f_l^1(t,x_t,y_t) \\ f_l^2(t,x_t,y_t) \end{bmatrix},$$

and

$$\widetilde{\phi}(t) = \begin{bmatrix} \phi(t) \\ \overline{\phi}(t) \end{bmatrix}.$$

5.1 Multiple delays

Assume that $W(t)$ is an infinite sequence of independent standard Brownian motions adapted to $\{\mathcal{F}_t\}_{t\geq 0}$, that is,

$$W(t) = \left(W^1(t), W^2(t), \ldots\right)^\top,$$

where each $W^k(t)$, $k \in \mathbb{N}$ is a one-dimensional $\{\mathcal{F}_t\}$-Brownian motion.

An $\mathbb{R}^n$-valued random variable is an $\mathcal{F}$-measurable mapping

$$x(t) : \Omega \longrightarrow \mathbb{R}^n.$$

For a fixed $t \geq 0$, the value $x(t,\omega)$ represents the state of the system at time t, corresponding to the outcome $\omega \in \Omega$.

A family of random variables

$$S = \left\{x(t,\omega) : \Omega \rightarrow \mathbb{R}^n;\ t \in J\right\}$$

indexed by a time interval is called a *stochastic process*. The process $\{x(t)\}_{t\in J}$ is said to be *adapted* if, for every $t \in J$, the random variable $x(t)$ is $\mathcal{F}_t$-measurable.

In the context of systems with multiple delays, we allow the evolution of the process $x(t)$ to depend not only on its current state but also on several past states $x(t-\tau_i)$, where $\tau_i > 0$ and $i = 1, \ldots, m$ denote distinct delay times. Such systems naturally arise in applications where memory or aftereffects play a significant role, for instance, in population dynamics, control theory, and viscoelasticity.

We assume that the stochastic process $x(t)$ has almost surely continuous sample paths and satisfies suitable integrability conditions, such as

$$\mathbb{E}\|x(t)\|^2 < \infty, \quad t \in J,$$

which ensure the well-posedness of the associated stochastic integrals.

5.2 Contraction principle

The classical Banach contraction principle was extended for contractive maps on spaces endowed with vector-valued metric spaces by Perov in 1964. The purpose of this section is to present the version of Schaefer's fixed-point theorem and the nonlinear alternative of the Leary–Schauder-type in generalized Banach spaces.

Theorem 5.1. *Let (X, d) be a complete generalized metric space with*

$$d : X \times X \longrightarrow \mathbb{R}^n,$$

and let

$$N : X \longrightarrow X,$$

be such that

$$d(N(x), N(y)) \le Md(x, y),$$

for all $x, y \in X$ and some square matrix M of nonnegative numbers. If the matrix M converges to zero, that is, $M^k \longrightarrow 0$ as $k \longrightarrow \infty$, then N has a unique fixed point $x_* \in X$

$$d(N^k(x_0), x_*) \le M^k(I - M)^{-1} d(N(x_0), x_0),$$

for every $x_0 \in X$ and $k \ge 1$.

Theorem 5.2 ([5]). *Let E be a generalized Banach space, $C \subset E$ be a nonempty closed convex subset of E, and $N : C \to C$ be a continuous operator with a relatively compact range. Then N has at least one fixed point in C.*

As a consequence of the Schauder fixed-point theorem, we present Schaefer's fixed-point theorem and the nonlinear alternative Leary–Schauder-type theorem in a generalized Banach space.

Theorem 5.3. *Let $(E, \|\cdot\|)$ be a generalized Banach space and $N : E \to E$ be a continuous compact mapping. Moreover, assume that the set*

$$\mathcal{A} = \{x \in E : x = \lambda N(x) \text{ for some } \lambda \in (0, 1)\},$$

is bounded. Then N has a fixed point.

Lemma 5.1. *Let*

$$f^i, g^i : \mathcal{D}_{\mathcal{F}_0} \times \mathcal{D}_{\mathcal{F}_0} \to \mathbb{R}^n,$$

be a continuous function. Then $z = (x, y)$ is the unique solution of the initial value problem

$$\begin{cases} dx(t) = \sum_{i=1}^{n_*} x(t-T_i) + \sum_{l=1}^{\infty} f_l^1(x_t,y_t)dW^l(t) + g^1(x_t,y_t)dt, & t \in J,\ t \neq t_k, \\ dy(t) = \sum_{i=1}^{n_*} y(t-T_i) + \sum_{l=1}^{\infty} f_l^2(x_t,y_t)dW^l(t) + g^2(x_t,y_t)dt, & t \in J,\ t \neq t_k, \\ \Delta x(t) = I_k(x(t_k)), & t = t_k,\ k = 1,2,\dots,m, \\ \Delta y(t) = \overline{I}_k(y(t_k)), & \\ x(t) = \phi(t) \in \mathcal{D}_{\mathcal{F}_0}, & t \in [-r,0], \\ y(t) = \overline{\phi}(t) \in \mathcal{D}_{\mathcal{F}_0}, & t \in [-r,0], \end{cases}$$

where $r = \max_{1\le i\le n_*} T_i$ *if and only if z is a solution of the impulsive integral functional differential equation*

$$\begin{cases} x(t) = \phi(t) \in \mathcal{D}_{\mathcal{F}_0}, \quad t \in [-r,0], \\ x(t) = \phi(0) + \sum_{i=1}^{n_*} \int_{T_i}^{0} \phi(s)ds + \sum_{i=1}^{n_*} \int_0^{t-T_i} y(s)ds + \sum_{l=1}^{\infty} \int_0^t f_l^1(x_s,y_s)dW^l(s) \\ \quad + \int_0^t g^1(x_s,y_s)ds + \sum_{0\le t_k\le t} I_k(x(t_k)), \quad t \in J, \\ y(t) = \overline{\phi}(t) \in \mathcal{D}_{\mathcal{F}_0}, \quad t \in [-r,0], \\ y(t) = \overline{\phi}(0) + \sum_{i=1}^{n_*} \int_{T_i}^{0} \overline{\phi}(s)ds + \sum_{i=1}^{n_*} \int_0^{t-T_i} y(s)ds + \sum_{l=1}^{\infty} \int_0^t f_l^2(x_s,y_s)dW^l(s) \\ \quad + \int_0^t g^2(x_s,y_s)ds + \sum_{0\le t_k\le t} \overline{I}_k(x(t_k)), \quad t \in J. \end{cases} \tag{5.2}$$

Proof. Let (x,y) be a possible solution to the problem (5.2). Then

$$z|_{[-r,t_1]} = (x|_{[-r,t_1]}, y|_{[-r,t_1]}),$$

is a solution to

$$dx(t) = \sum_{i=1}^{n_*} x(t-T_i) + \sum_{l=1}^{\infty} f_l^1(x_t,y_t)dW^l(t) + g^1(x_t,y_t)dt, \quad t \in J = [0,b].$$

Assume that

$$t_k < t \le t_{k+1}, \quad k = 1,\dots,m.$$

By integrating the above inequality, we have

$$x(t_1^-) - x(0) = \sum_{i=1}^{n_*} \int_0^{t_1} x(s-T_i)ds + \sum_{l=1}^{\infty} \int_0^{t_1} f_l^1(x_s,y_s)dW^l(s) + \int_0^{t_1} g^1(x_s,y_s)ds,$$

$$x(t_1^-) - x(0) = \sum_{i=1}^{n_*} \int_{-T_i}^{t_1-T_i} x(s)ds + \sum_{l=1}^{\infty} \int_0^{t_1} f_l^1(x_s,y_s)dW^l(s) + \int_0^{t_1} g^1(x_s,y_s)ds,$$

$$x(t_2^-) - x(t_1^+) = \sum_{i=1}^{n_*} \int_{t_1}^{t_2} x(s-T_i)ds + \sum_{l=1}^{\infty} \int_{t_1}^{t_2} f_l^1(x_s,y_s)dW^l(s) + \int_{t_1}^{t_2} g^1(x_s,y_s)ds,$$

$$x(t_2^-) - x(t_1^-) = I_1(x(t_1)) + \sum_{i=1}^{n_*} \int_{t_1-T_i}^{t_2} x(s)ds + \sum_{l=1}^{\infty} \int_{t_1}^{t_2} f_l^1(x_s, y_s)dW^l(s) + \int_{t_1}^{t_2} g^1(x_s, y_s)ds,$$

by recurrence

$$x(t_k^-) - x(t_{k-1}^+) = \sum_{i=1}^{n_*} \int_{t_{k-1}}^{t_k} x(s - T_i)ds + \sum_{l=1}^{\infty} \int_{t_{k-1}}^{t_k} f_l^1(x_s, y_s)dW^l(s) + \int_{t_{k-1}}^{t_k} g^1(x_s, y_s)ds,$$

$$x(t_k^-) - x(t_{k-1}^-) = I_k(x(t_k)) + \sum_{i=1}^{n_*} \int_{t_{k-1}-T_i}^{t_k-T_i} x(s)ds + \sum_{l=1}^{\infty} \int_{t_{k-1}}^{t_k} f_l^1(x_s, y_s)dW^l(s) + \int_{t_{k-1}}^{t_k} g^1(x_s, y_s)ds,$$

$$x(t) - x(t_k^-) = I_k(x(t_k)) + \sum_{i=1}^{n_*} \int_{t_k-T_i}^{t-T_i} x(s)ds + \sum_{l=1}^{\infty} \int_{t_k}^{t} f_l^1(x_s, y_s)dW^l(s) + \int_{t_k}^{t} g^1(x_s, y_s)ds.$$

Adding these together, we get

$$x(t) = x(0) + \sum_{0 \le t_k \le t} I_k(x(t_k)) + \sum_{i=1}^{n_*} \int_{-T_i}^{t-T_i} x(s)ds + \sum_{l=1}^{\infty} \int_0^t f_l^1(x_s, y_s)dW^l(s) + \int_0^t g^1(x_s, y_s)ds$$

$$x(t) = \phi(0) + \sum_{0 \le t_k \le t} I_k(x(t_k)) + \sum_{i=1}^{n_*} \int_{-T_i}^{0} x(s)ds + \sum_{i=1}^{n_*} \int_0^{t-T_i} x(s)ds + \sum_{l=1}^{\infty} \int_0^t f_l^1(x_s, y_s)dW^l(s)$$
$$+ \int_0^t g^1(x_s, y_s)ds.$$

Then x is a solution to the problem (5.2). Let

$$t \in [0, b] \backslash \{t_1, \dots, t_m\},$$

and

$$x(t) = \phi(0) + \sum_{0 \le t_k \le t} I_k(x(t_k)) + \sum_{i=1}^{n_*} \int_{-T_i}^{0} \phi(s)ds + \sum_{i=1}^{n_*} \int_0^{t-T_i} x(s)ds + \sum_{l=1}^{\infty} \int_0^t f_l^1(x_s, y_s)dW^l(s)$$
$$+ \int_0^t g^1(x_s, y_s)ds.$$

Similarly,

$$y(t) = \overline{\phi}(0) + \sum_{0\le t_k\le t} I_k(x(t_k)) + \sum_{i=1}^{n_*} \int_{-T_i}^{0} \overline{\phi}(s)ds + \sum_{i=1}^{n_*} \int_{0}^{t-T_i} y(s)ds + \sum_{l=1}^{\infty} \int_{0}^{t} f_l^2(x_s, y_s)dW^l(s)$$
$$+ \int_0^t g^2(x_s, y_s)ds. \qquad \square$$

Definition 5.1. Given $\phi \in \mathcal{D}_{\mathcal{F}_0}$, $\mathbb{R}^n$-valued stochastic process $z = (x, y)$ and

$$\{z(t),\ t \in [-r, b]\},$$

is called a mild solution of the problem (5.1) if $y(t)$ is measurable and $\mathcal{F}_t$-adapted, for each $t > 0$,

$$z(t) = \tilde{\phi}(t) \quad \text{on } [-r, 0],$$
$$\Delta z|_{t=t_k} = I_k(y(t_k^-)), \quad k = 1, 2, \ldots, m,$$

the restriction of $z(\cdot,\cdot)$ to

$$[0, b) - \{t_1, t_2, \ldots, t_m\},$$

is continuous, z satisfies the integral equation

$$\begin{cases} x(t) = \phi(t) \in \mathcal{D}_{\mathcal{F}_0}, \quad t \in [-r, 0], \\ x(t) = \phi(0) + \sum_{i=1}^{n_*} \int_{T_i}^{0} \phi(s)ds + \sum_{i=1}^{n_*} \int_0^{t-T_i} y(s)ds + \sum_{l=1}^{\infty} \int_0^t f_l^1(s, x_s, y_s)dW^l(s) \\ \quad + \int_0^t g^1(s, x_s, y_s)ds + \sum_{0\le t_k\le t} I_k(x(t_k)), \quad t \in J \\ y(t) = \overline{\phi}(t) \in \mathcal{D}_{\mathcal{F}_0}, \quad t \in [-r, 0], \\ y(t) = \overline{\phi}(0) + \sum_{i=1}^{n_*} \int_{T_i}^{0} \overline{\phi}(s)ds + \sum_{i=1}^{n_*} \int_0^{t-T_i} y(s)ds + \sum_{l=1}^{\infty} \int_0^t f_l^2(s, x_s, y_s)dW^l(s) \\ \quad + \int_0^t g^2(s, x_s, y_s)ds + \sum_{0\le t_k\le t} \overline{I}_k(x(t_k)), \quad t \in J. \end{cases}$$

First, we will list the following hypotheses that will be imposed:

(H_{10}) There exist functions $a_i,\ b_i \in L^1([0, b], \mathbb{R}^+)$ such that

$$\int_0^t \|f^i(s, x_s, y_s) - f^i(s, \overline{x}_s, \overline{y}_s)\|^2 ds \le \int_{-r}^t a_i(s)|x(s) - \overline{x}(s)|^2 + b_i(s)|y(s) - \overline{y}(s)|^2 ds,$$

for all $x, y, \overline{x}, \overline{y} \in \mathcal{D}_{\mathcal{F}_T}$.

(H_{11}) There exist functions $\alpha_i, \beta_i \in L^1([0, b], \mathbb{R}^+)$ such that

$$\int_0^t |g^i(s, x_s, y_s) - g^i(s, \overline{x}(s), \overline{y})(s)|^2 ds \le \int_{-r}^t \alpha_i(s)\|x(s) - \overline{x}(s)\|^2 + \beta_i(s)\|y(s) - \overline{y}(s)\|^2 ds,$$

for all $x, y, \overline{x}, \overline{y} \in \mathcal{D}_{\mathcal{F}_T}$, and a. e. $t \in [0, b]$.

(H_{12}) there exist constants $d_k \geq 0$ and $\overline{d}_k \geq 0$, $k = 1, \ldots, m$ such that

$$E(|I_k(x) - I_k(\overline{x})|)^2 \leq d_k E(|x - \overline{x}|)^2,$$
$$E(|\overline{I}_k(y) - \overline{I}_k(\overline{y})|)^2 \leq \overline{d}_k E(|y - \overline{y}|)^2,$$

for all $x, y, \overline{x}, \overline{y} \in \mathcal{D}_{\mathcal{F}_b}$.

For our main consideration of problem (5.1), a Perov fixed point is used to investigate the existence and uniqueness of solutions for a system of impulsive stochastic differential equations.

Theorem 5.4. *Assume that* (H_{10})–(H_{12}) *are satisfied, and the matrix*

$$M = \begin{pmatrix} \sqrt{\frac{1}{\tau} + 4\sum_{k=1}^{m} d_k} & \frac{1}{\sqrt{\tau}} \\ \frac{1}{\sqrt{\tau}} & \sqrt{\frac{1}{\tau} + 4\sum_{k=1}^{m} \overline{d}_k} \end{pmatrix},$$

where $\sqrt{\tau} > 2$*. If* M *converges to zero. Then the problem* (5.1) *has a unique solution on* $[-r, b]$.

Proof. Consider the operator

$$N : \mathcal{D}_{\mathcal{F}_T} \times \mathcal{D}_{\mathcal{F}_T} \to \mathcal{D}_{\mathcal{F}_T} \times \mathcal{D}_{\mathcal{F}_T},$$

defined by

$$N(x, y) = (N_1(x, y), N_2(x, y)), \quad (x, y) \in \mathcal{D}_{\mathcal{F}_T} \times \mathcal{D}_{\mathcal{F}_T},$$

where

$$N_1(x,y)(t) = \begin{cases} \phi(t) \in \mathcal{D}_{\mathcal{F}_0}, \quad t \in [-r, 0], \\ \phi(0) + \sum_{i=1}^{n_*} \int_{T_i}^{0} \phi(s) ds + \sum_{i=1}^{n_*} \int_0^{t-T_i} x(s) ds + \sum_{l=1}^{\infty} \int_0^t f_l^1(s, x(s), y(s)) dW^l(s) \\ \quad + \int_0^t g^1(s, x(s), y(s)) ds + \sum_{0<t_k<t} I_k(x(t_k^-)), \quad \text{if } t \in [0, b], \end{cases}$$

and

$$N_2(x,y)(t) = \begin{cases} \overline{\phi}(t) \in \mathcal{D}_{\mathcal{F}_0}, \quad t \in [-r, 0], \\ \overline{\phi}(0) + \sum_{i=1}^{n_*} \int_{T_i}^{0} \overline{\phi}(s) ds + \sum_{i=1}^{n_*} \int_0^{t-T_i} y(s) ds + \sum_{l=1}^{\infty} \int_0^t f_l^2(s, x_s, y_s) dW^l(s) \\ \quad + \int_0^t g^2(s, x_s, y_s) ds + \sum_{0<t_k<t} \overline{I}_k(y(t_k^-)), \quad \text{if } t \in [0, b]. \end{cases}$$

The operator N is well-defined; in other words, given

$$(x, y) \in \mathcal{D}_{\mathcal{F}_b} \times \mathcal{D}_{\mathcal{F}_b}.$$

We shall use Theorem 5.1 to prove that N has a fixed point. Indeed, let

$$(x,y),(\overline{x},\overline{y}) \in \mathcal{D}_{\mathcal{F}_b} \times \mathcal{D}_{\mathcal{F}_b}.$$

Then we have for each $t \in [0,b]$; for each $t \in [-r,b]$ we have

$$\begin{aligned}
&|N_1(x(t),y(t)) - N_1(\overline{x}(t),\overline{y}(t))|^2 \\
&= \Bigg|\sum_{l=1}^{\infty}\int_0^t f_l^1(s,x_s,y_s)dW^l(s) + \sum_{i=1}^{n_*}\int_0^{t-T_i} x(s)ds + \int_0^t g^1(s,x_s,y_s)ds + \sum_{0<t_k<t} I_k(x(t_k^-)) \\
&\quad - \sum_{l=1}^{\infty}\int_0^t f_l^1(s,\overline{x}_s,\overline{y}_s)dW^l(s) - \sum_{i=1}^{n_*}\int_0^{t-T_i} \overline{x}(s)ds - \int_0^t g^1(s,\overline{x}_s,\overline{y}_s)ds - \sum_{0<t_k<t} I_k(x(t_k^-))\Bigg|^2.
\end{aligned}$$

By the BDG inequality, we get

$$\begin{aligned}
E|N_1(x(t),y(t)) - N_1(\overline{x}(t),\overline{y}(t))|^2 \le{}& 4C_2\int_0^t E\|f^1(s,x_s,y_s) - f^1(s,\overline{x}_s,\overline{y}_s)\|^2 ds \\
&+ 4b\int_0^t E|g^1(s,x_s,y_s) - g^1(s,\overline{x}_s,\overline{y}_s)|^2 ds \\
&+ 4\sum_{k=1}^{m} E|I_k(x(t_k)) - I_k(\overline{x}(t_k))|^2 \\
&+ 4n_* b\int_0^t E|x(s) - \overline{x}(s)|^2 ds.
\end{aligned}$$

Therefore, by condition (H_{10})–(H_{12}),

$$\begin{aligned}
&E|N_1(x(t),y(t)) - N_1(\overline{x}(t),\overline{y}(t))|^2 \\
&\le \int_0^t (4C_2a_1(s) + 4ba_1(s))E(|x(s) - \overline{x}(s)|)^2 ds \\
&\quad + \int_0^t (4C_2b_1(s) + 4tb\beta_1(s))E(|y(s) - \overline{y}(s)|)^2 ds \\
&\quad + 4\sum_{k=1}^{m} d_k e^{\tau\hat{a}(s)}e^{-\tau\hat{a}(s)}E(|x(t_k) - \overline{x}(t_k)|)^2
\end{aligned}$$

and, therefore, since

$$(x(s),y(s)) = (\overline{x}(s),\overline{y}(s)),$$

over the interval $[-r, 0]$, by taking the supremum in the above inequality

$$\begin{aligned}
&\sup_{h\in[0,t]} E\left|N_1(x(h),y(h)) - N_1(\overline{x}(h),\overline{y}(h))\right|^2 \\
&\le \int_0^t a(s)e^{\tau\widehat{a}(s)}e^{-\tau\widehat{a}(s)} \sup_{\theta\in[0,s]} E\|x(\theta)-\overline{x}(\theta)\|^2 ds \\
&\quad + \int_0^t a(s)e^{\tau\widehat{a}(s)}e^{-\tau\widehat{a}(s)} \sup_{\theta\in[0,s]} E\|y(\theta)-\overline{y}(\theta)\|^2 ds \\
&\quad + 4\sum_{k=1}^m d_k e^{\tau\widehat{a}(s)}e^{-\tau\widehat{a}(s)} E\left(|x(t_k)-\overline{x}(t_k)|\right)^2 \\
&\le \int_0^t a(s)e^{\tau\widehat{a}(s)} ds\|x-\overline{x}\|_*^2 + \int_0^t a(s)e^{\tau\widehat{a}(s)} ds\|y-\overline{y}\|_*^2 \\
&\le \frac{1}{\tau}\int_0^t (e^{\tau\widehat{a}(s)})' ds\|x-\overline{x}\|_*^2 + \frac{1}{\tau}\int_0^t (e^{\tau\widehat{a}(s)})' ds\|y-\overline{y}\|_*^2 \\
&\le \left(\frac{1}{\tau} + 4\sum_{k=1}^m d_k\right)e^{\tau\widehat{a}(t)}\|x-\overline{x}\|_*^2 + \frac{1}{\tau}e^{\tau\widehat{a}(t)}\|y-\overline{y}\|_*^2.
\end{aligned}$$

Therefore,

$$e^{-\tau\widehat{a}(t)} \sup_{h\in[0,t]} E\left|N_1(x(h),y(h)) - N_1(\overline{x}(h),\overline{y}(h))\right|^2 \le \left(\frac{1}{\tau} + 4\sum_{k=1}^m d_k\right)\|x-\overline{x}\|_*^2 + \frac{1}{\tau}\|y-\overline{y}\|_*^2,$$

where

$$a(s) = \begin{cases} 0, & t \in [-r, 0], \\ \max\{4C_2 a_1(s) + 4ba_1(s), 4C_2 b_1(s) + 4b\beta_1(s)\}, & t \in [0, b], \end{cases}$$

and

$$\widehat{a}(t) = \int_0^t a(s)ds, \quad \|x\|_*^2 = \sup_{h\in[0,t]} E|x(h,.)|^2 e^{-\tau\widehat{a}(t)}, \quad \|.\|_*.$$

This yields

$$\left\|N_1(x,y) - N_1(\overline{x},\overline{y})\right\|_*^2 \le \left(\frac{1}{\tau} + 4\sum_{k=1}^m \overline{d}_k\right)\|x-\overline{x}\|_*^2 + \frac{1}{\tau}\|y-\overline{y}\|_*^2.$$

Using the fact that for all $a, b \ge 0$, we have

$$\sqrt{a+b} \leq \sqrt{a} + \sqrt{b}. \tag{5.3}$$

We conclude that

$$\|N_1(x,y) - N_1(\overline{x},\overline{y})\|_* \leq \left(\sqrt{\frac{1}{\tau} + 4\sum_{k=1}^{m} d_k}\right)\|x - \overline{x}\|_* + \frac{1}{\sqrt{\tau}}\|y - \overline{y}\|_*.$$

Similar computations for N_1 yield

$$\|N_2(x,y) - N_2(\overline{x},\overline{y})\|_* \leq \frac{1}{\sqrt{\tau}}\|x - \overline{x}\|_* + \left(\sqrt{\frac{1}{\tau} + 4\sum_{k=1}^{m} \overline{d}_k}\right)\|y - \overline{y}\|_*.$$

Thus,

$$\begin{aligned}\|N(x,y) - N(\overline{x},\overline{y})\|_* &= \begin{pmatrix} \|N_1((x,y) - N_1(\overline{x},\overline{y})\|_* \\ \|N_2(x,y) - N_2(\overline{x},\overline{y})\|_* \end{pmatrix} \\ &\leq \begin{pmatrix} \sqrt{\frac{1}{\tau} + 4\sum_{k=1}^{m} d_k} & \frac{1}{\sqrt{\tau}} \\ \frac{1}{\sqrt{\tau}} & \sqrt{\frac{1}{\tau} + 4\sum_{k=1}^{m} \overline{d}_k} \end{pmatrix} \begin{pmatrix} \|x - \overline{x}\|_* \\ \|y - \overline{y}\|_* \end{pmatrix}.\end{aligned}$$

Hence,

$$\|N(x,y) - N(\overline{x},\overline{y})\|_* \leq M \begin{pmatrix} \|x - \overline{x}\|_* \\ \|y - \overline{y}\|_* \end{pmatrix},$$

for all $(x,y), (\overline{x},\overline{y}) \in D_{t_0} \times D_{t_0}$, where

$$M = \begin{pmatrix} \sqrt{\frac{1}{\tau} + 4\sum_{k=1}^{m} d_k} & \frac{1}{\sqrt{\tau}} \\ \frac{1}{\sqrt{\tau}} & \sqrt{\frac{1}{\tau} + 4\sum_{k=1}^{m} \overline{d}_k} \end{pmatrix}.$$

We choose a suitable $\sqrt{\tau} > 2$ such that the matrix

$$\|M\| < 1,$$

then M converges to zero. As a consequence of Perov's fixed-point theorem, N has a unique fixed

$$(x,y) \in \mathcal{D}_{\mathcal{F}_b} \times \mathcal{D}_{\mathcal{F}_b},$$

which is the unique solution to problem (5.1) □

5.3 Existence with Nagumo-type growth conditions

In this section, we present the existence result under a nonlinearity f^i and g^i, $i = 1,2$ satisfying Nagumo type growth conditions:

(H_{13}) There exists a function $p_i \in L^1(J, \mathbb{R}^+)$ and a continuous nondecreasing function $\psi_i : [0,\infty) \to [0,\infty)$ for each $i = 1,2$, such that

$$\begin{cases} E(\|f^1(t,x,y)\|)^2 \le p_1(t)\psi_1(\|x\|^2_{\mathcal{D}_{\mathcal{F}_0}} + \|y\|^2_{\mathcal{D}_{\mathcal{F}_0}}), \\ E(\|f^2(t,x,y)\|)^2 \le p_2(t)\psi_2(\|x\|^2_{\mathcal{D}_{\mathcal{F}_0}} + \|y\|^2_{\mathcal{D}_{\mathcal{F}_b}}), \end{cases}$$

for all $x, y \in \mathcal{D}_{\mathcal{F}_b}$

(H_{14}) There exists a function $p_i \in L^1(J, \mathbb{R}^+)$ and a continuous nondecreasing function

$$\psi_i : [0,\infty) \to [0,\infty), \quad i = 3,4,$$

such that

$$\begin{cases} E(|g^1(t,x,y)|)^2 \le p_3(t)\psi_3(\|x\|^2_{\mathcal{D}_{\mathcal{F}_0}} + \|y\|^2_{\mathcal{D}_{\mathcal{F}_0}}), \\ E(|g^2(t,x,y)|)^2 \le p_4(t)\psi_4(\|x\|^2_{\mathcal{D}_{\mathcal{F}_0}} + \|y\|^2_{\mathcal{D}_{\mathcal{F}_0}}), \end{cases}$$

with

$$\int_{v(0)}^{v(t)} \frac{ds}{s + \Phi(s)} \le \int_0^b m(s)ds < \int_{v(0)}^{\infty} \frac{ds}{s + \Phi(s)},$$

where

$$p(t) = \sum_{i=1}^{4} p_i(t), \quad \text{and} \quad \Phi(t) = \sum_{i=1}^{4} \psi_i(t), \quad m(t) = \max\{bC_* p(t), c_4\}.$$

Here,

$$C_* = \max\{8b, 8C_2\},$$

and

$$v(0) = bC_1 + (1 - bC_4)(\|\phi\|^2_{\mathcal{D}_{\mathcal{F}_0}} + \|\overline{\phi}\|^2_{\mathcal{D}_{\mathcal{F}_0}}), \quad C_4 = 8n_*,$$

and

$$C_1 = 8(E|\phi(0)|^2 + E|\overline{\phi}(0)|^2) + 8n_* r(\|\phi\|^2_{\mathcal{D}_{\mathcal{F}_0}} + \|\overline{\phi}\|^2_{\mathcal{D}_{\mathcal{F}_0}}) + 8\left(\sum_{k=1}^{m} c_k + \sum_{k=1}^{m} \overline{c}_k\right),$$

for all $x, y \in \mathcal{D}_{\mathcal{F}_b}$.

(H_{15}) There exist positive constants $c_k, \widetilde{c}_k$, $k = 1, \ldots, m$, such that

$$E(|I_k(x_k)|)^2 \leq c_k,$$
$$E(|\bar{I}_k(y_k)|)^2 \leq \widetilde{c}_k,$$

for all $x, y \in \mathcal{D}_{\mathcal{F}_b}$.

Theorem 5.5. *Assume that* (H_{13})–(H_{15}) *holds. Then* (5.1) *possesses at least one solution on* $[-r, b]$.

Proof. Clearly, the fixed point of N is a solution to (5.1), where N is defined in Theorem 5.4. In order to apply Theorem (5.3), we first show that N is completely continuous. The proof will be given in several steps.

Step 1. $N = (N_1, N_2)$ is continuous.

Let (x_n, y_n) be a sequence such that

$$(x_n, y_n) \to (x, y) \in \mathcal{D}_{\mathcal{F}_b} \times \mathcal{D}_{\mathcal{F}_b},$$

as $n \to \infty$. Then

$$\begin{aligned}
&E|N_1(x_n(t), y_n(t)) - N_1(x(t), y(t))|^2 \\
&\leq 4E\left|\sum_{l=1}^{\infty} \int_0^t (f_l^1(s, (x_n)_s), (y_n)_s)) - f_l^1(s, x_s, y_s) dW^l(s)\right|^2 \\
&\quad + 4E\left|\int_0^t g^1(s, (x_n)_s), (y_n)_s)) - g^1(s, x_s, y_s) ds\right|^2 \\
&\quad + 4\sum_{k=1}^{m} E|I_k(x_n(t_k)) - I_k(x(t_k))|^2. \\
&\quad + 4\sum_{i=1}^{n_*} \int_0^{t-T_i} E|x_n(s) - x(s)|^2 ds.
\end{aligned}$$

We obtain

$$\begin{aligned}
&E|N_1(x_n(t), y_n(t)) - N_1(x(t), y(t))|^2 \\
&\leq 4C_2 \int_0^t E\|(f^1(s, (x_n)_s), (y_n)_s)) - f^1(s, x_s, y_s)\|^2 ds \\
&\quad + 4\int_0^t E|g^1(s, (x_n)_s), (y_n)_s)) - g^1(s, x_s, y_s)|^2 ds
\end{aligned}$$

$$+4\sum_{k=1}^{m} E|I_k(x_n(t_k)) - I_k(x(t_k))|^2.$$

$$+4\sum_{i=1}^{n_*}\int_0^{t-T_i} E|x_n(s) - x_n(s)|^2 ds.$$

Since f^i, g^i is a Carathéodory function and for each $i = 1, 2, I_k, \overline{I}_k$ are continuous functions. By the Lebesgue dominated convergence theorem, we get

$$\begin{aligned}
&\sup_{t\in J} E|N_1(x_n(t), y_n(t)) - N_1(x(t), y(t))|^2 \\
&\quad \le 4C_2 E\|f^1(., x_n, y_n) - f^1(., x, y)\|_{L^2}^2 \\
&\qquad + 4bE\|g^1(., x_n, y_n) - g^1(., x, y)\|_{L^2}^2 \\
&\qquad + 4\sum_{k=1}^{m} E|I_k(x_n(t_k)) - I_k(x(t_k))|^2 \\
&\qquad + 4\sum_{i=1}^{n_*}\int_0^{t-T_i} E|x_n(s) - x(s)|^2 ds \to 0 \quad \text{as } n \to \infty.
\end{aligned}$$

Similarly,

$$\begin{aligned}
&\sup_{t\in J} E|N_2(x_n(t), y_n(t)) - N_2(x(t), y(t))|^2 \\
&\quad \le 4C_2 E\|f^2(., x_n, y_n) - f^2(., x, y)\|_{L^2}^2 + 4bE\|g^2(., x_n, y_n) - g^2(., x, y)\|_{L^2}^2 \\
&\quad + 4\sum_{k=1}^{m} E|\overline{I}_k(y_n(t_k)) - \overline{I}_k(x(t_k))|^2 + 4\sum_{i=1}^{n_*}\int_0^{t-T_i} E|y_n(s) - y(s)|^2 ds \to 0 \quad \text{as } n \to \infty.
\end{aligned}$$

Thus, N is continuous.

Step 2. N maps bounded sets into bounded sets in $\mathcal{D}_{\mathcal{F}_b} \times \mathcal{D}_{\mathcal{F}_b}$. Indeed, it is enough to show that for any $q > 0$ there exists a positive constant l such that for each

$$(x, y) \in B_q = \{(x, y) \in \mathcal{D}_{\mathcal{F}_b} \times \mathcal{D}_{\mathcal{F}_b} : \|x\|_{\mathcal{D}_{\mathcal{F}_b}}^2 \le q,\ \|y\|_{\mathcal{D}_{\mathcal{F}_b}}^2 \le q\},$$

we have

$$\|N(x, y)\|_{\mathcal{D}_{\mathcal{F}_b}}^2 \le l = (l_1, l_2).$$

Then for each $t \in J$, we get

$$E|N_1(x(t), y(t))|^2 \le E|\phi(0)|^2 + \sum_{i=1}^{n_*}\int_0^{T_i} E|\phi(-s)|^2 ds + \sum_{i=1}^{n_*}\int_0^{t-T_i} E|x(s)|^2 ds$$

$$+E\left|\sum_{l=1}^{\infty}\int_0^t f_l^1(s,x(s),y(s))dW^l(s)\right|^2+\int_0^t E|g^1(s,x(s),y(s))|^2 ds$$
$$+\sum_{0<t_k<t} E|I_k(x(t_k^-)|^2.$$

This implies by (H_{13})–(H_{15}), for each $t\in J$ we have

$$E|N_1(x(t),y(t))|^2\le E|\phi(0)|^2+\sum_{i=1}^{n_*}\int_0^{T_i}E|\phi(-s)|^2ds+\sum_{i=1}^{n_*}\int_0^{t-T_i}E|x(s)|^2ds$$
$$+4C_2\|p_1\|_{L^1}\psi_1(2q)+4b\|p_3\|_{L^1}\psi_2(2q)ds+4\sum_{k=1}^m c_k.$$

Therefore,

$$E|N_1(x(t),y(t))|^2\le E|\phi(0)|^2+n_*r\|\phi\|^2_{\mathcal{D}_{\mathcal{F}_0}}+n_*qb+4C_2\|p_1\|_{L^1}\psi_1(2q)$$
$$+4b\|p_3\|_{L^1}\psi_3(2q)ds+4\sum_{k=1}^m c_k=l_1.$$

Similarly, we have

$$E|N_2(x(t),y(t))|^2\le E|\overline{\phi}(0)|^2+n_*rn_*r\|\overline{\phi}\|^2_{\mathcal{D}_{\mathcal{F}_0}}+n_*qb+4C_2\|p_2\|_{L^1}\psi_2(2q)$$
$$+4b\|p_4\|_{L^1}\psi_2(2q)ds+4\sum_{k=1}^m \overline{c}_k=l_2.$$

Step 3. N maps bounded sets into equi-continuous sets of $\mathcal{D}_{\mathcal{F}_b}\times\mathcal{D}_{\mathcal{F}_b}$.
Let B_q be a bounded set in $\mathcal{D}_{\mathcal{F}_b}\times\mathcal{D}_{\mathcal{F}_b}$ as in Step 2. Let $\tau_1,\tau_2\in J$, $\tau_1<\tau_2$ and $u\in B_q$. Thus, we have

$$E|N_1(x(\tau_2),y(\tau_2))-N_1(x(\tau_1),y(\tau_1))|^2$$
$$\le 4\sum_{i=1}^{n_*}\int_{\tau_1-T_i}^{\tau_2-T_i}E|x(s)|^2+4\int_{\tau_1}^{\tau_2}E|g^1(s,x(s),y(s))ds|^2+4\left|\sum_{l=1}^{\infty}\int_{\tau_1}^{\tau_2}f_l^1(s,x(s),y(s))dW^l(s)\right|^2$$
$$+4\sum_{\tau_1<t_k<\tau_2}E|I_k(x(t_k))|^2$$
$$\le 4\sum_{i=1}^{n_*}\int_{\tau_1-T_i}^{\tau_2-T_i}E|x(s)|^2+4C_2\int_{\tau_1}^{\tau_2}E\|f^1(s,x(s),y(s))\|^2ds$$
$$+4b\int_{\tau_1}^{\tau_2}E|g^1(s,x(s),y(s))|^2ds+4\sum_{\tau_1<t_k<\tau_2}E|I_k(x(t_k))|^2$$

$$\leq 4n_*(\tau_2 - \tau_1)q + 3C_2\psi_1(q)\int_{\tau_1}^{\tau_2} p_1(s)ds + 4b\psi_3(q)\int_{\tau_1}^{\tau_2} p_3(s)ds + 4\sum_{\tau_1<t_k<\tau_2} c_k.$$

Similarly, we have

$$E|N_2(x(\tau_2),y(\tau_2)) - N_2(x(\tau_1),y(\tau_1))|^2$$
$$\leq 4n_*(\tau_2 - \tau_1)q + 3C_2\psi_2(q)\int_{\tau_1}^{\tau_2} p_2(s)ds + 4b\psi_4(q)\int_{\tau_1}^{\tau_2} p_4(s)ds + 4\sum_{\tau_1<t_k<\tau_2} \bar{c}_k.$$

The right-hand term tends to zero as $|\tau_2 - \tau_1| \to 0$. As a consequence of Steps 1 to 3 together with the Arzelá–Ascoli, we conclude that N maps B_q into a precompact set in $\mathcal{D}_{\mathcal{F}_b} \times \mathcal{D}_{\mathcal{F}_b}$.

Step 4. $N(B_q)(t)$ is precompact in $\mathbb{R}^n \times \mathbb{R}^n$. As a consequence of Steps 2 to 3, together with the Arzelá–Ascoli theorem, it suffices to show that N maps B_q into a precompact set in $\mathbb{R}^n \times \mathbb{R}^n$. Let $0 < t < b$ be fixed and let ϵ be a real number satisfying $0 < \epsilon < t$. For $(x,y) \in B_q$, we define

$$N_1^\epsilon(x,y) = \overline{\phi}(0) + \sum_{i=1}^{n_*}\int_{-T_i}^{0} \phi(s)ds + \sum_{i=1}^{n_*}\int_{0}^{t-\epsilon-T_i} x(s)ds + \sum_{l=1}^{\infty}\int_{0}^{t-\epsilon} f_l^1(s,x_s,y_s)dW^l(s)$$
$$+ \int_0^{t-\epsilon} g^1(s,x_s,y_s)ds + \sum_{0<t_k<t-\epsilon} \overline{I}_k(y(t_k^-)),$$
$$H_\epsilon = \{N^\epsilon(x,y)(t) = (N_1^\epsilon(x,y)(t), N_2^\epsilon(x,y)(t)) \ (x,y) \in B_q\},$$

is precompact in $\mathbb{R}^n \times \mathbb{R}^n$ for every ϵ and $0 < \epsilon < t$ Moreover, for every $(x,y) \in B_q$, we have

$$E\|N_1(x,y) - N_1^\epsilon(x,y)\|^2$$
$$\leq 4n_*\epsilon q + 3C_2\psi_1(q)\int_{t-\epsilon}^{t} p_1(s)ds + 4b\psi_3(q)\int_{t-\epsilon}^{t} p_3(s)ds + 4\sum_{t-\epsilon<t_k<t} c_k.$$

Similarly,

$$E\|N_2(x,y) - N_2^\epsilon(x,y)\|^2$$
$$\leq 4n_*\epsilon q + 3C_2\psi_2(q)\int_{t-\epsilon}^{t} p_2(s)ds + 4b\psi_4(q)\int_{t-\epsilon}^{t} p_4(s)ds + 4\sum_{t-\epsilon<t_k<t} \bar{c}_k.$$

Therefore, there are precompact sets arbitrarily close to the set

$$H_\epsilon = \{N^\epsilon(x,y)(t) = (N_1^\epsilon(x,y)(t), N_2^\epsilon(x,y)(t)) \ (x,y) \in B_q\}.$$

Hence, the set

$$H = \{N(x,y)(t) = (N_1(x,y)(t), N_2(x,y)(t)) \ (x,y) \in B_q\},$$

is precompact in $\mathbb{R}^n \times \mathbb{R}^n$ and the right-hand term tends to 0 uniformly in t as $\epsilon \to 0^+$. Hence, we concluded the relative compactness of $N(B_q)(t)$ for $t \geq 0$. By the Arzelá–Ascoli theorem, we conclude that

$$N : \mathcal{D}_{\mathcal{F}_b} \times \mathcal{D}_{\mathcal{F}_b} \to \mathcal{D}_{\mathcal{F}_b} \times \mathcal{D}_{\mathcal{F}_b},$$

is a completely continuous operator. N maps B_q into a precompact set in $PC \times PC$.

Step 5. A priori bounds.
Now it remains to show that the set

$$\mathcal{A} = \{(x,y) \in \mathcal{D}_{\mathcal{F}_b} \times \mathcal{D}_{\mathcal{F}_b} : (x,y) = \lambda N(x,y), \lambda \in (0,1)\},$$

is bounded.
Let $(x,y) \in \mathcal{A}$. Then $x = \lambda N_1(x,y)$ and $y = \lambda N_2(x,y)$ for some $0 < \lambda < 1$. Thus, for $t \in J$, we have

$$\begin{aligned} x(t) = \lambda(\phi(0) &+ \sum_{i=1}^{n_*} \int_{-T_i}^{0} \phi(s)ds + \sum_{i=1}^{n_*} \int_{0}^{t-T_i} x(s)ds \\ &+ \sum_{l=1}^{\infty} \int_0^t f_l^1(s, x_s, y_s) dW^l(s) \\ &+ \int_0^t g^1(s, x_s, y_s)ds + \sum_{0<t_k<t} I_k(x(t_k^-)), \end{aligned}$$

and

$$\phi(t) \in \mathcal{D}_{\mathcal{F}_0}, \quad t \in [-r, 0].$$

This implies for each $t \in J$,

$$\begin{aligned} E|x(t)|^2 \leq 8E|\phi(0)|^2 &+ 8n_* r\|\phi\|^2_{\mathcal{D}_{\mathcal{F}_0}} + 8n_* \int_0^t E|x(s)|^2 ds \\ &+ 8C_8 \int_0^t p_1(s)\psi_1(\|x_s\|^2_{\mathcal{D}_{\mathcal{F}_0}} + \|y_s\|^2_{\mathcal{D}_{\mathcal{F}_0}})ds \end{aligned}$$

$$+ 8b \int_0^t p_3(s)\psi_3(\|x_s\|^2_{\mathcal{D}_{\mathcal{F}_0}} + \|y_s\|^2_{\mathcal{D}_{\mathcal{F}_0}})ds + 8\sum_{k=1}^m c_k,$$

and

$$E|y(t)|^2 \le 8E|\overline{\phi}(0)|^2 + 8n_* r\|\overline{\phi}\|^2_{\mathcal{D}_{\mathcal{F}_0}} + 8n_* \int_0^t E|y(s)|^2 ds$$
$$+ 8C_2 \int_0^t p_2(s)\psi_2(\|x_s\|^2_{\mathcal{D}_{\mathcal{F}_0}} + \|y_s\|^2_{\mathcal{D}_{\mathcal{F}_0}})ds$$
$$+ 8b \int_0^t p_4(s)\psi_4(\|x_s\|^2_{\mathcal{D}_{\mathcal{F}_0}} + \|y_s\|^2_{\mathcal{D}_{\mathcal{F}_0}})ds + 8\sum_{k=1}^m \overline{c}_k.$$

Consider functions $\mu, \overline{\mu}$ defined on J by

$$\mu(t) = \sup\{E|x(s)|^2 : 0 \le s \le t\}, \quad t \in J,$$

and

$$\overline{\mu}(t) = \sup\{E|y(s)|^2 : 0 \le s \le t\}, \quad t \in J.$$

Since

$$\|x_s\|_{\mathcal{D}_{\mathcal{F}_0}} = \int_{-r}^0 E|x(s+\theta)|^2 d\theta = \int_{-r+s}^s E|x(\theta)|^2 d\theta,$$

then

$$\|x_s\|_{\mathcal{D}_{\mathcal{F}_0}} \le \int_{-r}^0 E|x(\theta)|^2 d\theta + \int_0^s E|x(\theta)|^2 d\theta \le \int_{-r}^0 E|x(\theta)|^2 d\theta + b\mu(t),$$

and

$$\|y_s\|_{\mathcal{D}_{\mathcal{F}_0}} \le \int_{-r}^0 E|y(\theta)|^2 d\theta + b\overline{\mu}(t).$$

Hence,

$$\|x_s\|_{\mathcal{D}_{\mathcal{F}_0}} \le \|\phi\|^2_{\mathcal{D}_{\mathcal{F}_0}} + b\mu(t), \quad \|y_s\|_{\mathcal{D}_{\mathcal{F}_0}} \le \|\overline{\phi}\|^2_{\mathcal{D}_{\mathcal{F}_0}} + b\overline{\mu}(t).$$

Then, for $t \in J$ we have

$$\mu(t) \le 8E|\phi(0)|^2 + 8n_* r\|\phi\|^2_{\mathcal{D}_{\mathcal{F}_0}} + 8n_* \int_0^t \mu(s)ds$$
$$+ 8C_2 \int_0^t p_1(s)\psi_1(\|\phi\|^2_{\mathcal{D}_{\mathcal{F}_0}} + b\mu(s) + \|\overline{\phi}\|^2_{\mathcal{D}_{\mathcal{F}_0}} + b\overline{\mu}(s))ds$$
$$+ 8b \int_0^t p_3(s)\psi_3(\|\phi\|^2_{\mathcal{D}_{\mathcal{F}_0}} + b\mu(t) + \|\overline{\phi}\|^2_{\mathcal{D}_{\mathcal{F}_0}} + b\overline{\mu}(t))ds + 8\sum_{k=1}^m c_k.$$

Similarly,

$$\overline{\mu}(t) \le 8E|\overline{\phi}(0)|^2 + 8n_* r\|\overline{\phi}\|^2_{\mathcal{D}_{\mathcal{F}_0}} + 8n_* \int_0^t \overline{\mu}(s)ds$$
$$+ 8C_2 \int_0^t p_2(s)\psi_2(\|\phi\|^2_{\mathcal{D}_{\mathcal{F}_0}} + b\mu(s) + \|\overline{\phi}\|^2_{\mathcal{D}_{\mathcal{F}_0}} + b\overline{\mu}(s))ds$$
$$+ 8b \int_0^t p_4(s)\psi_4(\|\phi\|^2_{\mathcal{D}_{\mathcal{F}_0}} + b\mu(t) + \|\overline{\phi}\|^2_{\mathcal{D}_{\mathcal{F}_0}} + b\overline{\mu}(t))ds + 8\sum_{k=1}^m \overline{c}_k.$$

Thus, we have

$$\mu(t) + \overline{\mu}(t) \le 8(E|\phi(0)|^2 + E|\overline{\phi}(0)|^2) + 8n_* r(\|\phi\|^2_{\mathcal{D}_{\mathcal{F}_0}} + \|\overline{\phi}\|^2_{\mathcal{D}_{\mathcal{F}_0}})$$
$$+ 8\left(\sum_{k=1}^m c_k + \sum_{k=1}^m \overline{c}_k\right) + 8n_* \int_0^t (\mu(s) + \overline{\mu}(s))ds$$
$$+ 8C_2 \int_0^t p_1(s)\psi_1(\|\phi\|^2_{\mathcal{D}_{\mathcal{F}_0}} + b\mu(s) + \|\overline{\phi}\|^2_{\mathcal{D}_{\mathcal{F}_0}} + b\overline{\mu}(s))ds$$
$$+ 8b \int_0^t p_3(s)\psi_3(\|\phi\|^2_{\mathcal{D}_{\mathcal{F}_0}} + b\mu(t) + \|\overline{\phi}\|^2_{\mathcal{D}_{\mathcal{F}_0}} + b\overline{\mu}(t))ds$$
$$+ 8C_2 \int_0^t p_2(s)\psi_2(\|\phi\|^2_{\mathcal{D}_{\mathcal{F}_0}} + b\mu(s) + \|\overline{\phi}\|^2_{\mathcal{D}_{\mathcal{F}_0}} + b\overline{\mu}(s))ds$$
$$+ 8b \int_0^t p_4(s)\psi_4(\|\phi\|^2_{\mathcal{D}_{\mathcal{F}_0}} + b\mu(t) + \|\overline{\phi}\|^2_{\mathcal{D}_{\mathcal{F}_0}} + b\overline{\mu}(t))ds.$$

We denote

$$p(t) = \sum_{i=1}^4 p_i(t), \quad \text{and} \quad \Phi = \sum_{i=1}^4 \psi_i(t),$$

$$\mu(t)+\overline{\mu}(t)\leq C_1+C_4\left(\int_0^t \mu(s)+\overline{\mu}(s)\right)ds$$
$$+C_*\int_0^t p(s)\Phi(b(\mu(s)+\overline{\mu}(s))+\|\overline{\phi}\|^2_{\mathcal{D}_{\mathcal{F}_0}}+\|\phi\|^2_{\mathcal{D}_{\mathcal{F}_0}})ds,$$

where

$$C_1=8(E|\phi(0)|^2+E|\overline{\phi}(0)|^2)+8n_*r(\|\phi\|^2_{\mathcal{D}_{\mathcal{F}_0}}+\|\overline{\phi}\|^2_{\mathcal{D}_{\mathcal{F}_0}})+8\left(\sum_{k=1}^m c_k+\sum_{k=1}^m \overline{c}_k\right),$$

and

$$C_*=\max\{8b,8C_2\},\quad C_4=8n_*.$$

Thus, we have

$$\begin{aligned}
&b(\mu(t)+\overline{\mu}(t))+\|\phi\|^2_{\mathcal{D}_{\mathcal{F}_0}}+\|\overline{\phi}\|^2_{\mathcal{D}_{\mathcal{F}_0}}\\
&\quad\leq bC_1+(1-bC_4)(\|\phi\|^2_{\mathcal{D}_{\mathcal{F}_0}}+\|\overline{\phi}\|^2_{\mathcal{D}_{\mathcal{F}_0}})\\
&\quad+C_4\left(\int_0^t b(\mu(s)+\overline{\mu}(s))+\|\phi\|^2_{\mathcal{D}_{\mathcal{F}_0}}+\|\overline{\phi}\|^2_{\mathcal{D}_{\mathcal{F}_0}}\right)ds\\
&\quad+bC_*\int_0^t p(s)\Phi(b(\mu(s)+\overline{\mu}(s))+\|\overline{\phi}\|^2_{\mathcal{D}_{\mathcal{F}_0}}+\|\phi\|^2_{\mathcal{D}_{\mathcal{F}_0}})ds.
\end{aligned}$$

Let us take the right-hand side of the above inequality as $v(t)$. Then we have

$$v(0)=bC_1+(1-bC_4)(\|\phi\|^2_{\mathcal{D}_{\mathcal{F}_0}}+\|\overline{\phi}\|^2_{\mathcal{D}_{\mathcal{F}_0}}),$$

and

$$b(\mu(t)+\overline{\mu}(t))+\|\phi\|^2_{\mathcal{D}_{\mathcal{F}_0}}+\|\overline{\phi}\|^2_{\mathcal{D}_{\mathcal{F}_0}}\leq v(t)\quad\text{for a.e. } t\in J,$$

and

$$\begin{aligned}
v'(t)&=bC_*p(t)\Phi(b(\mu(t)+\overline{\mu}(t))+\|\overline{\phi}\|^2_{\mathcal{D}_{\mathcal{F}_0}}+\|\phi\|^2_{\mathcal{D}_{\mathcal{F}_0}})\\
&\quad+C_4(b(\mu(t)+\overline{\mu}(t))+\|\phi\|^2_{\mathcal{D}_{\mathcal{F}_0}}+\|\overline{\phi}\|^2_{\mathcal{D}_{\mathcal{F}_0}}).
\end{aligned}$$

Using the increasing character of Φ, we obtain

$$v'(t)\leq bC_*p(t)\Phi(v(t))+C_4v(t)\leq m(t)(v(t)+\Phi(v(t))\quad\text{for a.e. } t\in J,$$

where

$$m(t) = \max\{bC_* p(t), c_4\}.$$

This implies that for each $t \in [0, b]$,

$$\int_{v(0)}^{v(t)} \frac{ds}{s + \Phi(s)} \le \int_0^b m(s)ds < \int_{v(0)}^{\infty} \frac{ds}{s + \Phi(s)}.$$

Consequently, there exists a constant K such that

$$b(\mu(t) + \overline{\mu}(t)) + \|\phi\|^2_{\mathcal{D}_{\mathcal{F}_0}} + \|\overline{\phi}\|^2_{\mathcal{D}_{\mathcal{F}_0}} \le v(t) < K.$$

Now, from the definition of $\mu, \overline{\mu}$ it follows that

$$E|x(t)|^2 + E|y(t)|^2 \le \mu(t) + \overline{\mu}(t) \le \frac{K}{b}, \quad \text{for each } t \in J.$$

Consequently,

$$\|x\|_{\mathcal{D}_{\mathcal{F}_b}} \le \frac{K}{b},$$

and

$$\|y\|_{\mathcal{D}_{\mathcal{F}_b}} \le \frac{K}{b}.$$

This shows that $\mathcal{A}$ is bounded. As a consequence of Theorem 5.3 we deduce that N has a fixed point (x, y), which is a solution to the problem (5.1). □

6 Impulsive stochastic differential equations

In this chapter, we investigate the following first-order stochastic impulsive equation driven by fractional Brownian motion with the Hurst index $H > 1/2$:

$$\begin{cases} dx(t) = (Ax(t) + f^1(t, x(t), y(t))dt + \sum_{l=1}^{\infty} \sigma_l^1(t, x(t)), y(t))dB_l^H(t), & t \in [0, T],\ t \neq t_k, \\ dy(t) = (Ay(t) + f^2(t, x(t), y(t)))dt + \sum_{l=1}^{\infty} \sigma_l^2(t, x(t), y(t))dB_l^H(t), & t \in J,\ t \neq t_k, \\ \Delta x(t) = I_k(x(t_k)), & t = t_k,\ k = 1, 2, \dots, m, \\ \Delta y(t) = \bar{I}_k(y(t_k)), & t = t_k,\ k = 1, 2, \dots, m, \\ x(0) = x_0, \\ y(t) = y_0, \end{cases} \tag{6.1}$$

where X is a real separable Hilbert space with inner product $\langle \cdot, \cdot \rangle$ and induced norm $\|\cdot\|$, and

$$A : D(I) \subset X \longrightarrow X$$

is the infinitesimal generator of a strongly continuous semigroup of bounded linear operators $(S(t))_{t \geq 0}$ in X, and

$$f^1, f^2 : [0, T] \times X \times X \longrightarrow X$$

are given functions; B_l^H is an infinite sequence of mutually independent fractional Brownian motions $l = 1, 2, \dots$, with a Hurst parameter H, and

$$\begin{aligned} I_k, \bar{I}_k &\in C(X, X) \quad (k = 1, 2, \dots, m), \\ \sigma_l^1, \sigma_l^2 &: J \times X \times X \to L_Q^0(Y, X). \end{aligned}$$

Here, $L_Q^0(Y, X)$ denotes the space of all Q-Hilbert–Schmidt operators from Y into X, which will also be defined in the next section. Moreover, the fixed times t_k satisfy

$$0 < t_1 < t_2 < \cdots < t_m < T,$$

and $y(t_k^-)$ and $y(t_k^+)$ denote the left and right limits of $y(t)$ at $t = t_k$.

$$\begin{cases} \sigma(t, x, y) = (\sigma_1(t, x, y), \sigma_2(t, x, y), \dots), \\ \|\sigma(t, x, y)\|^2 = \sum_{j=1}^{\infty} \|\sigma_j(t, x, y)\|_{L_Q^0}^2 < \infty, \end{cases} \tag{6.2}$$

where $\sigma(\cdot, \cdot, \cdot) \in \ell^2$, and ℓ^2 is given by

$$\ell^2 = \left\{ \phi = (\phi_j)_{j \geq 1} : [0, T] \times X \times X \to L_Q^0(Y, X) : \|\phi(t, x, y)\|^2 = \sum_{j=1}^{\infty} \|\phi_j(t, x, y)\|_{L_Q^0}^2 < \infty \right\}.$$

https://doi.org/10.1515/9783112251898-006

It is obvious that the system (6.1) can be seen as a fixed-point problem for the model

$$\begin{cases} dz(t) = A_* z(t) + f(t, z(t))dt + \sum_{l=1}^{\infty} \sigma_l(t, z))dB_l^H(t), & t \in [0, T],\ t \neq t_k, \\ \Delta z(t) = I_k^*(z(t_k)), & t = t_k,\ k = 1, 2, \dots, m, \\ z(0) = z_0, \end{cases}$$

where

$$z(t) = \left[x(t)y(t)\right],$$

$$A_* = \begin{bmatrix} A & 0 \\ 0 & A \end{bmatrix},$$

$$f(t, z) = \begin{bmatrix} f^1(t, x(t), y(t)) \\ f^2(t, x(t), y(t)) \end{bmatrix},$$

$$\sigma_l(t, z) = \begin{bmatrix} \sigma_l^1(t, x, y) \\ \sigma_l^2(t, x, y) \end{bmatrix}$$

and

$$z_0 = \begin{bmatrix} x_0 \\ y_0 \end{bmatrix}.$$

6.1 Mild solution

Let

$$J_k = (t_k, t_{k+1}], \quad k = 1, 2, \dots, m.$$

In order to define a solution for problem (4.2), consider the following space of piecewise continuous functions:

$$PC = \Big\{x : \Omega \times [0, T] \longrightarrow X,\ x \in C(J_k, L^2(\Omega, X)),\ k = 1, \dots, m \text{ such that } x(t_k^+, .) \text{ and } x(t_k^-, .) \text{ exist with } x(t_k^-, .) = x(t_k, .) \text{ and } \sup_{t \in [0,T]} E\|x(t, .)\|^2 < \infty \text{ almost surely}\Big\},$$

endowed with the norm

$$\|x\|_{PC} = \sup_{s \in [0,T]} \left(E\|x(s, .)\|^2\right)^{\frac{1}{2}}.$$

It is not difficult to check that PC is a Banach space with norm $\| \cdot \|_{PC}$.

First, we will list the following hypotheses that will be imposed in our main theorem. In this section, we assume that there exists $M > 0$, such that

$$\|S(t)\| \le M, \quad \text{for every } t \in [0,T].$$

(H_{16}) There exist functions $a_i,\ b_i \in L^1([0,T],\mathbb{R}^+)$, such that

$$|f^i(t,x,y) - f^i(t,\overline{x},\overline{y})|_X^2 \le a_i(t)|x-\overline{x}|_X^2 + b_i(t)|y-\overline{y}|_X^2, \quad i = 1,2,$$

for all $x, y, \overline{x}, \overline{y} \in X$.

(H_{17}) There exist functions $\alpha_i, \beta_i \in L^1([0,T],\mathbb{R}^+)$, such that

$$\|\sigma^i(t,x,y) - \sigma^i(t,\overline{x},\overline{y})\|^2 \le \alpha_i(t)|x-\overline{x}|_X^2 + \beta_i(t)|y-\overline{y}|_X^2.$$

The function

$$\sigma : J \times X \times X \longrightarrow L_Q^0(Y,X),$$

satisfies

$$\sum_{l=1}^{\infty} \int_0^T \|\sigma_l^i(s,x(s),y(s))\|_{L_Q^0}^2 ds < \infty. \tag{6.3}$$

for all $x, y, \overline{x}, \overline{y} \in X$, and a. e. $t \in J$.

Now, we will first define the concept of a mild solution to our problem.

Definition 6.1. An $X-$ valued stochastic process

$$u = (x,y) \in PC \times PC,$$

is said to be a solution of (6.1) with respect to the probability space $(\Omega, \mathcal{F}, \mathbb{P})$, if:

1. $u(t)$ is $\mathcal{F}_t$-adapted for all $t \in J_k = (t_k, t_{k+1}] \quad k = 1,2,\dots,m$.
2. $u(t)$ is right continuous and has a limit on the left almost surely.
3. $u(0) = (x_0, y_0)$.
4. $u(t)$ satisfies for all $t \in [0,T]$ and almost surely that

$$\begin{cases} x(t) = S(t)x_0 + \int_0^t S(t-s)f^1(s,x(s),y(s))ds \\ \quad + \sum_{l=1}^{\infty} \int_0^t S(t-s)\sigma_l^1(t,x(s),y(s))dB_l^H(s) \\ \quad + \sum_{0<t_k<t} S(t-t_k)I_k(x(t_k)), \quad \mathbb{P}\text{-a.s.}, \ t \in J \\ y(t) = S(t)y_0 + \int_0^t S(t-s)f^2(s,x(s),y(s))ds \\ \quad + \sum_{l=1}^{\infty} \int_0^t S(t-s)\sigma_l^2(t,x(s),y(s))dB_l^H(s) \\ \quad + \sum_{0<t_k<t} S(t-t_k)I_k(y(t_k)), \quad \mathbb{P}\text{-a.s.}, \ t \in J. \end{cases} \tag{6.4}$$

Notice that, thanks to (6.3) and the fact that $H \in (1/2, 1)$, (2.50) holds, which implies that the stochastic integrals in (6.4) are well-defined since $S(\cdot)$ is a strongly continuous semigroup for every $t \in [0, T]$, and that this concept of solution can be considered more general than the classical concept of a solution to equation (4.2). A continuous solution of (6.4) is called a mild solution of (6.1).

Let us now prove the existence and uniqueness of a mild solution for (6.1) by using Perov's fixed-point theorem.

Theorem 6.1. *Assume that* (H_{16}) *and* (H_{17}) *hold. Then problem* (6.1) *possesses a unique mild solution on* $[0, T]$.

Proof. The proof will be split into several steps.
Step 1. Consider the problem

$$\begin{cases} dx(t) = (Ax(t) + f^1(t, x(t), y(t))dt + \sum_{l=1}^{\infty} \sigma_l^1(t, x(t)), y(t))dB_l^H(t), & t \in [0, t_1], \\ dy(t) = (Ay(t) + f^2(t, x(t), y(t)))dt + \sum_{l=1}^{\infty} \sigma_l^2(t, x(t), y(t))dB_l^H(t), & t \in [0, t_1], \\ x(0) = x_0, \\ y(0) = y_0. \end{cases} \tag{6.5}$$

Let

$$D_{t_0} = \left\{x : \Omega \times [0, t_1] \longrightarrow X,\ x \in C(J_1, L^2(\Omega, X)) : \sup_{t\in[0,t_1]} E|x(t, \cdot)|_X^2 < \infty\right\}.$$

Consider the operator

$$P^0 : D_{t_0} \times D_{t_0} \to D_{t_0} \times D_{t_0},$$

defined by

$$P^0(x, y) = (P_1^0(x, y), P_2^0(x, y)), \quad (x, y) \in D_{t_0} \times D_{t_0},$$

where

$$\begin{cases} P_1^0(x, y) = S(t)x_0 + \int_0^t S(t-s)f^1(s, x(s), y(s))ds \\ \quad + \sum_{l=1}^{\infty} \int_0^t S(t-s)\sigma_l^1(t, x(s), y(s))dB_l^H(s), \quad t \in [0, t_1]. \\ P_2^0(x, y) = S(t)y_0 + \int_0^t S(t-s)f^2(s, x(s), y(s))ds \\ \quad + \sum_{l=1}^{\infty} \int_0^t S(t-s)\sigma_l^2(t, x(s), y(s))dB_l^H(s), \quad t \in [0, t_1]. \end{cases} \tag{6.6}$$

The operators in (6.6) are well-defined. In other words, given

$$(x, y) \in D_{t_0} \times D_{t_0},$$

we see that

$$P^0(x,y) \in D_{t_0} \times D_{t_0},$$

as well. We will use Theorem 3.11 to prove that P^0 possesses a fixed point. Let

$$(x,y),(\overline{x},\overline{y}) \in D_{t_0} \times D_{t_0},$$

then, for each $t \in [0,t_1]$, by Lemma 2.13 and assumptions (H_{16}), (H_{17}), imply

$$\begin{aligned}
&E|P_1^0(x(t),y(t)) - P_1^0(\overline{x}(t),\overline{y}(t))|_X^2 \\
&\leq 2M^2 t_1 \int_0^t a_1(s)E|x(s)-\overline{x}(s)|_X^2 + b_1(s)E|y(s)-\overline{y}(s)|_X^2 ds \\
&2c_H H(2H-1)t_1^{2H-1}\int_0^t M^2 ds \int_0^t \alpha_1(s)E|x(s)-\overline{x}(s)|_X^2 ds \\
&\quad + 2c_H H(2H-1)t_1^{2H-1}\int_0^t M^2 ds \int_0^t \beta_1(s)E|y(s)-\overline{y}(s)|_X^2 ds \\
&\leq \int_0^t \alpha(s)e^{\tau\hat{a}(s)}e^{-\tau\hat{a}(s)} \sup_{s\in[0,t_1]} E|x(s)-\overline{x}(s)|_X^2 ds \\
&\quad + \int_0^t \alpha(s)e^{\tau\hat{a}(s)}e^{-\tau\hat{a}(s)} \sup_{s\in[0,t_1]} E|y(s)-\overline{y}(s)|_X^2 ds \\
&\leq \int_0^t \alpha(s)e^{\tau\hat{a}(s)}ds\|x-\overline{x}\|_*^2 + \int_0^t \alpha(s)e^{\tau\hat{a}(s)}ds\|y-\overline{y}\|_*^2 \\
&\leq \frac{1}{\tau}\int_0^t (e^{\tau\hat{a}(s)})' ds\|x-\overline{x}\|_*^2 + \frac{1}{\tau}\int_0^t (e^{\tau\hat{a}(s)})' ds\|y-\overline{y}\|_*^2 \\
&\leq \frac{1}{\tau}e^{\tau\hat{a}(t)}\|x-\overline{x}\|_*^2 + \frac{1}{\tau}e^{\tau\hat{a}(t)}\|y-\overline{y}\|_*^2.
\end{aligned}$$

Therefore,

$$e^{-\tau\hat{a}(t)}E|L_1(x(t),y(t)) - L_1(\overline{x}(t),\overline{y}(t))|_X^2 \leq \frac{1}{\tau}\|x-\overline{x}\|_*^2 + \frac{1}{\tau}\|y-\overline{y}\|_*^2,$$

where $\|\cdot\|_*$ is the Bielecki-type norm on D_{t_0} defined by

$$\|x\|_*^2 = \sup_{t\in[0,t_1]} E|x(t,.)|_X^2 e^{-\tau\hat{a}(t)},$$

where

$$\widehat{\alpha}(t) = \int_0^t \alpha(s)ds, \quad t \in [0, t_1],$$

and

$$\alpha(s) = \max\{2M^2 t_1 a_1(s) + 2c_H H(2H-1)t_1^{2H} M^2 \alpha_1(s), 2M^2 t_1 b_1(s) \\ + 2c_H H(2H-1)t_1^{2H} M^2 \beta_1(s)\}.$$

Hence,

$$\|P_1^0(x,y) - P_1^0(\overline{x},\overline{y})\|_*^2 \le \frac{1}{\tau}\|x-\overline{x}\|_*^2 + \frac{1}{\tau}\|y-\overline{y}\|_*^2.$$

Using (5.3), we conclude that

$$\|P_1^0(x,y) - P_1^0(\overline{x},\overline{y})\|_* \le \frac{1}{\sqrt{\tau}}\|x-\overline{x}\|_* + \frac{1}{\sqrt{\tau}}\|y-\overline{y}\|_*.$$

Similar computations for N_1 yield

$$\|P_2^0(x,y) - P_2^0(\overline{x},\overline{y})\|_* \le \frac{1}{\sqrt{\tau}}\|x-\overline{x}\|_* + \frac{1}{\sqrt{\tau}}\|y-\overline{y}\|_*.$$

Thus,

$$\begin{aligned}\|P^0(x,y) - P^0(\overline{x},\overline{y})\|_* &= \begin{pmatrix}\|P_1^0((x,y) - P_1^0(\overline{x},\overline{y})\|_* \\ \|P_2^0(x,y) - P_2^0(\overline{x},\overline{y})\|_*\end{pmatrix} \\ &\le \frac{1}{\sqrt{\tau}}\begin{pmatrix}1 & 1\\ 1 & 1\end{pmatrix}\begin{pmatrix}\|x-\overline{x}\|_* \\ \|y-\overline{y}\|_*\end{pmatrix}.\end{aligned}$$

Hence,

$$\|P^0(x,y) - P^0(\overline{x},\overline{y})\|_* \le \frac{1}{\sqrt{\tau}} M_{\alpha,\beta}\begin{pmatrix}\|x-\overline{x}\|_* \\ \|y-\overline{y}\|_*\end{pmatrix},$$

for all $(x,y), (\overline{x},\overline{y}) \in D_{t_0} \times D_{t_0}$, where

$$M_{\alpha,\beta} = \begin{pmatrix}1 & 1\\ 1 & 1\end{pmatrix}.$$

If we choose a suitable $\sqrt{\tau} > 2$, such that the matrix

$$\frac{\|M_{\alpha,\beta}\|}{\sqrt{\tau}} < 1,$$

is nonnegative,

$$\frac{M_{\alpha,\beta}}{\tau},$$

is nonsingular, and

$$I - \frac{\|M_{\alpha,\beta}\|}{\tau},$$

is nonsingular,

$$\left(I - \frac{M_{\alpha,\beta}}{\sqrt{\tau}}\right)^{-1} = I + \frac{M_{\alpha,\beta}}{\sqrt{\tau}} + \frac{M_{\alpha,\beta}^2}{\tau} + \cdots$$

From Lemma 3.5, we obtain that $\frac{M_{\alpha,\beta}}{\sqrt{\tau}}$ converges to zero. As a consequence of Perov's fixed-point theorem, P^0 has a unique fixed point

$$(x,y) \in D_{t_0} \times D_{t_0},$$

which is the unique solution to problem (6.5). Let us denote this solution by (x_1, y_1).

Step 2. Now consider the problem

$$\begin{cases} dx(t) = (Ax(t) + f^1(t,x(t),y(t))dt + \sum_{l=1}^{\infty} \sigma_l^1(t,x(t)),y(t))dB_l^H(t), & t \in (t_1,t_2], \\ dy(t) = (Ay(t) + f^2(t,x(t),y(t)))dt + \sum_{l=1}^{\infty} \sigma_l^2(t,x(t),y(t))dB_l^H(t), & t \in (t_1,t_2], \\ x(t_1^+) = x_1(t_1^-) + I_1(x_1(t_1)), \\ y(t_1^+) = y_1(t_1^-) + \bar{I}_1(y_1(t_1)). \end{cases} \tag{6.7}$$

Let

$$D_{t_1} = \Big\{x \in C((t_1,t_2], L^2(\Omega, X)), \text{ such that } x(t_1^+,.) \text{ and } x(t_1^-,.) \\ \text{exist with } x(t_1^-,.) = x(t_1,.) \text{ and } \sup_{t\in(t_1,t_2]} E|x(t,.)|_X^2 < \infty\Big\}.$$

Set

$$C_1 = D_{t_0} \cap D_{t_1}.$$

Consider the operator $P : C_1 \times C_1 \to C_1 \times C_1$ defined by

$$P^1(x,y) = (P_1^1(x,y), P_2^1(x,y)), \quad (x,y) \in C_1 \times C_1,$$

where

$$\begin{cases} P_1^1(x,y) = S(t-t_1)[x_1(t_1) + I_1(x_1(t_1^-))] + \int_{t_1}^t S(t-s)f^1(s,x(s),y(s))ds \\ \quad + \sum_{l=1}^{\infty} \int_{t_1}^t S(t-s)\sigma_l^1(t,x(s),y(s))dB_l^H(s), \quad t \in [t_1,t_2], \\ P_2^1(x,y) = S(t-t_1)[y_1(t_1) + I_1(y_1(t_1^-))] + \int_{t_1}^t S(t-s)f^2(s,x(s),y(s))ds \\ \quad + \sum_{l=1}^{\infty} \int_{t_1}^t S(t-s)\sigma_l^2(t,x(s),y(s))dB_l^H(s), \quad t \in [t_1,t_2]. \end{cases}$$

As in Step 1, we can show that P^1 is well-defined and the integral equation possesses a unique fixed point (x,y), which is a solution to problem (6.7). Denote this solution by (x_2,y_2).

Step 3. We continue this process, taking into account that

$$(x_m,y_m) = (x|_{[t_m,T]},y|_{[t_m,T]}),$$

is a solution to the problem

$$\begin{cases} dx(t) = (Ax(t) + f^1(t,x(t),y(t))dt + \sum_{l=1}^{\infty} \sigma_l^1(t,x(t)),y(t))dB_l^H(t), & t \in (t_m,T], \\ dy(t) = (Ay(t) + f^2(t,x(t),y(t)))dt + \sum_{l=1}^{\infty} \sigma_l^2(t,x(t),y(t))dB_l^H(t), & t \in (t_m,T], \\ x(t_m^+) = x_m(t_m^-) + I_m(x_m(t_m)), \\ y(t_m^+) = y_m(t_m^-) + \bar{I}_m(y_m(t_1)). \end{cases}$$

Let

$$D_{t_m} = \Big\{x \in C((t_m,T],L^2(\Omega,X)) \text{ such that } x(t_m^+,.) \text{ and } x(t_m^-,.) \text{ exist with} \\ x(t_m^-,.) = x(t_m,.) \text{ and } \sup_{t\in(t_m,T]} E|x(t,.)|_X^2 < \infty \text{ almost surely}\Big\}.$$

Set

$$C_m = \bigcap_{k=0}^{m} D_{t_k}.$$

Then there exists a fixed point $(x_m(t),y_m(t))$ of

$$P^m = (P_1^m,P_2^m) : C_m \times C_m \to C_m \times C_m.$$

The unique solution (x,y) of problem (6.1) is then defined by

$$\begin{cases} P_1^m(x,y) = S(t-t_m)[x_m(t_m) + I_m(x_1(t_m^-))] + \int_{t_m}^t S(t-s)f^1(s,x(s),y(s))ds \\ \quad + \sum_{l=1}^{\infty} \int_{t_m}^t S(t-s)\sigma_l^1(t,x(s),y(s))dB_l^H(s), \quad t \in [t_m,T], \\ P_2^m(x,y) = S(t-t_m)[y_m(t_m) + I_m(y_1(t_m^-))] + \int_{t_m}^t S(t-s)f^2(s,x(s),y(s))ds \\ \quad + \sum_{l=1}^{\infty} \int_{t_m}^t S(t-s)\sigma_l^2(t,x(s),y(s))dB_l^H(s), \quad t \in [t_m,T]. \end{cases}$$

$$(x(t),y(t)) = \begin{cases} (x_1(t),y_1(t)), & \text{if } t \in [0,t_1], \\ (x_2(t),y_2(t)), & \text{if } t \in (t_1,t_2], \\ \vdots & \\ (x_{m+1}(t),y_{m+1}(t)), & \text{if } t \in (t_m,T] \end{cases}$$

and the proof is complete. □

The second result in this section, dealing with the existence of solutions to our problem, will be obtained by applying the Leary–Schauder fixed-point theorem. To this end, we first need to introduce the following hypotheses:

(H_{18}) There exist functions

$$\overline{a}_i, \overline{b}_i, c_i \in L^1([0,T], \mathbb{R}^+),$$

such that each

$$|f^i(t,x,y)|_X^2 \le \overline{a}_i(t)|x|_X^2 + \overline{b}_i(t)|y|_X^2 + c_i(t), \quad i = 1,2.$$

for all $x, y \in X$, and a. e. $t \in J$.

(H_{19}) There exist functions

$$\overline{\alpha}_i \in L^1([0,T], \mathbb{R}^+),$$

and

$$\overline{\beta}_i, \overline{c}_i \in L^1([0,T], \mathbb{R}^+),$$

such that

$$\|\sigma^i(t,x,y)\|^2 \le \overline{\alpha}_i(t)|x|_X^2 + \overline{\beta}_i(t)|y|_X^2 + \overline{c}_i(t), \quad i = 1,2,$$

for all $x, y \in X$, and a. e. $t \in J$.

(H_{20}) The semigroup $\{S(t)\}_{t>0}$ is compact in X.

(H_{21}) There exist constants $d_k, \lambda_k \ge 0$ and $\overline{d}_k, \overline{\lambda}_k \ge 0$, $k = 1, \dots, m$, such that

$$|I_k(x)|_X^2 \le d_k|x|_X^2 + \lambda_k,$$
$$|\overline{I}_k(y)|^2 \le \overline{d}_k|y|_X^2 + \overline{\lambda}_k,$$

for all $x, y \in X$.

(H_{22}) f^i, g^i are L^2-Carathéodory maps.

We now prove the second goal of this section.

Theorem 6.2. *Assume conditions* (H_{18})–(H_{22}) *hold. Then problem* (6.1) *has at least one solution.*

Proof. We transform problem (6.5) into a fixed-point problem (6.5). Consider the operator

$$N : PC \times PC \to PC \times PC,$$

defined by

$$N(x,y) = (N_1(x,y), N_2(x,y)), \quad (x,y) \in PC \times PC,$$

where

$$\begin{aligned} N_1(x,y) &= S(t)x_0 + \int_0^t S(t-s)f^1(s,x(s),y(s))ds + \sum_{l=1}^{\infty}\int_0^t S(t-s)\sigma_l^1(t,x(s),y(s))dB_l^H(s) \\ &\quad + \sum_{0<t_k<t} S(t-t_k)I_k(x(t_k)), \quad t \in [0,T], \\ N_2(x,y) &= S(t)y_0 + \int_0^t S(t-s)f^2(s,x(s),y(s))ds + \sum_{l=1}^{\infty}\int_0^t S(t-s)\sigma_l^2(t,x(s),y(s))dB_l^H(s) \\ &\quad + \sum_{0<t_k<t} S(t-t_k)\bar{I}_k(y(t_k)), \quad t \in [0,T]. \end{aligned}$$

Clearly, the fixed points of N are solutions to (6.1). In order to apply Theorem 5.3, we first show that N is completely continuous. The proof will be carried out in several steps.

Step 1. N is continuous.

Let (x_n, y_n) be a sequence such that

$$(x_n, y_n) \to (x,y) \in PC \times PC,$$

as $n \to \infty$, and observe that thanks to (H_{18})–(H_{22}), $I_k, \bar{I}_k$, and $k = 1, 2, \ldots, m$, they are continuous.

Then

$$\begin{aligned} &|N_1(x_n(t), y_n(t)) - N_1(x(t), y(t))|_X^2 \\ &\leq 3\left|\sum_{l=1}^{\infty}\int_0^t S(t-s)(\sigma_l^1(s,x_n(s),y_n(s))\sigma_l^1(s,x(s),y(s)))dB_l^H(s)\right|_X^2 \\ &\quad + 3\left|\int_0^t S(t-s)(f^1(s,x_n(s),y_n(s)) - f^1(s,x(s),y(s)))ds\right|_X^2 \end{aligned}$$

$$+3\left|\sum_{0<t_k<t} S(t-t_k)I_k(x_n(t_k)) - \sum_{0<t_k<t} S(t-t_k)I_k(x(t_k))\right|_X^2.$$

Thus, we deduce

$$\begin{aligned}
&E|N_1(x_n(t),y_n(t)) - N_1(x(t),y(t))|_X^2 \\
&\leq 3M^2 c_2(H)H(2H-1)T^{2H-1}\int_0^t E\|\sigma^1(s,x_n(s),y_n(s)) - \sigma^1(s,x(s),y(s))\|^2 ds \\
&\quad + 3M^2\int_0^t E|f^1(s,x_n(s),y_n(s)) - f^1(s,x(s),y(s))|_X^2 ds \\
&\quad + 3M^2\sum_{0<t_k<t} E|I_k(x_n(t_k)) - I_k(x(t_k))|_X^2.
\end{aligned}$$

By the Lebesgue dominated convergence theorem, we have

$$\begin{aligned}
&\sup_{t\in J} E|N_1(x_n(t),y_n(t)) - N_1(x(t),y(t))|_X^2 \\
&\leq \bar{c}_H\int_0^T E\|\sigma^1(s,x_n(s),y_n(s)) - \sigma^1(s,x(s),y(s))\|^2 ds \\
&\quad + 3M^2\int_0^T E|f^1(s,x_n(s),y_n(s)) - f^1(s,x(s),y(s))|_X^2 ds \\
&\quad + 3M^2\sum_{k=1}^m E|I_k(x_n(t_k)) - I_k(x(t_k))|_X^2 \to 0 \quad \text{as } n\to\infty,
\end{aligned}$$

where

$$\bar{c}_H = 3M^2 c_2(H)H(2H-1)T^{2H-1}.$$

Similarly,

$$\begin{aligned}
&\sup_{t\in J} E|N_2(x_n(t),y_n(t)) - N_1(x(t),y(t))|_X^2 \\
&\leq \bar{c}_H\int_0^T E\|\sigma^2(s,x_n(s),y_n(s)) - \sigma^2(s,x(s),y(s))\|^2 ds \\
&\quad + 2M^2\int_0^T E|f^2(s,x_n(s),y_n(s)) - f^2(s,x(s),y(s))|_X^2 ds \\
&\quad + 3M^2\sum_{k=1}^m E|\bar{I}_k(y_n(t_k)) - \bar{I}_k(y(t_k))|_X^2 \to 0 \text{ as } n\to\infty.
\end{aligned}$$

Therefore, N is continuous.

Step 2. N maps bounded sets into bounded sets in $PC \times PC$. Indeed, it is enough to show that for any $q > 0$, there exists a positive constant κ such that for each

$$(x,y) \in B_q = \{(x,y) \in PC \times PC : \|x\|_{PC} \leq q,\ \|y\| \leq q\},$$

we have

$$\|N(x,y)\|_{PC} \leq \kappa = (\kappa_1, \kappa_2).$$

Then, for each $t \in J$, and thanks to Lemma 2.12,

$$\begin{aligned}
E\big|N_1(x(t),y(t))\big|_X^2 &\leq 4M^2E|x_0|_X^2 + 4M^2\|c_1\|_{L^1} + 4M^2c_2(H)H(2H-1)T^{2H-1}\|\overline{c}_1\|_{L^1} \\
&\quad + 4M^2\Big(\sum_{0<t_k<t}(d_kE|x|_X^2 + \lambda_k)\Big) \\
&\quad + 4M^2\int_0^t (c_2(H)H(2H-1)T^{2H-1}\overline{a}_1(s) + \overline{a}_1(s))E\big|x(s)\big|_X^2 ds \\
&\quad + 4M^2\int_0^t (c_2(H)H(2H-1)T^{2H-1}\overline{\beta}_1(s) + \overline{b}_1(s))E\big|y(s)\big|_X^2 ds \\
&\leq 4M^2E|x_0|_X^2 + 4M^2\|c_1\|_{L^1} + 4M^2c_2(H)H(2H-1)T^{2H-1}\|\overline{c}_1\|_{L^1} \\
&\quad + 4M^2\sum_{0<t_k<t}\lambda_k \\
&\quad + 4M^2\int_0^t (c_2(H)H(2H-1)T^{2H-1}\overline{a}_1(s) + \overline{a}_1(s))q ds \\
&\quad + 4M^2\int_0^t (c_2(H)H(2H-1)T^{2H-1}\overline{\beta}_1(s) + \overline{b}_1(s))q ds \\
&\quad + 4M^2\sum_{0<t_k<t} d_k q = \kappa_1.
\end{aligned}$$

Similarly, we have

$$\begin{aligned}
E\big|N_2(x,y)\big|_X^2 &\leq 4M^2E|y_0|_X^2 + 4M^2\|c_2\|_{L^1} + 4M^2c_2(H)H(2H-1)T^{2H-1}\|\overline{c}_2\|_{L^1} \\
&\quad + 4M^2\sum_{0<t_k<t}\overline{\lambda}_k \\
&\quad + 4M^2\int_0^t (c_2(H)H(2H-1)T^{2H-1}\overline{a}_2(s) + \overline{a}_2(s))q\, ds
\end{aligned}$$

$$+ 4M^2 \int_0^t (c_2(H)H(2H-1)T^{2H-1}\overline{\beta}_2(s) + \overline{b}_2(s))q\,ds$$
$$+ 4M^2 \sum_{0<t_k<t} \overline{d}_k q = \kappa_2.$$

Step 3. N maps bounded sets into equicontinuous sets of $PC \times PC$.

Let B_q be a bounded set in $PC \times PC$ as in Step 2. Let $\tau_1, \tau_2 \in J$, $\tau_1 < \tau_2$ and $u \in B_q$. Then, for $i = 1, 2$, we obtain

$$
\begin{aligned}
&E|N_i(\tau_2, x(\tau_2), y(\tau_2)) - N_i(\tau_1, x(\tau_1), y(\tau_1))|_X^2 \\
&\quad \le 7\|(S(\tau_2) - S(\tau_1))\|^2 E|x_0|_X^2 \\
&\qquad + 7E\left|\int_0^{\tau_1} (S(\tau_2 - s) - S(\tau_1 - s))f^i(s, x(s), y(s))ds\right|_X^2 \\
&\qquad + 7E\left|\int_{\tau_1}^{\tau_2} S(\tau_2 - s)f^i(s, x(s), y(s))ds\right|_X^2 \\
&\qquad + 7E\left|\sum_{l=1}^{\infty}\int_0^{\tau_1} (S(\tau_2 - s) - S(\tau_1 - s))\sigma_l^i(s, x(s), y(s))dB_l^H(s)\right|_X^2 \\
&\qquad + 7E\left|\sum_{l=1}^{\infty}\int_{\tau_1}^{\tau_2} S(\tau_2 - s)\sigma_l^i(s, x(s), y(s))dB_l^H(s)\right|_X^2 \\
&\qquad + 7E\left|\sum_{0<t_k<\tau_2} (S(\tau_1 - t_k) - S(\tau_2 - t_k))I_k(x(t_k))\right|_X^2 \\
&\qquad + 7E\left|\sum_{\tau_1<t_k<\tau_2} (S(\tau_2 - t_k))I_k(x(t_k))\right|_X^2.
\end{aligned}
$$

From Lemma 2.12, we deduce

$$
\begin{aligned}
&E|N_i(\tau_2, x(\tau_2), y(\tau_2)) - N_i(\tau_1, x(\tau_1), y(\tau_1))|_X^2 \\
&\quad \le 7\|(S(\tau_2 - \tau_1) - Id)\|^2 E|x_0|_X^2 \\
&\qquad + 7\tau_2 \int_0^{\tau_1} \|(S(\tau_2 - s) - S(\tau_1 - s))\|^2 E|f^i(s, x(s), y(s))|_X^2 ds \\
&\qquad + 7(\tau_2 - \tau_1)\int_{\tau_1}^{\tau_2} \|S(\tau_2 - s)\|^2 E|f^i(s, x(s), y(s))|_X^2 ds \\
&\qquad + 7c_2(H)H(2H-1)\tau_2^{2H-1}\int_0^{\tau_1} \|S(\tau_2 - s) - S(\tau_1 - s)\|^2 E\|\sigma^i(s, x(s), y(s))\|^2 ds
\end{aligned}
$$

$$+ 7c_2(H)H(2H-1)|\tau_2-\tau_1|^{2H-1}\int_{\tau_1}^{\tau_2}\|S(\tau_1-s)\|^2E\|\sigma^i(s,x(s),y(s))\|^2ds$$
$$+ 7\sum_{0<t_k<\tau_2}\|(S(\tau_2-\tau_1)-I_d)\|^2E|I_k(x(t_k))|_X^2$$
$$+ 7\sum_{\tau_1<t_k<\tau_2}\|S(\tau_2-t_k)\|^2E|I_k(x(t_k))|_X^2.$$

From (H_{18})–(H_{19}) and (H_{21}), we have

$$\begin{aligned}
&E|N_i(\tau_2,x(\tau_2),y(\tau_2))-N_i(\tau_1,x(\tau_1),y(\tau_1))|_X^2\\
&\quad\le 7\|S(\tau_2-\tau_1)-Id\|^2E|x_0|_X^2\\
&\quad+7\tau_2\int_0^{\tau_2}\|S(\tau_2-s)-S(\tau_1-s)\|^2(\overline{a}_i(s)E|x(s)|_X^2+\overline{b}_i(s)E|y(s)|_X^2+c_i(s))ds\\
&\quad+7(\tau_2-\tau_1)\int_{\tau_1}^{\tau_2}\|S(\tau_1-s)\|^2(\overline{a}_i(s)E|x(s)|_X^2+\overline{b}_i(s)E|y(s)|_X^2+c_i(s))ds\\
&\quad+7c_2(H)H(2H-1)\tau_2^{2H-1}\\
&\quad\times\int_0^{\tau_2}\|S(\tau_2-s)-S(\tau_1-s)\|^2(\overline{\alpha}_i(s)E|x(s)|_X^2+\overline{\beta}_i(s)E|y(s)|_X^2+\overline{c}_i(s))ds\\
&\quad+7c_2(H)H(2H-1)(\tau_2-\tau_1)^{2H-1}\\
&\quad\times\int_{\tau_1}^{\tau_2}\|S(\tau_1-s)\|^2(\overline{\alpha}_i(s)E|x(s)|_X^2+\overline{\beta}_i(s)E|y(s)|_X^2+\overline{c}_i(s))ds\\
&\quad+7\sum_{0<t_k<\tau_2}\|S((\tau_2-\tau_1)-I_d)\|^2(d_kE|x(t_k)|_X^2+\lambda_k)\\
&\quad+7\sum_{\tau_1<t_k<\tau_2}\|S(\tau_2-t_k)\|^2(d_kE|x(t_k)|_X^2+\lambda_k).
\end{aligned}$$

Now, it is straightforward to see that the right-hand side tends to zero as $\tau_2-\tau_1\to 0$ since the compactness of $S(t)$ for $t>0$ implies continuity in the uniform operator topology. This proves the equicontinuity.

Step 4. $N(B_q)(t)$ is precompact in $X\times X$.
As a consequence of Steps 2 to 3, and in light of the Arzelá–Ascoli theorem, it is sufficient to show that N maps B_q into a precompact set in $X\times X$. Let $0<t<b$ be fixed, and let ϵ be a real number satisfying $0<\epsilon<t$. For $(x,y)\in B_q$, we define

$$N_i^\epsilon(x,y)=S(t)z_0^i+\int_0^{t-\epsilon}S(t-s)f^i(s,x(s),y(s))ds$$

$$+\sum_{l=1}^{\infty}\int_{0}^{t-\epsilon} S(t-s)\sigma_l^i(t,x(s),y(s))dB_l^H(s)$$
$$+\sum_{0<t_k<t-\epsilon} S(t-t_k)I_k^i(z(t_k))$$
$$= S(\epsilon)S(t-\epsilon)z_0^i + S(\epsilon)\int_{0}^{t-\epsilon} S(t-s-\epsilon)f^i(s,x(s),y(s))ds$$
$$+ S(\epsilon)\sum_{l=1}^{\infty}\int_{0}^{t-\epsilon} S(t-s-\epsilon)\sigma_l^i(t,x(s),y(s))dB_l^H(s)$$
$$+\sum_{0<t_k<t-\epsilon} S(t-t_k)I_k^i(z^i(t_k)).$$

Since $S(t)$ is a compact operator, the set

$$H_\epsilon = \{N^\epsilon(x,y)(t) = (N_1^\epsilon(x,y)(t), N_2^\epsilon(x,y)(t)),\ (x,y)\in B_q\}$$

is precompact in $X\times X$ for every ϵ such that $0<\epsilon<t$ Moreover, for every $(x,y)\in B_q$ and $i=1,2$, we have

$$E\|N_i(x,y)-N_i^\epsilon(x,y)\|^2 \le 3\int_{t-\epsilon}^{t}(\overline{a}_i(s)q+\overline{b}_i(s)q+c_i(s))ds$$
$$+3M^2c_2(H)H(2H-1)T^{2H-1}\int_{t-\epsilon}^{t}(\overline{\alpha}_i(s)q+\overline{\beta}_i(s)q+\overline{c}_i(s))ds$$
$$+3M^2\sum_{t-\epsilon<t_k<t}(d_kq+\lambda_k).$$

Therefore, there are precompact sets arbitrarily close to the set

$$H_\epsilon = \{N^\epsilon(x,y)(t) = (N_1^\epsilon(x,y)(t), N_2^\epsilon(x,y)(t)),\ (x,y)\in B_q\}.$$

Hence, the set

$$H = \{N(x,y)(t) = (N_1(x,y)(t), N_2(x,y)(t)),\ (x,y)\in B_q\},$$

is precompact in $X\times X$, and the right-hand side tends to 0 uniformly in t as $\epsilon\to 0^+$. Hence, we can conclude the relative compactness of $N(B_q)(t)$ for $t\ge 0$. By the Arzelá–Ascoli theorem, we conclude that

$$N: PC\times PC \to PC\times PC,$$

is a completely continuous operator.

Step 5. A priori bounds on solutions.

The set

$$U = \{(x,y) \in PC \times PC : x = \lambda N_1(x,y) \text{ and } y = \lambda N_2(x,y) \text{ for some } 0 < \lambda < 1\}$$

is bounded.

Let $(x,y) \in PC \times PC$ be a solution of the abstract nonlinear equation $x = \lambda N_1(x,y)$ and $y = \lambda N_2(x,y)$ for some $0 < \lambda < 1$. Thus, for $t \in [0,t_1]$,

$$\begin{aligned} E|x(t)|_X^2 &\leq 3M^2E|x_0|_X^2 + 3M^2\int_0^t (\overline{a}_1(s)E|x(s)|_X^2 + \overline{b}_1(s)E|y(s)|_X^2 + c_1(s))ds \\ &\quad + 3M^2c_2(H)H(2H-1)t_1^{2H-1}\int_0^t (\overline{\alpha}_1(s)E|x(s)|_X^2 + \overline{\beta}_1(s)E|y(s)|_X^2 + \overline{c}_1(s))ds \\ &\leq 3M^2E|x_0|^2 + 3M^2\|c_1\|_{L^1} + 3M^2c_2(H)H(2H-1)t_1^{2H-1}\|\overline{c}_1\|_{L^1} \\ &\quad + 3M^2\int_0^t (c_2(H)H(2H-1)^2t_1^{2H-1}\overline{\alpha}_1(s) + \overline{a}_1(s))E|x(s)|_X^2 ds \\ &\quad + 3M^2\int_0^t (c_2(H)H(2H-1)^2t_1^{2H-1}\overline{\beta}_1(s) + \overline{b}_1(s))E|y(s)|_X^2 ds, \end{aligned}$$

$$E|x(t)|_X^2 \leq A_1 + \int_0^t B_1(s)E|x(s)|_X^2 ds + \int_0^t C_1(s)E|y(s)|_X^2\, ds,$$

and similarly,

$$E|y(t)|_X^2 \leq A_2 + \int_0^t B_2(s)E|x(s)|_X^2 ds + \int_0^t C_2(s)E|y(s)|_X^2,$$

where for each $j = 1,2$,

$$\begin{aligned} A_j &= 3M^2E|z_0^j|^2 + 3M^2\|c_j\|_{L^1} + 3M^2c_2(H)H(2H-1)t_1^{2H-1}\|\overline{c}_j\|_{L^1}, \\ B_j(s) &= 3M^2(c_2(H)H(2H-1)^2t_1^{2H-1}\overline{\alpha}_i(s) + \overline{a}_i(s)), \quad s \in [0,t_1], \end{aligned}$$

and

$$C_j(s) = 3M^2(c_2(H)H(2H-1)^2t_1^{2H-1}\overline{\beta}_j(s) + \overline{b}_j(s)),$$

and $z_0^1 = x_0$ and $z_0^2 = y_0$.

From these previous expressions, we can easily deduce

$$\sup_{r\in[0,t]}(E|y(r)|_X^2 + E|x(r)|_X^2) \le K_1 + \int_0^t K_2(s) \sup_{\eta\in[0,s]}(E|x(\eta)|_X^2 + E|y(\eta)|_X^2)ds$$

where

$$K_1 = A_1 + A_2, \quad K_2(s) = \max\{B_1(s) + B_2(s), c_1(s) + C_2(s)\}.$$

Using the Gronwall inequality now,

$$\sup_{r\in[0,t]}(E|y(r)|_X^2 + E|x(r)|_X^2) \le K_1 \exp\left(\int_0^{t_1} K_2(s)ds\right) = \overline{M}_0,$$

whence

$$\sup_{t\in[0,t_1]} E|x(t)|_X^2 \le \overline{M}_0, \quad \text{and} \quad \sup_{t\in[0,t_1]} E|y(t)|_X^2 \le \overline{M}_0,$$

where $\overline{M}_0$ depends only on t_1. For $t \in (t_1, t_2]$, we have

$$x(t) = S(t-t_1)[x(t_1) + I_1(x_1(t_1^-))] + \int_{t_1}^t S(t-s)f^1(s, x(s), y(s))ds$$
$$+ \sum_{l=1}^{\infty} \int_{t_1}^t S(t-s)\sigma_l^1(t, x(s), y(s))dB_l^H(s).$$

Then, by a similar argument, we obtain

$$E|x(t)|_X^2 = E\Bigg|S(t-t_1)[x_1(t_1) + I_1(x_1(t_1^-))] + \int_{t_1}^t S(t-s)f^1(s, x(s), y(s))ds$$
$$+ \sum_{l=1}^{\infty} \int_{t_1}^t S(t-s)\sigma_l^1(t, x(s), y(s))dB_l^H(s)\Bigg|_X^2$$
$$\le \|S(t-t_1)\|^2(E|x_1(t_1)|_X^2 + E|I_1(x_1(t_1^-))|_X^2)$$
$$+ E\left|\int_{t_1}^t S(t-s)f^1(s, x(s), y(s))ds\right|_X^2$$
$$+ E\left|\sum_{l=1}^{\infty} \int_{t_1}^t S(t-s)\sigma_l^1(t, x(s), y(s))dB_l^H(s)\right|_X^2.$$

Consequently,

$$\begin{aligned}E|x(t)|_X^2 &\le 4M^2(\overline{M}^2 + d_1 E|x(t_1|_X^2 + \lambda)\\ &\quad + 4M^2\int_{t_1}^{t}(\overline{a}_1(s)E|x(s)|_X^2 + \overline{b}_1(s)E|y(s)|_X^2 + c_1(s))ds\\ &\quad + 4M^2 c_2(H)H(2H-1)t_2^{2H-1}\int_{t_1}^{t}(\overline{a}_1(s)E|x(s)|_X^2 + \overline{\beta}_1(s)E|y(s)|_X^2 + \overline{c}_1(s))ds\\ &\le A_3 + \int_{t_1}^{t} B_3(s)E|x(s)|_X^2 ds + \int_{t_1}^{t} C_3(s)E|y(s)|_X^2 ds.\end{aligned}$$

Then we deduce

$$E|y(t)|_X^2 \le A_4 + \int_{t_1}^{t} B_4(s)E|x(s)|_X^2 ds + \int_{t_1}^{t} C_4(s)E|y(s)|_X^2,$$

where

$$A_3 = 4M^2(\overline{M}^2 + \lambda) + 4M^2\|c_1\|_{L^1} + 4M^2 c_2(H)H(2H-1)t_2^{2H-1}\|\overline{c}_1\|_{L^1},$$

and

$$A_4 = 4M^2(\overline{M}^2 + \overline{\lambda}) + 4M^2\|c_2\|_{L^1} + 4M^2 c_2(H)H(2H-1)t_2^{2H-1}\|\overline{c}_2\|_{L^1},$$

and

$$\begin{aligned}B_3(s) &= 4M^2 d_1 + 4M^2(c_2(H)H(2H-1)^2 t_2^{2H-1}\overline{a}_1(s) + \overline{a}_1(s))\\ B_4(s) &= 4M^2 + (c_2(H)H(2H-1)^2 t_2^{2H-1}\overline{a}_2(s) + \overline{a}_2(s)), \quad s \in (t_1, t_2],\end{aligned}$$

and

$$\begin{aligned}C_3(s) &= 4M^2(c_2(H)H(2H-1)^2 t_2^{2H-1}\overline{\beta}_1(s) + \overline{b}_1(s)), \quad s \in (t_1, t_2],\\ C_4(s) &= 4M^2\overline{d}_1 + 4M^2(c_2(H)H(2H-1)^2 t_2^{2H-1}\overline{\beta}_1(s) + \overline{b}_2(s)), \quad s \in (t_1, t_2].\end{aligned}$$

Combining $E|x(t)|_X^2$ and $E|y(t)|_X^2$,

$$\sup_{r\in[t_1,t]}(E|y(r)|_X^2 + E|x(r)|_X^2) \le K_3 + \int_{t_1}^{t} K_4(s)\sup_{\eta\in[t_1,s]}(E|x(\eta)|_X^2 + E|y(\eta)|_X^2)ds$$

where

$$K_3 = A_3 + A_4,$$
$$K_4(s) = \max\{B_3(s) + B_4(s), c_3(s) + C_4(s)\}.$$

Using once more the Gronwall inequality,

$$\sup_{t\in(t_1,t_2]} (E|y(t)|_X^2 + E|x(t)|_X^2) \leq K_3 \exp\left(\int_{t_1}^{t_2} K_4(s)ds\right) = \overline{M}_1.$$

Consequently, there exists a constant $\overline{M}_1$, which only depends on t_1, t_2, such that

$$\sup_{t\in(t_1,t_2]} E|x(t)|_X^2 \leq \overline{M}_1,$$

and

$$\sup_{t\in(t_1,t_2]} E|y(t)|_X^2 \leq \overline{M}_1.$$

If we iterate this procedure in every interval $(t_i, t_{i+1}]$, we can prove that there exists a positive constant $\overline{M}_i$, which only depends on t_i, t_{i+1}, such that

$$\sup_{t\in(t_i,t_{i+1}]} E|x(t)|_X^2 \leq \overline{M}_i,$$

and

$$\sup_{t\in(t_i,t_{i+1}]} E|y(t)|_X^2 \leq \overline{M}_i, \quad i = 0, 1, 2, \dots, m-1.$$

Set

$$U = \{(x,y) \in PC \times PC : \|x\|_{PC} < M^*, \ \|y\|_{PC} < M^*\},$$

where

$$M^* = \sup\{\overline{M}_i : i = 0, 1, 2, \dots, m-1\}.$$

Then U is an open subset of $PC \times PC$, and it is straightforward to see that there is no $(x,y) \in \partial U$, such that $x = \lambda N_1(x,y)$ and $y = \lambda N_2(x,y)$ for some $0 < \lambda < 1$. By Theorem 5.3, N possesses at least one fixed point (x,y). □

Remark 6.1. We can replace conditions (H_{18}) and (H_{20}) with the following hypotheses:

$(\widetilde{H})$ There exist integrable functions

$$\eta : [0, T] \longrightarrow \mathbb{R}^+,$$

and

$$\psi : [0, \infty) \longrightarrow (0, \infty),$$

that are continuous and nondecreasing such that, for all $t \in [0, T]$ and $x, y \in X$,

$$E\|f^i(t,x,y)\|^2 \le \eta(t)\psi(E\|x\|^2 + E\|y\|^2), \quad E\|\sigma^i(t,x,y)\|^2 \le \eta(t)\psi(E|x|^2 + E|y|^2),$$

where

$$\sigma^i(t,x,y) = \sum_{l=1}^{\infty} \sigma_l^i(t,x,y), \quad i = 1, 2.$$

6.2 Weak solutions

The mild solutions to the system (6.1) are also proven to be weak solutions.

Definition 6.2. An X-valued stochastic process

$$u(t) = (x(t), y(t)), \quad t \in [0, T],$$

is called a weak solution of (6.1) if for each $\varphi \in D(A^*)$,

$$\begin{aligned}\langle x(t), \varphi\rangle &= \langle x_0, \varphi\rangle + \int_0^t \langle x(s), a^*\varphi\rangle d\tau + \int_0^t \langle f^1(\tau, x(\tau), y(\tau)), \varphi\rangle d\tau \\ &\quad + \sum_{l=1}^{\infty} \int_0^t \langle \sigma_l^1(\tau, x(\tau), y(\tau)), \varphi\rangle dB_l^H(\tau) + \sum_{0<t_k<t} \langle I_k(x(t_k)), \varphi\rangle, \quad t \in J,\end{aligned}$$

and

$$\begin{aligned}\langle y(t), \varphi\rangle &= \langle y_0, \varphi\rangle + \int_0^t \langle y(s), a^*\varphi\rangle d\tau + \int_0^t \langle f^2(\tau, x(\tau), y(\tau)), \varphi\rangle d\tau \\ &\quad + \sum_{l=1}^{\infty} \int_0^t \langle \sigma_l^2(\tau, x(\tau), y(\tau)), \varphi\rangle dB_l^H(\tau) + \sum_{0<t_k<t} \langle \bar{I}_k(y(t_k)), \varphi\rangle, \quad t \in J.\end{aligned}$$

Theorem 6.3. *Assume conditions* (H_{18})–(H_{21}) *hold. The mild solution of* (6.1) *is also a weak solution.*

Proof. Let $(x(t), y(t))$ be a mild solution to (6.1). Then, for each $\varphi \in D(A^*)$ and $t \in [0, t_1]$, we obtain

$$E\left|\int_0^t \langle x(s), a^*\varphi\rangle ds - \int_0^t \langle S(s)x_0, a^*\varphi\rangle ds - \int_0^t \int_0^s \langle S(s-\tau)f^1(\tau, x(\tau), y(\tau)), a^*\varphi\rangle d\tau ds\right.$$

$$
\begin{aligned}
&\quad - \sum_{l=1}^{\infty}\int_0^t\int_0^s \langle S(s-\tau)\sigma_l^1(\tau,x(\tau),y(\tau)),a^*\varphi\rangle dB_l^H(\tau)ds\Bigg| \\
&\le \int_0^t E\Bigg|\langle x(s),a^*\varphi\rangle - \langle S(t)x_0,a^*\varphi\rangle - \int_0^s \langle S(s-\tau)f^1(\tau,x(\tau),y(\tau)),a^*\varphi\rangle d\tau \\
&\quad - \sum_{l=1}^{\infty}\int_0^s \langle S(s-\tau)\sigma_l^1(\tau,x(\tau),y(\tau)),a^*\varphi\rangle dB_l^H(\tau)ds\Bigg| ds \\
&= \int_0^t E\Bigg|\Bigg\langle x(s) - S(s)x_0 - \int_0^s S(s-\tau)f^1(\tau,x(\tau),y(\tau))d\tau \\
&\quad - \sum_{l=1}^{\infty}\int_0^s S(s-\tau)\sigma_l^1(\tau,x(\tau),y(\tau))dB_l^H(\tau)ds)ds, a^*\varphi\Bigg\rangle\Bigg| ds \\
&= 0.
\end{aligned}
$$

Thus,

$$
\begin{aligned}
\int_0^t \langle x(s),a^*\varphi\rangle ds &= \int_0^t \langle S(t)x_0,a^*\varphi\rangle ds + \int_0^t\int_0^s \langle S(s-\tau)f^1(\tau,x(\tau),y(\tau)),a^*\varphi\rangle d\tau ds \\
&\quad + \sum_{l=1}^{\infty}\int_0^t\int_0^s \langle S(s-\tau)\sigma_l^1(\tau,x(\tau),y(\tau)),a^*\varphi\rangle dB_l^H(\tau)ds \\
&= \sum_{j=1}^{3}\widetilde{P}_j.
\end{aligned}
$$

Similarly,

$$
\begin{aligned}
\int_0^t \langle y(s),a^*\varphi\rangle ds &= \int_0^t \langle S(t)y_0,a^*\varphi\rangle ds + \int_0^t\int_0^s \langle S(s-\tau)f^2(\tau,x(\tau),y(\tau)),a^*\varphi\rangle d\tau ds \\
&\quad + \sum_{l=1}^{\infty}\int_0^t\int_0^s \langle S(s-\tau)\sigma_l^2(\tau,x(\tau),y(\tau)),a^*\varphi\rangle dB_l^H(\tau)ds \\
&= \sum_{j=1}^{3}\widehat{P}_j.
\end{aligned}
$$

Now, for $\varphi \in D(A^*)$ and $t \in [0,t_1]$, we use the fact that

$$
\frac{d}{dt}S^*(t)\varphi = S^*(t)A^*\varphi,
$$

which yields

$$\widetilde{P}_1 = \int_0^t \langle S(s)x_0, a^*\varphi\rangle ds = \int_0^t \langle x_0, S^*(s)A^*\varphi\rangle ds$$
$$= \int_0^t \left\langle x_0, \frac{d}{ds}S^*(s)\varphi \right\rangle ds$$
$$= \langle S(t)x_0 - x_0, \varphi\rangle,$$

and analogously,

$$\widehat{P}_1 = \int_0^t \langle S(s)y_0, a^*\varphi\rangle = \langle S(t)y_0 - y_0, \varphi\rangle.$$

In addition, using Fübini's theorem for $t \in [0, t_1]$,

$$\widetilde{P}_2 = \int_0^t \int_0^s \langle S(s-\tau)f^1(\tau, x(\tau), y(\tau)), a^*\varphi\rangle d\tau ds$$
$$= \int_0^t \int_\tau^t \langle 1_{(0,s]}f^1(\tau, x(\tau), y(\tau)), S^*(s-\tau)A^*\varphi\rangle d\tau ds$$
$$= \int_0^t \langle S(t-\tau)f^1(\tau, x(\tau), y(\tau)) - f^1(\tau, x(\tau), y(\tau)), \varphi\rangle d\tau$$
$$= \int_0^t \langle S(t-\tau)f^1(\tau, x(\tau), y(\tau)), y(\tau)), \varphi\rangle d\tau - \int_0^t \langle f^1(\tau, x(\tau), y(\tau)), \varphi\rangle d\tau,$$

and

$$\widehat{P}_2 = \int_0^t \int_0^s \langle S(s-\tau)f^2(\tau, x(\tau), y(\tau)), a^*\varphi\rangle d\tau ds$$
$$= \int_0^t \langle S(t-\tau)f^2(\tau, x(\tau), y(\tau)), y(\tau)), \varphi\rangle d\tau - \int_0^t \langle f^2(\tau, x(\tau), y(\tau)), \varphi\rangle d\tau.$$

As for the estimates of the third terms, we have for $t \in [0, t_1]$,

$$\widetilde{P}_3 = \sum_{l=1}^{\infty} \int_0^t \int_0^s \langle S(s-\tau)\sigma_l^1(\tau, x(\tau), y(\tau)), a^*\varphi\rangle dB_l^H(\tau) ds$$

$$= \sum_{l=1}^{\infty} \int_0^t \int_\tau^t \langle 1_{(0,s]}\sigma_l^1(\tau, x(\tau), y(\tau)), S^*(s-\tau)A^*\varphi\rangle dB_l^H(\tau) ds$$
$$= \sum_{l=1}^{\infty} \int_0^t \langle S(t-\tau)\sigma_l^1(\tau, x(\tau), y(\tau)), \varphi\rangle dB_l^H(\tau) - \sum_{l=1}^{\infty} \int_0^t \langle \sigma_l^1(\tau, x(\tau), y(\tau)), \varphi\rangle dB_l^H(\tau),$$

and

$$\widehat{P}_3 = \sum_{l=1}^{\infty} \int_0^t \int_0^s \langle S(s-\tau)\sigma_l^2(\tau, x(\tau), y(\tau)), a^*\varphi\rangle dB_l^H(\tau) ds$$
$$= \sum_{l=1}^{\infty} \int_0^t \langle S(t-\tau)\sigma_l^2(\tau, x(\tau), y(\tau)), \varphi\rangle dB_l^H(\tau) - \sum_{l=1}^{\infty} \int_0^t \langle \sigma_l^2(\tau, x(\tau), y(\tau)), \varphi\rangle dB_l^H(\tau).$$

Now, taking into account all the previous estimates, we can write

$$\int_0^t \langle x(s), a^*\varphi\rangle ds$$
$$= \int_0^t \langle Ax(s), \varphi\rangle ds$$
$$= \langle S(t)x_0 - x_0, \varphi\rangle + \int_0^t \langle S(t-\tau)f^1(\tau, x(\tau), y(\tau)), y(\tau)), \varphi\rangle d\tau - \int_0^t \langle f^1(\tau, x(\tau), y(\tau)), \varphi\rangle d\tau$$
$$+ \sum_{l=1}^{\infty} \int_0^t \langle S(t-\tau)\sigma_l^1(\tau, x(\tau), y(\tau)), \varphi\rangle dB_l^H(\tau) - \sum_{l=1}^{\infty} \int_0^t \langle \sigma_l^1(\tau, x(\tau), y(\tau)), \varphi\rangle dB_l^H(\tau)$$
$$= \langle x(t) - x_0, \varphi\rangle - \int_0^t \langle f^1(\tau, x(\tau), y(\tau)), \varphi\rangle d\tau - \sum_{l=1}^{\infty} \int_0^t \langle \sigma_l^1(\tau, x(\tau), y(\tau)), \varphi\rangle dB_l^H(\tau).$$

Finally,

$$\langle x(t), \varphi\rangle = \langle x_0, \varphi\rangle + \int_0^t \langle x(s), a^*\varphi\rangle d\tau + \int_0^t \langle f^1(\tau, x(\tau), y(\tau)), \varphi\rangle d\tau$$
$$+ \sum_{l=1}^{\infty} \int_0^t \langle \sigma_l^1(\tau, x(\tau), y(\tau)), \varphi\rangle dB_l^H(\tau),$$

and similar computations for y and $t \in [0, t_1]$ imply

$$\langle y(t), \varphi \rangle = \langle y_0, \varphi \rangle + \int_0^t \langle y(s), a^* \varphi \rangle d\tau + \int_0^t \langle f^2(\tau, x(\tau), y(\tau)), \varphi \rangle d\tau$$
$$+ \sum_{l=1}^{\infty} \int_0^t \langle \sigma_l^2(\tau, x(\tau), y(\tau)), \varphi \rangle dB_l^H(\tau).$$

Therefore, the mild solution is also a weak solution in the interval $[0, t_1]$.

We repeat this scheme in every subinterval $(t_{m-1}, t_m])$ until we reach the points $t \in (t_m, T]$. As the arguments are the same as in the previous case, we prefer to omit the details. □

6.3 Exponential stability

In this section, we are interested in the exponential decay to zero in mean square of the mild solutions to (6.1). We will assume that solutions are defined globally in time

$$PC = \Big\{x : \Omega \times [0, T] \longrightarrow X,\ x \in C(J_k, L^2(\Omega, X)),\ k = 1, \dots, m \text{ such that}$$
$$x(t_k^+, .) \text{ and } x(t_k^-, .) \text{ exist with } x(t_k^-, .) = x(t_k, .) \text{ and}$$
$$\sup_{t \in [0,T]} E\|x(t, .)\|^2 < \infty \text{ almost surely}\Big\},$$

endowed with the norm

$$\|x\|_{PC} = \sup_{s \in [0,T]} \left(E\|x(s, .)\|^2\right)^{\frac{1}{2}}.$$

It is not difficult to check that PC is a Banach space with norm $\|\cdot\|_{PC}$.

Consider the Banach space

$$PC_b = \{y \in PC_*(\mathbb{R}^+ \times \Omega, X) : y \text{ is bounded almost surely}\},$$

where

$$PC_*(\mathbb{R}^+ \times \Omega, X) = \{y\colon [0, \infty) \times \Omega \to X,\ y_k \in C((t_k, t_{k+1}), L^2(\Omega, X)),\ k = 0, 1, 2 \dots,$$
$$y(t_k^-) \text{ and } y(t_k^+) \text{ exist and satisfy } y(t_k) = y(t_k^-) \text{ for } k = 1, \dots\}$$

and $y_k = y_{|(t_k, t_{k+1}]}$. Endowed with the norm

$$\|y\|_b^2 = \sup\{E|y(t)|_X^2 : t \in [0, \infty)\},$$

PC_b is a Banach space.

For the study of this problem, we first introduce the following hypotheses:

(H_{23}) There exist $M > 0$ and $\gamma > 0$, such that

$$\|S(t)\| \le Me^{-\gamma t},$$

for all $t > 0$.

(H_{24}) There exist nonnegative numbers a_i and b_i for each $i \in \{1,2\}$, and for all $0 \le \delta \le \gamma$ it holds

$$\int_0^t e^{\delta s} E|f^i(s,x(s),y(s)) - f^i(s,\overline{x}(s),\overline{y}(s))|_X^2 ds \le a_i \int_0^t e^{\delta s} E|x(s) - \overline{x}(s)|_X^2 ds + b_i \int_0^t e^{\delta s} E|y(s) - \overline{y}(s)|_X^2,$$

and

$$\int_0^\infty e^{\gamma s} |f^i(s,0,0)|_X^2 ds < \infty.$$

for all $x, y, \overline{x}, \overline{y} \in PC_b$.

(H_{25}) There exist functions

$$\alpha_i, \beta_i \in C(\mathbb{R}^+, \mathbb{R}^+), \quad i = 1,2,$$

such that, for all $0 \le \delta \le \gamma$,

$$x, y, \overline{x}, \overline{y} \in PC_b,$$

and $t \in \mathbb{R}^+$, we have

$$\int_0^t e^{\delta s} E\|\sigma^i(s,x(s),y(s)) - \sigma^i(s,\overline{x}(s),\overline{y}(s))\|^2 ds \le \alpha_i(t) \int_0^t e^{\delta s} E|x(s) - \overline{x}(s)|_X^2 ds + \beta_i(t) \int_0^t e^{\delta s} E|y(s) - \overline{y}(s)|_X^2 ds,$$

and

$$\int_0^\infty e^{\gamma s} E\|\sigma^i(s,0,0)\|^2 ds < \infty,$$

$$\sup_{t \in \mathbb{R}^+} t^{2H-1} \alpha_i(t) < \infty$$

$$\sup_{t \in \mathbb{R}^+} t^{2H-1} \beta_i(t) < \infty.$$

(H_{26}) There exist constants $d_k \geq 0$ and $\overline{d}_k \geq 0$, $k = 1, \ldots,$ such that

$$E|I_k(x) - I_k(\overline{x})|_X^2 \leq d_k E|x - \overline{x}|_X^2$$
$$E|\overline{I}_k(y) - \overline{I}_k(\overline{y})|_X^2 \leq \overline{d}_k E|y - \overline{y}|_X^2,$$

and

$$\sum_{k=1}^{\infty} e^{\gamma t_k} E|I_k(0)|_X^2 < \infty, \quad 4M^2 \sum_{k=1}^{\infty} d_k < 1,$$

and

$$\sum_{k=1}^{\infty} e^{\gamma t_k} E|\overline{I}_k(0)|_X^2 < \infty, \quad 4M^2 \sum_{k=1}^{\infty} \overline{d}_k < 1,$$

for all $x, y, \overline{x}, \overline{y} \in PC_b$.

Theorem 6.4. *Assume* (H_{23})–(H_{26}) *hold and that* $\gamma > A_2 > 0$ *where*

$$A_2 = \max\left\{\frac{4M^2 c_2(H) H(2H-1)(\overline{c}_2 + \overline{c}_1)}{\widetilde{k}}, \frac{4\gamma^{-1} M^2 (c_2 + c_1)}{\widetilde{k}}\right\},$$
$$c_i = \max\{a_i, b_i\},$$
$$\overline{c}_i = \max\left\{\sup_{t \in \mathbb{R}_+} t^{2H-1} \alpha_i(t), \sup_{t \in \mathbb{R}_+} t^{2H-1} \beta_i(t)\right\},$$

and

$$\bar{k}_* = 4M^2 \sum_{k=1}^{\infty} \overline{d}_k,$$
$$k_* = 4M^2 \sum_{k=1}^{\infty} d_k,$$
$$\frac{1}{\widetilde{k}} = \max\left(\frac{1}{1 - k_*}, \frac{1}{1 - \bar{k}_*}\right).$$

Then there exists a unique mild solution (x, y) *to problem* (6.1) *that converges to zero in mean square, that is,*

$$\lim_{t \to +\infty} \begin{pmatrix} E\|x(t)\|^2 \\ E\|y(t)\|^2 \end{pmatrix} = \begin{pmatrix} 0 \\ 0 \end{pmatrix}.$$

Proof. It is clear that each mild solution to problem (6.1) is a fixed point of the operator N defined in Theorem 6.1. By using (H_{23})–(H_{26}), we can easily prove that

$$N(PC_b) \subset PC_b,$$

and from Perov's fixed-point theorem, there exists a unique

$$(x, y) \in PC_b \times PC_b,$$

which is a fixed point of N. Now, we show that

$$\begin{aligned}\lim_{t\to+\infty} E|N(x,y)|_X^2 &= \lim_{t\to+\infty} (E|N_1(x,y)|_X^2, E|N_2(x,y)|_X^2)\\ &= \lim_{t\to+\infty} (E|x(t)|_X^2, E|y(t)|_X^2)\\ &= (0,0).\end{aligned}$$

First, observe that we have

$$\begin{aligned}E|x(t)|_X^2 \le{}& 4E|S(t)x_0|_X^2 + 4E\left|\int_0^t S(t-s)f^1(s,x(s),y(s))ds\right|_X^2\\ &+ 4E\left|\sum_{l=1}^{\infty}\int_0^t S(t-s)\sigma_l^1(s,x(s),y(s))dB_l^H(s)\right|_X^2\\ &+ 4E\left|\sum_{0<t_k<t} S(t-t_k)I_k(x(t_k))\right|_X^2,\end{aligned}$$

and

$$\begin{aligned}E|y(t)|_X^2 \le{}& 4E|S(t)y_0|_X^2 + 4E\left|\int_0^t S(t-s)f^1(s,x(s),y(s))ds\right|_X^2\\ &+ 4E\left|\sum_{l=1}^{\infty}\int_0^t S(t-s)\sigma_l^1(s,x(s),y(s))dB_l^H(s)\right|_X^2\\ &+ 4E\left|\sum_{0<t_k<t} S(t-t_k)\bar{I}_k(y(t_k))\right|_X^2.\end{aligned}$$

Therefore, by Lemma 2.52 and (H_{23}),

$$E|x(t)|_X^2 \le 4M^2e^{-2\gamma t}E|x_0|_X^2 + 4\gamma^{-1}M^2\int_0^t e^{-\gamma(t-s)}E|f^1(s,x(s),y(s))|_X^2 ds$$

$$
\begin{aligned}
&+ 4M^2 c_2(H)H(2H-1)t^{2H-1}\sum_{l=1}^{\infty}\int_0^t e^{-2\gamma(t-s)}\|\sigma_l^1(s,x(s),y(s))\|^2 ds \\
&+ 4M^2 \sum_{0<t_k<t} e^{-2\gamma(t-t_k)} E|I_k(x(t_k))|_X^2,
\end{aligned}
$$

and, consequently,

$$
\begin{aligned}
e^{\gamma t}E|x(t)|_X^2 \le{}& 4M^2E|x_0|_X^2 + 4\gamma^{-1}M^2\int_0^t e^{\gamma s}E|f^1(s,x(s),y(s))|_X^2 ds \\
&+ 4M^2 c_2(H)H(2H-1)t^{2H-1}\sum_{l=1}^{\infty}\int_0^t e^{\gamma s}\|\sigma_l^1(s,x(s),y(s))\|^2 ds \\
&+ 4M^2 \sum_{0<t_k<t} e^{\gamma t_k} E|I_k(x(t_k))|_X^2.
\end{aligned}
$$

Thanks to the fact that $A_2 - \gamma < 0$, we can choose $\theta > 0$, such that

$$
\eta = A_2 - \gamma + \theta < 0,
$$

and for these constants, we have

$$
\begin{aligned}
e^{(\gamma-\theta)t}E|x(t)|_X^2 \le{}& 4M^2e^{-\theta t}E|x_0|_X^2 + 4e^{-\theta t}\gamma^{-1}M^2\int_0^t e^{\gamma s}E|f^1(s,x(s),y(s))|_X^2 ds \\
&+ 4M^2 c_2(H)H(2H-1)t^{2H-1}e^{-\theta t}\sum_{l=1}^{\infty}\int_0^t e^{\gamma s}\|\sigma_l^1(s,x(s),y(s))\|^2 ds \\
&+ 4M^2e^{-\theta t} \sum_{0<t_k<t} e^{\gamma t_k} E|I_k(x(t_k))|_X^2 \\
\le{}& 4M^2e^{-\theta t}E|x_0|_X^2 + 4\gamma^{-1}M^2\int_0^t e^{(\gamma-\theta)s}E|f^1(s,x(s),y(s))|_X^2 ds \\
&+ 4M^2 c_H H(2H-1)t^{2H-1}\int_0^t e^{(\gamma-\theta)s}E\|\sigma^1(s,x(s),y(s))\|^2 ds \\
&+ 4M^2e^{-\theta t} \sum_{0<t_k<t} e^{\gamma t_k} E|I_k(x(t_k))|_X^2 \\
={}& 4\sum_{i=1}^{4} P_i.
\end{aligned}
$$

Similarly,

$$
\begin{aligned}
e^{(\gamma-\theta)t}E|y(t)|_X^2 \le\ & 4M^2e^{-\theta t}E|y_0|_X^2 + 4\gamma^{-1}M^2\int_0^t e^{(\gamma-\theta)s}E|f^2(s,x(s),y(s))|_X^2 ds \\
& + 4M^2c_2(H)H(2H-1)t^{2H-1}\int_0^t e^{(\gamma-\theta)s}E\|\sigma^2(s,x(s),y(s))\|^2 ds \\
& + 4M^2e^{-\theta t}\sum_{0<t_k<t} e^{\gamma t_k}E|\bar{I}_k(y(t_k))|_X^2 \\
=\ & 4\sum_{i=1}^4 \bar{P}_i.
\end{aligned}
$$

Now it is easy to see, for any $t \ge 0$, that

$$
\begin{aligned}
P_2 &= \int_0^t e^{(\gamma-\theta)s}E|f^1(s,x(s),y(s))|_X^2 ds \\
&\le \int_0^t e^{(\gamma-\theta)s}E|f^1(s,x(s),y(s)) - f^1(s,0,0) + f^1(s,0,0)|_X^2 ds \\
&\le \int_0^t e^{(\gamma-\theta)s}E|f^1(s,x(s),y(s)) - f^1(s,0,0)|_X^2 ds + \int_0^t e^{(\gamma-\theta)s}E|f^1(s,0,0)|_X^2 ds \\
&\le a_1\int_0^t e^{(\gamma-\theta)s}E|x(s)|_X^2 ds + b_1\int_0^t e^{(\gamma-\theta)s}E|y(s)|_X^2 ds + \int_0^t e^{(\gamma-\theta)s}E|f^1(s,0,0)|_X^2 ds,
\end{aligned}
$$

and similarly,

$$
\begin{aligned}
\bar{P}_2 &= \int_0^t e^{(\gamma-\theta)s}E|f^2(s,x(s),y(s))|_X^2 ds \\
&\le a_2\int_0^t e^{(\gamma-\theta)s}E|x(s)|_X^2 ds + b_2\int_0^t e^{(\gamma-\theta)s}E|y(s)|_X^2 ds + \int_0^t e^{(\gamma-\theta)s}E|f^2(s,0,0)|_X^2 ds.
\end{aligned}
$$

By condition (H_{24}), we deduce the existence of positive constants $K_1, \bar{K}_1$, such that

$$
4\gamma^{-1}M^2\int_0^t e^{(\gamma-\theta)s}E|f^1(s,0,0)|_X^2 ds \le K_1,
$$

and

$$
4\gamma^{-1}M^2\int_0^t e^{(\gamma-\theta)s}E|f^2(s,0,0)|_X^2 ds \le \bar{K}_1.
$$

As for the term P_3, we have

$$\begin{aligned} P_3 &= \int_0^t e^{(\gamma-\theta)s} E\|\sigma^1(s,x(s),y(s))\|^2 ds \\ &\le \int_0^t e^{(\gamma-\theta)s} E\|\sigma^1(s,x(s),y(s)) - \sigma^1(s,0,0) + \sigma^1(s,0,0)\|^2 ds \\ &\le \int_0^t e^{(\gamma-\theta)s} E\|\sigma^1(s,x(s),y(s)) - \sigma^1(s,0,0)\|^2 ds + \int_0^t e^{(\gamma-\theta)s} E\|\sigma^1(s,0,0)\|^2 ds \\ &\le \alpha_1(t) \int_0^t e^{(\gamma-\theta)s} E|x(s)|_X^2 ds + \beta_1(t) \int_0^t e^{(\gamma-\theta)s} E|y(s)|_X^2 ds + \int_0^t e^{(\gamma-\theta)s} E\|\sigma^1(s,0,0)\|^2 ds, \end{aligned}$$

and similarly,

$$\begin{aligned} \overline{P}_3 &= \int_0^t e^{(\gamma-\theta)s} E\|\sigma^2(s,x(s),y(s))\|^2 ds \\ &\le \alpha_2(t) \int_0^t e^{(\gamma-\theta)s} E|x(s)|_X^2 ds + \beta_2(t) \int_0^t e^{(\gamma-\theta)s} E|y(s)|_X^2 ds + \int_0^t e^{(\gamma-\theta)s} E\|\sigma^1(s,0,0)\|^2 ds. \end{aligned}$$

From (H_{25}), we deduce the existence of positive constants $K_2, \overline{K}_2$, such that

$$4M^2 c_2(H)H(2H-1)t^{2H-1} \int_0^t e^{(\gamma-\theta)s} E\|\sigma^1(s,0,0)\|^2 ds \le K_2,$$

and

$$4M^2 c_2(H)H(2H-1)t^{2H-1} \int_0^t e^{(\gamma-\theta)s} E\|\sigma^2(s,0,0)\|^2 ds \le \overline{K}_2,$$

and for the last terms

$$\begin{aligned} P_4 &= e^{-\theta t} \sum_{0<t_k<t} e^{\gamma t_k} E|I_k(x(t_k))|_X^2 \\ &\le e^{-\theta t} \sum_{0<t_k<t} e^{\gamma t_k} E|I_k(x(t_k)) - I_k(0) + I_k(0)|_X^2 \\ &\le e^{-\theta t} \sum_{0<t_k<t} e^{\gamma t_k} E|I_k(x(t_k)) - I_k(0)|_X^2 + e^{-\theta t} \sum_{0<t_k<t} e^{\gamma t_k} E|I_k(0)|_X^2 \\ &\le e^{-\theta t} \sum_{0<t_k<t} e^{\gamma t_k} d_k E|x(t_k)|_X^2 + e^{-\theta t} \sum_{0<t_k<t} e^{\gamma t_k} E|I_k(0)|_X^2 \end{aligned}$$

$$\leq e^{(\gamma-\theta)t} \sum_{0<t_k<t} e^{\gamma(t_k-t)} d_k E|x(t_k)|_X^2 + e^{-\theta t} \sum_{0<t_k<t} e^{\gamma t_k} E|I_k(0)|_X^2,$$

and analogously,

$$\begin{aligned}\overline{P}_4 &= e^{-\theta t} \sum_{0<t_k<t} e^{\gamma t_k} E|\overline{I}_k(y(t_k))|_X^2 \\ &\leq e^{(\gamma-\theta)t} \sum_{0<t_k<t} e^{\gamma(t_k-t)} d_k E|x(t_k)|_X^2 + e^{-\theta t} \sum_{0<t_k<t} e^{\gamma t_k} E|\overline{I}_k(0)|_X^2.\end{aligned}$$

Now, thanks to (H_{26}), we have the existence of positive constants $K_3, \overline{K}_3$ such that, for all $t \geq 0$,

$$e^{-\theta t} \sum_{0<t_k<t} e^{\gamma t_k} E|I_k(0)|_X^2 \leq K_3,$$

and

$$e^{-\theta t} \sum_{0<t_k<t} e^{\gamma t_k} E|\overline{I}_k(0)|_X^2 \leq \overline{K}_3,$$

which further imply

$$\begin{aligned}e^{(\gamma-\theta)t} E|x(t)|_X^2 \leq{}& 4M^2 e^{-\theta t} E|x_0|_X^2 + \sum_{i=1}^{3} K_i + 4\gamma^{-1} M^2 a_1 \int_0^t e^{(\gamma-\theta)s} E|x(s)|_X^2 ds \\ &+ 4\gamma^{-1} M^2 b_1 \int_0^t e^{(\gamma-\theta)s} E|y(s)|_X^2 ds \\ &+ 4M^2 c_2(H) H(2H-1) t^{2H-1} \alpha_1 \int_0^t e^{(\gamma-\theta)s} E|x(s)|_X^2 ds \\ &+ 4M^2 c_2(H) H(2H-1) t^{2H-1} \beta_1 \int_0^t e^{(\gamma-\theta)s} E|y(s)|_X^2 ds \\ &+ 4M^2 e^{(\gamma-\theta)t} \sum_{0<t_k<t} e^{\gamma(t_k-t)} d_k E|x(t_k)|_X^2,\end{aligned}$$

and similarly,

$$\begin{aligned}e^{(\gamma-\theta)t} E|y(t)|_X^2 \leq{}& 4M^2 e^{-\theta t} E|y_0|_X^2 + \sum_{i=1}^{3} \overline{K}_i + 4\gamma^{-1} M^2 a_2 \int_0^t e^{(\gamma-\theta)s} E|x(s)|_X^2 ds \\ &+ 4\gamma^{-1} M^2 b_2 \int_0^t e^{(\gamma-\theta)s} E|y(s)|_X^2 ds\end{aligned}$$

$$+ 4M^2 c_2(H)H(2H-1)t^{2H-1}\alpha_2 \int_0^t e^{(\gamma-\theta)s} E|x(s)|_X^2 ds$$
$$+ 4M^2 c_2(H)H(2H-1)t^{2H-1}\beta_2 \int_0^t e^{(\gamma-\theta)s} E|y(s)|_X^2 ds$$
$$+ 4M^2 e^{(\gamma-\theta)t} \sum_{0<t_k<t} e^{\gamma(t_k-t)} \overline{d}_k E|y(t_k)|_X^2.$$

Consider functions $\mu, \overline{\mu}$ defined on $[0, +\infty)$ by

$$\mu(t) = \sup\{E|x(s)|_X^2 : 0 \leq s \leq t\},$$
$$\overline{\mu}(t) = \sup\{E|y(s)|_X^2 : 0 \leq s \leq t\}.$$

Together with (H_{25}), it follows that, for each $t > 0$, the constant

$$k_*(t) = 4M^2 \sum_{0<t_k<t} e^{\gamma(t_k-t)} d_k \leq k_* < 1.$$

Then

$$e^{(\gamma-\theta)t}\mu(t) \leq \frac{4M^2}{1-k_*} e^{-\theta t} E|x_0|_X^2 + \sum_{i=1}^{3} \frac{K_i}{1-k_*} + \frac{4\gamma^{-1}M^2 a_1}{1-k_*} \int_0^t e^{(\gamma-\theta)s}\mu(s)ds$$
$$+ 4\frac{\gamma^{-1}M^2 b_1}{1-k_*} \int_0^t e^{(\gamma-\theta)s}\overline{\mu}(s)ds + \frac{4M^2 c_2(H)H(2H-1)t^{2H-1}\alpha_1(t)}{1-k_*} \int_0^t e^{(\gamma-\theta)s}\mu(s)ds$$
$$+ \frac{4M^2 c_2(H)H(2H-1)t^{2H-1}\beta_1(t)}{1-k_*} \int_0^t e^{(\gamma-\theta)s}\overline{\mu}(s)ds.$$

Analogously, the constant

$$\bar{k}_*(t) = 4M^2 \sum_{0<t_k<t} e^{\gamma(t_k-t)} \overline{d}_k \leq \bar{k}_* < 1,$$

is present.

$$e^{(\gamma-\theta)t}\overline{\mu}(t) \leq \frac{4M^2}{1-\bar{k}_*} e^{-\theta t} E|y_0|_X^2 + \sum_{i=1}^{3} \frac{\overline{K}_i}{1-\bar{k}_*} + \frac{4\gamma^{-1}M^2 a_2}{1-\bar{k}_*} \int_0^t e^{(\gamma-\theta)s}\mu(s)ds$$
$$+ 4\frac{\gamma^{-1}M^2 b_2}{1-\bar{k}_*} \int_0^t e^{(\gamma-\theta)s}\overline{\mu}(s)ds + \frac{4M^2 c_2(H)H(2H-1)t^{2H-1}\alpha_2}{1-\bar{k}_*} \int_0^t e^{(\gamma-\theta)s}\mu(s)ds$$

$$+\frac{4M^2c_2(H)H(2H-1)t^{2H-1}\beta_2}{1-\bar{k}_*}\int_0^t e^{(\gamma-\theta)s}\overline{\mu}(s)ds.$$

Therefore,

$$\begin{aligned}
&e^{(\gamma-\theta)t}(\mu(t)+\overline{\mu}(t))\\
&\quad\le\frac{4M^2}{1-k_*}e^{-\theta t}E|x_0|_X^2+\frac{4M^2}{1-\bar{k}_*}e^{-\theta t}E|y_0|_X^2+\sum_{i=1}^3\frac{K_i}{1-k_*}+\sum_{i=1}^3\frac{\overline{K}_i}{1-\bar{k}_*}\\
&\quad+\frac{4\gamma^{-1}M^2}{1-k_*}\left(a_1\int_0^t e^{(\gamma-\theta)s}\mu(s)ds+b_1\int_0^t e^{(\gamma-\theta)s}\overline{\mu}(s)ds\right)\\
&\quad+\frac{4M^2c_2(H)H(2H-1)}{1-k_*}\left(t^{2H-1}\alpha_1(t)\int_0^t e^{(\gamma-\theta)s}\mu(s)ds+t^{2H-1}\beta_1(t)\int_0^t e^{(\gamma-\theta)s}\overline{\mu}(s)ds\right)\\
&\quad+\frac{4\gamma^{-1}M^2}{1-\bar{k}_*}\left(a_2\int_0^t e^{(\gamma-\theta)s}\mu(s)ds+b_2\int_0^t e^{(\gamma-\theta)s}\overline{\mu}(s)ds\right)\\
&\quad+\frac{4M^2c_2(H)H(2H-1)}{1-\bar{k}_*}\left(t^{2H-1}\alpha_2(t)\int_0^t e^{(\gamma-\theta)s}\mu(s)ds+t^{2H-1}\beta_2(t)\int_0^t e^{(\gamma-\theta)s}\overline{\mu}(s)ds\right),
\end{aligned}$$

the maximum being taken componentwise for each $i=1,2$, and

$$\begin{aligned}
c_i&=\max\{a_i,b_i\},\\
\overline{c}_i&=\max\{t^{2H-1}\alpha_i(t),t^{2H-1}\beta_i(t)\},\\
\frac{1}{\widetilde{k}}&=\max\left\{\frac{1}{1-\bar{k}_*},\frac{1}{1-k_*}\right\}.
\end{aligned}$$

$$\begin{aligned}
e^{(\gamma-\theta)t}(\mu(t)+\overline{\mu}(t))&\le\frac{4M^2}{\widetilde{k}}e^{-\theta t}(E|x_0|_X^2+E|y_0|_X^2)+\sum_{i=1}^3\frac{K_i+\overline{K}_i}{\widetilde{k}}\\
&+\frac{4\gamma^{-1}M^2c_1}{\widetilde{k}}\int_0^t e^{(\gamma-\theta)s}(\mu(s)+\overline{\mu}(s))ds\\
&+\frac{4M^2c_2(H)H(2H-1)\overline{c}_1}{\widetilde{k}}\int_0^t e^{(\gamma-\theta)s}(\mu(s)+\overline{\mu}(s))ds\\
&+\frac{4\gamma^{-1}M^2c_2}{\widetilde{k}}\int_0^t e^{(\gamma-\theta)s}(\mu(s)+\overline{\mu}(s))ds\\
&+\frac{4M^2c_2(H)H(2H-1)\overline{c}_2}{\widetilde{k}}\int_0^t e^{(\gamma-\theta)s}(\mu(s)+\overline{\mu}(s))ds,
\end{aligned}$$

where

$$A_2 = \max\left\{\frac{4M^2c_2(H)H(2H-1)(\overline{c}_2+\overline{c}_1)}{\widetilde{k}}, \frac{4\gamma^{-1}M^2(c_2+c_1)}{\widetilde{k}}\right\},$$

and

$$A_1 = \frac{4M^2}{\widetilde{k}}e^{-\theta t}(E|x_0|_X^2 + E|y_0|_X^2) + \sum_{i=1}^{3}\frac{K_i+\overline{K}_i}{\widetilde{k}},$$

$$e^{(\gamma-\theta)t}(\mu(t)+\overline{\mu}(t)) \le A_1 + A_2\int_0^t e^{(\gamma-\theta)s}(\mu(s)+\overline{\mu}(s))ds,$$

where A_1 is a suitable positive constant. Gronwall's lemma leads us to

$$e^{(\gamma-\theta)t}(\mu(t)+\overline{\mu}(t)) \le A_1 e^{A_2 t},$$

and, consequently,

$$\mu(t)+\overline{\mu}(t) \le A_1 e^{(A_2-\gamma+\theta)t} = A_1 e^{\eta t}.$$

Then

$$\lim_{t\to\infty}[\mu(t)+\bar{\mu}(t)] = 0 \quad \Rightarrow \quad \lim_{t\to\infty} E|x(t)|_X^2 = 0, \quad \lim_{t\to\infty} E|y(t)|_X^2 = 0.$$

The proof is therefore complete. □

6.4 Example

We consider finite or infinite fractional Brownian motion.

Example 6.1. Consider the following coupled stochastic partial differential equation with impulsive effects:

$$\begin{cases} du(t,\xi) = \frac{\partial^2}{\partial\xi^2}u(t,\xi) + F(t,u(t,\xi),v(t,\xi)) + \sigma(t)\frac{dB_Q^H}{dt}, & t\ge 0,\ t\ne t_k,\ 0\le\xi\le\pi, \\ dv(t,\xi) = \frac{\partial^2}{\partial\xi^2}v(t,\xi) + G(t,u(t,\xi),v(t,\xi)) + \sigma(t)\frac{dB_Q^H}{dt}, & t\ge 0,\ t\ne t_k,\ 0\le\xi\le\pi, \\ u(t_k^+,\xi) - u(t_k^-,\xi) = a_k u(t_k^-,\xi), & k=1,\dots,m, \\ v(t_k^+,\xi) - v(t_k^-,\xi) = \bar{a}_k v(t_k^-,\xi), & k=1,\dots,m, \\ u(t,0) = u(t,\pi) = 0, & t\ge 0, \\ v(t,0) = v(t,\pi) = 0, & t\ge 0, \\ u(0,\xi) = u_0(\xi), & 0\le\xi\le\pi, \\ v(0,\xi) = v_0(\xi), & 0\le\xi\le\pi, \end{cases} \tag{6.8}$$

where $a_k > 0$, B_Q^H denotes a fractional Brownian motion,

$$G, F : [0,\pi] \times \mathbb{R} \times \mathbb{R} \to \mathbb{R},$$

and are continuous functions.

Let

$$x(t)(\xi) = u(t,\xi), \quad y(t)(\xi) = v(t,\xi), \quad t \in J,\ \xi \in [0,\pi],$$
$$I_k\big(x(t_k)\big)(\xi) = a_k u\big(t_k^-,\xi\big), \quad \bar{I}_k\big(y(t_k)\big)(\xi) = a_k v\big(t_k^-,\xi\big), \quad \xi \in [0,\pi],\ k = 1,\dots,m,$$
$$f\big(t,x(t),y(t)\big)(\xi) = F\big(t,u(t,\xi),v(t,\xi)\big), \quad \xi \in [0,\pi],$$
$$g\big(t,x(t),y(t)\big)(\xi) = G\big(t,u(t,\xi),v(t,\xi)\big), \quad \xi \in [0,\pi],$$
$$u_0(\xi) = u(0,\xi), \quad v_0(\xi) = v(0,\xi), \quad \xi \in [0,\pi],$$

Take

$$\mathcal{K} = \mathcal{H} = L^2\big([0,\pi]\big).$$

We define the operator A by $Au = u''$, with domain

$$D(I) = \big\{u \in \mathcal{H}, u', u'' \in \mathcal{H} \text{ and } u(0) = u(\pi) = 0\big\}.$$

Then it is well known that

$$Az = -\sum_{n=1}^{\infty} e^{-n^2 t}\langle z, e_n\rangle e_n, \quad z \in \mathcal{H},$$

and A is the infinitesimal generator of an analytic semigroup

$$\big\{S(t)\big\}_{t\geq 0},$$

on $\mathcal{H}$, which is given by

$$S(t)u = \sum_{n=1}^{\infty} e^{-n^2 t}\langle u, e_n\rangle e_n, \quad u \in \mathcal{H},$$

and

$$e_n(u) = (2/\pi)^{1/2}\sin(nu), \quad n = 1,2,\dots,$$

are the orthogonal set of eigenvectors of A. The analytic semigroup

$$\big\{S(t)\big\}_{t>0}, \quad t \in J,$$

is compact, and there exists a constant $M \geq 1$ such that

$$\big\|S(t)\big\|^2 \leq M.$$

In order to define the operator

$$Q : \mathcal{K} \longrightarrow \mathcal{K},$$

we choose a sequence

$$\{\sigma_n\}_{n\geq 1} \subset \mathbb{R}^+.$$

Set $Qe_n = \sigma_n e_n$, and assume that

$$\operatorname{tr}(Q) = \sum_{n=1}^{\infty} \sqrt{\sigma_n} < \infty.$$

Define the process $B_Q^H(s)$ by

$$B_Q^H = \sum_{n=1}^{\infty} \sqrt{\sigma_n} \gamma_n^H(t) e_n,$$

where $H \in (1/2, 1)$, and $\{\gamma_n^H\}_{n\in\mathbb{N}}$ is a sequence of two-sided, one-dimensional mutually independent fractional Brownian motions. Assume now that:

1. There exists certain positive number $d_k, \bar{d}_k$ $k \in \{1, \dots, m\}$, such that

   $$|I_k(\xi)| \le d_k, \quad |\bar{I}_k(\xi)| \le \bar{d}_k$$

   for any $\xi \in \mathbb{R}$.
2. The functions

   $$f, g : [0, T] \times \mathcal{H} \longrightarrow \mathcal{H},$$

   defined by

   $$f(t, u)(.) = F(t, u(.)), \quad g(t, u)(.) = G(t, u(.)),$$

   are continuous, and we impose suitable conditions on F and G to verify assumption (H_{17}).
3. Assume that there exists an integrable function

   $$\eta : [0, T] \longrightarrow \mathbb{R}^+,$$

   such that

   $$|F(t, x, y)|^2 \le \eta(t)\psi(|x|^2 + |y|^2),$$
   $$|G(t, x, y)|^2 \le \eta(t)\psi(|x|^2 + |y|^2)$$

   for any $t \in [0, T]$ and $x, y \in \mathbb{R}$, where

   $$\psi : [0, \infty) \longrightarrow (0, \infty),$$

   is continuous, nondecreasing, and concave with

   $$\int_1^{\infty} \frac{ds}{\psi(s)} = +\infty.$$

4. The function

   $$\sigma : [0, T] \longrightarrow L_Q^2(\mathcal{K}, \mathcal{H}),$$

   is bounded, that is, there exists a positive constant L, such that

   $$\int_0^T \|\sigma(s)\|_{L_Q^2}^2 ds < L, \quad \forall T > 0.$$

Thus, problem (6.9) can be written in abstract form

$$\begin{cases} dx(t) = [Ax(t) + f(t,x,y)]dt + \sigma(t)dB_Q^H(t), & t \in J = [0,T], \\ dy(t) = [Ay(t) + g(t,x,y)]dt + \sigma(t)dB_Q^H(t), & t \in J = [0,T], \\ x(t_k^+) - x(t_k) = I_k(x(t_k)), & k = 1,\dots,m; \\ y(t_k^+) - y(t_k) = I_k(y(t_k)), & k = 1,\dots,m; \\ x(0) = x_0, \\ y(0) = y_0. \end{cases} \tag{6.9}$$

Thanks to these assumptions, it is straightforward to check that (H_{16})–(H_{22}) holds true; therefore, the assumptions in Theorem 6.2 are fulfilled, and we can conclude that the system (6.9) possesses a mild solution on $[0,T]$.

In the case that we consider the problem for $t \in [0,\infty)$, we observe that:

1. $\|S(t)\| \le e^{-\pi^2 t}$.
2. The functions

$$f,g : [0,\infty) \times \mathcal{H} \longrightarrow \mathcal{H},$$

are continuous, and under suitable conditions on F, assumption (H_{16}) holds in $[0,+\infty)$.

3. The function

$$\sigma : [0,+\infty) \longrightarrow L_Q^2(\mathcal{K},\mathcal{H}),$$

is bounded, that is, there exists a positive constant L, such that

$$\int_0^\infty e^{\gamma s} \|\sigma(s)\|_{L_Q^0}^2 ds < L.$$

4. There exists a positive number d_k, $k \in \{1,\dots,m,\dots\}$, such that

$$|I_k(\xi)| \le d_k,$$
$$|\bar{I}_k(\xi)| \le \bar{d}_k,$$

and

$$\sum_{k=1}^\infty d_k < \infty,$$
$$\sum_{k=1}^\infty \bar{d}_k < \infty$$

for any $\xi \in \mathbb{R}$.

Thus, the problem (6.9) can be written in the abstract form

$$\begin{cases} dx(t) = [Ax(t) + f(t,x,y)]dt + \sigma(t)dB_Q^H(t), & t \in J = [0,\infty), \\ dy(t) = [Ay(t) + g(t,x,y)]dt + \sigma(t)dB_Q^H(t), & t \in J = [0,\infty), \\ x(t_k^+) - x(t_k) = I_k(x(t_k)), & k = 1,2,3\dots, \\ y(t_k^+) - y(t_k) = I_k(y(t_k)), & k = 1,2,3\dots, \\ x(0) = x_0, \\ y(0) = y_0. \end{cases} \tag{6.10}$$

Corollary 6.1. *Let*

$$\sigma_1, \sigma_2 : [0,\pi] \times L^2([0,\pi]) \times L^2([0,\pi]) \to L^2([0,\pi]),$$

be defined by

$$\sigma^1(t,x,y) = \sum_{k=1}^{\infty} \sigma_k^1(t,x,y), \quad \sigma^2(t,x,y) = \sum_{k=1}^{\infty} \sigma_k^2(t,x,y),$$

where

$$\sigma_k^1(t,x,y)v(s) = a_{2k+1} \int_0^t [\sin k^2 x(t)] f(s) ds + a_{2k} \int_0^t [\cos k^2 y(t)] f(s) ds, \quad v \in L^2([0,\pi])$$

and

$$\sigma_k^2(t,x,y)v(s) = b_{2k+1} \int_0^t [\sin k^2 x(s)] f(s) ds + b_{2k} \int_0^t [\cos k^2 y(s)] f(s) ds, \quad v \in L^2([0,\pi]).$$

If

$$\sum a_k^2 < \infty, \quad \sum b_k^2 < \infty,$$

then for each $k \in \mathbb{N}$, we have

$$\sigma_k^1, \sigma_k^2 \in L_Q^0(L^2([0,\pi], L^2([0,\pi])),$$

and

$$\sum_{l=1}^{\infty} \|\sigma_l^i(.,x,y)\|_{L_Q^0}^2 < \infty, \quad i = 1,2.$$

Proof. Clearly, for every $k \in \mathbb{N}$, σ_k^1, σ_k^2 are linear continuous operators. Let $(e_i)_{i\in\mathbb{N}}$ be an orthonormal basis of $L^2([0,\pi])$,

$$\begin{aligned} &\|\sigma_k^1(t,x,y)\|_{L_Q}^2 \\ &= \sum_{k=1}^{\infty} \int_0^{\pi} \left(\left| \sqrt{\sigma_n} a_{2k+1} \int_0^t [\sin k^2 x(s)] e_n(s) ds + \sqrt{\sigma_n} a_{2k} \int_0^t [\cos k^2 y(s)] e_n(s) ds \right|^2 \right) dt \\ &\le 2\pi^2 (a_{2k+1}^2 + a_{2k}^2) \sum_{n=1}^{\infty} \sigma_n. \end{aligned}$$

Therefore,

$$\|\sigma_k^1(t,x,y)\|_{L_Q^0} \le \pi\sqrt{2(a_{2k+1}^2 + a_{2k}^2)} \sum_{n=1}^{\infty} \sqrt{\sigma_n} < \infty,$$

$$\|\sigma_k^1(t,x,y)\|_{L_Q^0}^2 \le L_*(a_{2k+1}^2 + a_{2k}^2), \quad L_* = 2\pi^2 \sum_{n=1}^{\infty} \sqrt{\sigma_n}.$$

We deduce

$$\begin{aligned}\|\sigma^1(t,x,y)\|^2 &= \sum_{k=1}^{\infty} \|\sigma_l^1\|_{L_Q^0} \\ &\le L_* \sum_{k=1}^{\infty} (a_{2k+1}^2 + a_{2k}^2) \\ &\le L_* \sum_{k=1}^{\infty} a_k^2 < \infty,\end{aligned}$$

and

$$\begin{aligned}\|\sigma^2(t,x,y)\|^2 &= \sum_{k=1}^{\infty} \|\sigma_l^2\|_{L_Q^0}^2 \\ &\le L_* \sum_{k=1}^{\infty} b_k^2 < \infty.\end{aligned}$$

Hence,

$$E\|\sigma_1(t,x,y)\|^2 \le L_* \sum_{k=1}^{\infty} a_k^2 + E(\|x\|_{L^2}^2 + \|y\|_{L^2}^2) \quad \text{for all } x,y \in L^2([0,\pi]),$$

and

$$E\|\sigma_2(t,x,y)\|^2 \le L_* \sum_{k=1}^{\infty} b_k^2 + E(\|x\|_{L^2}^2 + \|y\|_{L^2}^2) \quad \text{for all } x,y \in L^2([0,\pi]). \qquad \square$$

Corollary 6.2. *Let*

$$K : [0,\pi] \times [0,\pi] \times \mathbb{R} \times \mathbb{R} \to \mathbb{R},$$

be a continuous function such that there exists $a, b \ge 0$, such that

$$|K(t,s,x,y) - K(t,s,\bar{x},\bar{y})| \le a|x-\bar{x}| + b|y-\bar{y}|, \quad \text{for all } x,\bar{x},y,\bar{y} \in \mathbb{R}.$$

Then

$$K_* : [0,\pi] \times L^2([0,\pi]) \times L^2([0,\pi]) \to L_Q^0(L^2([0,\pi]), L^2([0,\pi])),$$

defined by

$$K_*(t,x,y)f(t) = \int_0^t K(t,s,x(s),y(s))f(s)ds, \quad f \in L^2([0,\pi]),$$

is a Hilbert–Schmidt operator, and there exists $\alpha, \beta \geq 0$, such that

$$\|K_*(t,x,y) - K(t,s,\bar{x},\bar{y})\|_{L_Q^0} \leq \alpha\|x-\bar{x}\|_{L^2} + \beta\|y-\bar{y}\|_{L^2} \quad \textit{for each } x,\bar{x},y,\bar{y} \in L^2([0,\pi]).$$

Proof. It is clear that $K_*(t,x,y)$ is a bounded linear operator. Let $(e_i)_{i\in\mathbb{N}}$ be an orthonormal basis of $L^2([0,\pi])$,

$$\begin{aligned}
&\|K_*(t,x,y)\|^2_{L_Q^0}\\
&= \sum_{k=1}^{\infty}\int_0^{\pi}\left(\left|\sqrt{\sigma_n}\int_0^t [K(t,s,x(s),y(s))]e_n(s)ds\right|^2\right)dt\\
&\leq \sum_{n=1}^{\infty}\int_0^{\pi}\left(\sigma_n\left(\int_0^{\pi}[a|x(s)| + b|y(s)|]|e_n(s)|ds + \int_0^{\pi}|K(t,s,0,0)||e_n(s)|ds\right)^2\right)dt\\
&\leq 4\pi a^2\sum_{n=1}^{\infty}\sigma_n\|x\|^2_{L^2} + 4\pi b^2\sum_{n=1}^{\infty}\sigma_n\|y\|^2_{L^2} + 2\pi^2\sup_{(t,s)\in[0,\pi]\times[0;\pi]}|K(t,s,0,0)|\sum_{n=1}^{\infty}\sigma_n.
\end{aligned}$$

Thus,

$$\|K_*(t,x,y)\|_{L_Q^0} \leq \alpha_1\|x\|_{L^2} + \beta_1\|y\|_{L^2} + \gamma_1 < \infty,$$

where

$$\begin{aligned}
\alpha_1^2 &= 4\pi a^2\sum_{n=1}^{\infty}\sigma_n,\\
\beta_1^2 &= 4\pi b^2\sum_{n=1}^{\infty}\sigma_n,\\
\gamma_1^2 &= 2\pi^2\sup_{(t,s)\in[0,\pi]\times[0,\pi]}|K(t,s,0,0)|\sum_{n=1}^{\infty}\sigma_n.
\end{aligned}$$

We can easily prove

$$\|K_*(t,x,y) - K(t,s,\bar{x},\bar{y})\|^2_{L_Q} \leq \alpha\|x-\bar{x}\|^2_{L^2} + \beta\|y-\bar{y}\|^2_{L^2} \quad \text{for each } x,\bar{x},y,\bar{y} \in L^2([0,\pi]),$$

where

$$\alpha = \alpha_1^2, \quad \beta = \beta_1^2.$$

□

Corollary 6.3. *Let*

$$\{K_n\}_{n\in\mathbb{R}} : [0,\pi]\times[0,\pi]\times\mathbb{R}\times\mathbb{R}\to\mathbb{R},$$

be a summability kernel on $[0,\pi]$*, that is, it holds*

$$\sum_{n=1}^{\infty}\|K_n(t,.,x,y)\|_{L^2}^2<\infty,$$

and K is a continuous function such that there exists $a_n, b_n \geq 0$*, such that*

$$|K(t,s,x,y)-K(t,s,\bar{x},\bar{y})|\leq a_n|x-\bar{x}|+b_n|y-\bar{y}|,\quad \textit{for all } x,\bar{x},y,\bar{y}\in\mathbb{R}.$$

If

$$\sum_{n=1}^{\infty}(a_n^2+b_n^2)<\infty.$$

Then

$$\bar{K}_n : [0,\pi]\times L^2([0,\pi])\times L^2([0,\pi])\to L_Q^0(L^2([0,\pi]),L^2([0,\pi]))$$

is defined by

$$\bar{K}_n(t,x,y)f(t)=\int_0^t K_n(t,s,x(s),y(s))f(s)ds,\quad f\in L^2([0,\pi]).$$

It is a Hilbert–Schmidt operator and there exists $\alpha,\beta\geq 0$*, such that*

$$\sum_{n=1}^{\infty}\|\bar{K}_n(t,x,y)-\bar{K}_n(t,s,\bar{x},\bar{y})\|_{L_Q^0}^2\leq\alpha\|x-\bar{x}\|_{L^2}^2+\beta\|y-\bar{y}\|_{L^2}^2\quad \textit{for each } x,\bar{x},y,\bar{y}\in L^2([0,\pi]).$$

Proof. To prove the result, we use the same method as in the proof of Corollary 6.2. □

Lemma 6.1. *Assume that* σ *satisfies the conditions of Corollary* 6.1*, and that* (3) *or F, G are Lipschitz functions and* I_k*,* $\bar{I}_k$ *satisfies* (1) *or* (4)*. Then problem* (6.9) *has at least one solution, and problem* (6.1) *has exponentially stable mild solutions.*

Proof. If we use that F, G are Lipschitz functions and the assumptions in Corollary 6.2 hold, then by Theorem 6.1, problem (6.9) possesses a unique mild solution. By the same method, we can prove that all the conditions of Theorem 6.2 or Theorem 6.4 are satisfied. □

7 Coupled impulsive neutral functional differential equations

The results of this chapter extend the first two sections of Chapter 6. We prove some results on the existence and uniqueness of mild solutions for a system of semilinear impulsive differential equations with infinite fractional Brownian motions and the Wiener process. Our approach is based on a new version of the fixed-point theorem due to Krasnoselskii in generalized Banach spaces. On the existence of solutions for a system of stochastic impulsive differential equations of the following type:

$$\begin{cases} d[x(t) - h^1(t,x_t,y_t)] = [A_1x(t) + f^1(t,x_t,y_t)]dt + \sum_{l=1}^{\infty} \sigma_l^1(t)dB_l^H(t) + g^1(t)dW(t), \\ \quad t \in J = [0,b], \\ d[y(t) - h^2(t,x_t,y_t)] = [A_2y(t) + f^2(t,x_t,y_t)]dt + \sum_{l=1}^{\infty} \sigma_l^2(t)dB_l^H(t) + g^2(t)dW(t), \\ \quad t \in J = [0,b], \end{cases} \tag{7.1}$$

$$\begin{cases} \Delta x|_{t=t_k} = x(t_k^+) - x(t_k) = I_k(x(t_k),y(t_k)), \quad k = 1,\dots,m, \\ \Delta y|_{t=t_k} = y(t_k^+) - y(t_k) = \bar{I}_k(x(t_k),y(t_k)), \quad k = 1,\dots,m, \end{cases} \tag{7.2}$$

$$\begin{cases} x(t) = x_0(t) = \phi(t), \quad \text{for a. e. } t \in J_0 = (-\infty,0], \\ y(t) = y_0(t) = \bar{\phi}(t), \quad \text{for a. e. } t \in J_0 = (-\infty,0]. \end{cases} \tag{7.3}$$

Here, $x(\cdot), y(\cdot)$ takes the value in the separable Hilbert space X with the inner product $\langle\cdot,\cdot\rangle$ induced by the norm $\|\cdot\|$;

$$A_i : D(A_i) \subset X \longrightarrow X$$

is the infinitesimal generator of a strongly continuous semigroup of bounded linear operators

$$(S_i(t))_{t\geq 0},$$

in X for each $i = 1, 2$. B_l^H is an infinite sequence of independent fractional Brownian motions on a real and separable Hilbert space Y, $l = 1, 2, \dots$, with a Hurst parameter $H \in (\frac{1}{2}, 1)$. Let

$$\{W(t), t \in \mathbb{R}\},$$

be a standard cylindrical Wiener process with values in Y and with respect to a complete probability space $(\Omega, \mathcal{F}, \mathcal{F}_t, \mathbb{P})$ furnished with a family of right continuous and increasing σ-algebras

$$\{\mathcal{F}_t, t \in J\},$$

https://doi.org/10.1515/9783112251898-007

satisfying $\mathcal{F}_t \subset \mathcal{F}$. Here, $\mathcal{D}_{\mathcal{F}_0}$ is a linear space of a family of $\mathcal{F}_0$-measurable functions from $(-\infty, 0]$ into X, which will also be defined in the next section. The impulse times t_k satisfy

$$0 < t_1 < t_2 < \cdots < t_m < b.$$

As for x_t, we mean the segment solution, which is defined in the usual way, that is, if

$$x(\cdot,\cdot) : (-\infty, b] \times \Omega \to X,$$

then for any $t \geq 0$,

$$x_t(\cdot,\cdot) : (-\infty, 0] \times \Omega \to X,$$

is given by

$$x_t(\theta, \omega) = x(t+\theta, \omega), \quad \text{for } \theta \in (-\infty, 0],\ \omega \in \Omega,$$

belong to some abstract phase space $\mathcal{D}_{\mathcal{F}_0}$, that is, a phase space defined axiomatically. Thus, $\mathcal{D}_{\mathcal{F}_0}$ is a linear space of functions mapping $[0, b]$ into X endowed with a seminorm $\|\cdot\|_{\mathcal{D}_{\mathcal{F}_{b_1}}}$. Consider the following space of piecewise continuous functions:

$$\begin{aligned}\mathcal{D}_{\mathcal{F}_{b_1}} = \Big\{ & x\colon [0,b] \times \Omega \to X \text{ is random processes, } x|_{J_k}(\cdot, w) \in C(J_k, X) \\ & \text{for } k = 1, \dots m, \text{ and there exist } x(t_k^-, .) \text{ and } x(t_k^+, .) \text{ with } x(t_k, w) = x(t_k^-, w), \\ & k = 1, \dots, m, \text{ and } \sup_{t \in [0,b]} \mathbb{E}(|x(t)|^2) < \infty \Big\},\end{aligned}$$

where x_k denotes the restriction of x to

$$J_k = (t_{k-1}, t_k], \quad k = 1, 2, \dots, m,$$

and

$$J_0 = (-\infty, 0], \quad i = 1, 2.$$

We endow $\mathcal{D}_{\mathcal{F}_{b_1}}$ with a norm $\|\cdot\|_{\mathcal{F}_{b_1}}$ on $\mathcal{D}_{\mathcal{F}_{b_1}}$ defined by

$$\|x\|_{\mathcal{F}_{b_1}} = \sup_{0 \leq s \leq b} \left(\sqrt{\mathbb{E}\|x(s)\|^2} \right),$$

We will use an axiomatic definition of the phase space $\mathcal{D}_{\mathcal{F}_0}$.

Definition 7.1. $\mathcal{D}_{\mathcal{F}_0}$ is a linear space of a family of $\mathcal{F}_0$-measurable functions that are random from $(-\infty, 0]$ into X, endowed with the seminorm $\|\cdot\|_{\mathcal{D}_{\mathcal{F}_0}}$, which satisfies the following axioms:

1. If $x : (-\infty, b] \longrightarrow X$, $b > 0$ is such that $x|_{[0,b]} \in \mathcal{D}_{\mathcal{F}_0}$ and $x_0 \in \mathcal{D}_{\mathcal{F}_0}$, then for every $t \in [0, b)$, the following conditions hold:
   a. $x_t = x(t + \cdot) \in \mathcal{D}_{\mathcal{F}_0}$.
   b. $\|x(t)\| \leq L\|x_t\|_{\mathcal{D}_{\mathcal{F}_0}}$.
   c. $\|x_t\|_{\mathcal{D}_{\mathcal{F}_0}} \leq K(t)\sup\{\|x(s)\| : 0 \leq s \leq t\} + N(t)\|x_0\|_{\mathcal{D}_{\mathcal{F}_0}}$, where $L > 0$ is a constant;

   $$K, N : [0, \infty) \longrightarrow [0, \infty),$$

   K is continuous, N is locally bounded, and K, N, and L are independent of $x(\cdot)$. Denote

   $$\widehat{K} = \sup\{K(t) : t \in J\},$$

   and

   $$\widehat{N} = \sup\{N(t) : t \in J\}.$$

2. For the function $x(\cdot)$ in (i), x_t is a $\mathcal{D}_{\mathcal{F}_0}$-valued function $[0, b)$.
3. The space $\mathcal{D}_{\mathcal{F}_0}$ is complete.

Remark 7.1. Let $\phi \in \mathcal{D}_{\mathcal{F}_0}$ and $t \leq 0$. The notation ϕ_t represents the function defined by

$$\phi_t = \phi(t + \theta).$$

Consequently, if the function $x(\cdot)$ in axiom (i) is such that $x_0 = \phi$, then $x_t = \phi_t$. We observe that ϕ_t is well-defined for $t < 0$ since the domain of ϕ is $(-\infty, 0]$.

Remark 7.2. In retarded functional differential equations without impulses, the axioms of the abstract phase space $\mathcal{D}_{\mathcal{F}_0}$ include the continuity of the function $t \to x_t$. Due to the impulsive effect, this property is not satisfied in impulsive delay systems; for this reason, it has been eliminated in our abstract description of $\mathcal{D}_{\mathcal{F}_0}$.

Example 7.1. Let $(X, \|\cdot\|_X)$ be a Banach space and

$$\Psi \in C\big((-\infty, 0], X\big).$$

Assume that

$$h : (-\infty, 0] \to [0, +\infty),$$

is a continuous function with

$$l = \int_{-\infty}^{0} h(s)ds < \infty.$$

Define

$$\|\Psi\|_X = \sup_{\theta\le 0} \sqrt{\left(\mathbb{E}|\Psi(\theta)|^2\right)} < \infty,$$

$$\mathcal{C}_{\mathcal{F}_0} = \Big\{ x\colon (-\infty, 0] \times \Omega \to X,\ \text{for any, } a > 0 \ \ \sqrt{\mathbb{E}\left(|\Psi(\theta)|^2\right)}$$

$$\text{is bounded and measurable function on } [-a, 0]$$

$$\text{and } \int_{-\infty}^{0} h(s) \sup_{s\in[0,\theta]} \sqrt{\mathbb{E}\left(|\Psi(\theta)|^2\right)} < \infty \Big\}.$$

If $\mathcal{C}_{\mathcal{F}_0}$ is endowed with the norm,

$$\|\phi\|_{\mathcal{D}_{\mathcal{F}_0}} = \int_{-\infty}^{0} h(s) \sup_{s\in[0,\theta]} \sqrt{\mathbb{E}\left(|\Psi(\theta)|^2\right)}.$$

Then $(\mathcal{C}_{\mathcal{F}_0}, \|\cdot\|_{\mathcal{D}_{\mathcal{F}_0}})$ is a Banach space that satisfies the axioms in Definition 7.1 with $\hat{K} = l, \hat{N} = 1$.

Now, for a given $b > 0$, we define

$$\mathcal{D}_{\mathcal{F}_b} = \{x\colon (-\infty, b] \times \Omega \to X,\ x \in \mathcal{D}_{\mathcal{F}_0} \cap \mathcal{D}_{\mathcal{F}_{b_1}}\}.$$

Endowed with the norm

$$\|x\|_{\mathcal{D}_{\mathcal{F}_b}} = \|x_0\|_{\mathcal{D}_{\mathcal{F}_0}} + \sup_{0\le s\le b} \left(\sqrt{\mathbb{E}\|x(s)\|^2}\right), \quad x \in \mathcal{D}_{\mathcal{F}_b}.$$

Then we will consider our initial data

$$(x_0, y_0) = \phi, \bar{\phi} \in \mathcal{D}_{\mathcal{F}_0}.$$

Let Y be another real separable Hilbert space, and suppose that $B_Q^H = B^H$ is a Y-valued fractional Brownian motion with increment covariance given by a non-negative trace class operator Q_i (see the next section for more details), and let us denote by $\mathcal{L}(Y, X)$ the space of all bounded, continuous, and linear operators from Y into X.

Then we assume that

$$h^i : J \times \mathcal{D}_{\mathcal{F}_0} \times \mathcal{D}_{\mathcal{F}_0} \to X,$$
$$f^i : J \times \mathcal{D}_{\mathcal{F}_0} \times \mathcal{D}_{\mathcal{F}_0} \to X,$$

and

$$\sigma_l^i : J \to L_{Q_i}^0(Y, X), \quad l = 1, \dots,$$
$$g^i : J \to L^0(Y, X).$$

Here, $L_{Q_i}^0(Y, X)$ denotes the space of all Q_i-Hilbert–Schmidt operators from Y into X, and

$$L^0(Y,X) = L_2(Q^{1/2}Y,X),$$

is the space of all Hilbert–Schmidt operators from $Q^{1/2}Y$ to X with the inner product

$$\langle \psi, \varphi \rangle_{L^0} = \mathrm{tr}(\psi Q \varphi^*),$$

and

$$L^0_{Q_i} \neq L^0.$$

As for the impulse functions, we will assume that

$$I_k, \bar{I}_k \in C(X \times X, X) \quad (k = 1, \dots, m),$$

and

$$\begin{aligned} \Delta x|_{t=t_k} &= x(t_k^+) - x(t_k^-), \\ x(t_k^+) &= \lim_{h \to 0^+} x(t_k + h), \end{aligned}$$

and

$$x(t_k^-) = \lim_{h \to 0^+} x(t_k - h).$$

Put

$$\begin{cases} \sigma^i(t) = (\sigma_1^i(t), \sigma_2^i(t), \dots, \sigma_l^i(t), \dots), \\ \|\sigma^i(t)\|^2 = \sum_{l=1}^{\infty} \|\sigma_l^i(t)\|^2_{L^0_{Q_i}} < \infty, \end{cases}$$

where $\sigma^i(\cdot) \in l^2$, $i = 1, 2$ and l^2 are given as

$$l^2 = \left\{ \phi^i = (\phi_l)_{l \geq 1}^i : [0,T] \to L^0_{Q_i}(Y,X) : \|\phi^i(t)\|^2 = \sum_{l=1}^{\infty} \|\phi_l^i(t)\|^2_{L^0_{Q_i}} < \infty,\ i = 1, 2 \right\}.$$

It is obvious that the system (7.1)–(7.3) can be seen as a fixed-point problem

$$\begin{cases} d(z(t) - h(t, z_t)) = A_* z(t) + f(t, z_t)dt + g(t, z_t)dW(t) + \sum_{l=1}^{\infty} \sigma_l(t) dB_l^H(t), \\ \quad t \in [0, b],\ t \neq t_k, \\ \Delta z(t) = I_k^*(z(t_k)), \quad t = t_k,\ k = 1, 2, \dots, m \\ z(t) = z_0(t) \in \mathcal{D}_{\mathcal{F}_0} \times \mathcal{D}_{\mathcal{F}_0}, \\ \quad \text{for a. e. } t \in J_0 = (-\infty, 0], \end{cases} \tag{7.4}$$

where

$$z_t = \begin{bmatrix} x_t \\ y_t \end{bmatrix},$$

$$A_* = \begin{bmatrix} A_1 & 0 \\ 0 & A_2 \end{bmatrix},$$

$$f(t, z_t) = \begin{bmatrix} f^1(t, x_t, y_t) \\ f^2(t, x_t, y_t) \end{bmatrix},$$

$$\sigma_l(t) = \begin{bmatrix} \sigma_l^1(t) \\ \sigma_l^2(t) \end{bmatrix},$$

and

$$z_0(t) = \begin{bmatrix} \phi(t) \\ \bar{\phi}(t) \end{bmatrix},$$

$$h(t, z_t) = \begin{bmatrix} h^1(t, x_t, y_t) \\ h^2(t, x_t, y_t) \end{bmatrix},$$

$$(t) = \begin{bmatrix} g^1(t) \\ g^2(t) \end{bmatrix},$$

$$I_k^*(z(t_k)) = \begin{bmatrix} I_k(x(t_k), y(t_k)) \\ \bar{I}_k(x(t_k), y(t_k))) \end{bmatrix}.$$

7.1 Existence and solution set

Assume $B^H(t)$ is an infinite sequence of independent fractional Brownian motions defined on $(\Omega, \mathcal{F}, \mathbb{P})$, that is,

$$B^H(t) = (B_1^H(t), b_2^H(t), \dots)^T.$$

An X-valued random variable is an $\mathcal{F}$-measurable function $x(t, \cdot) : \Omega \to X$, and the collection of random variables

$$S = \{x(t, .) : \Omega \to X \mid t \in J\},$$

is called a stochastic process. Generally, we just write $x(t)$ instead of $x(t, \omega)$.

For a stochastic process

$$x(\cdot, \cdot) : [0, T] \times \Omega \to X,$$

we will write $x(t)$ (or simply x when no confusion is possible) instead of $x(t, \omega)$.

Lemma 7.1. *For any*

$$\sigma_l^i : [0, T] \to L_{Q_i}^0(Y, X),$$

such that (2.50) holds, and for any $\alpha, \beta \in [0,T]$ with $\alpha > \beta$,

$$\mathbb{E}\left|\int_{\alpha}^{\beta} \sigma_l^i(s)dB_l^H(s)\right|_X^2 \leq c_H H(2H-1)(\alpha-\beta)^{2H-1}\sum_{n=1}^{\infty}\int_{\alpha}^{\beta}|\sigma_l^i(s)Q_i^{1/2}e_n|_X^2 ds.$$

If, in addition,

$$\sum_{n=1}^{\infty}|\sigma_l^i Q_i^{1/2}e_n|_X \quad \text{is uniformly convergent for } t \in [0,T],$$

then

$$\mathbb{E}\left|\int_{\alpha}^{\beta} \sigma_l^i(s)dB_l^H(s)\right|_X^2 \leq c_H H(2H-1)(\alpha-\beta)^{2H-1}\int_{\alpha}^{\beta}\|\sigma_l^i(s)\|_{L_{Q_i}^0}^2 ds. \tag{7.5}$$

Corollary 7.1. *Let $(\mathcal{F}_t)_{t\geq 0}$ be a right-continuous complete σ^i-algebra filtration generated by B_l^H, $l = 1,2,\ldots$. Given sequences of measurable $(\mathcal{F}_t)$-adapted processes σ_l^i set*

$$M^k = \sum_{l=1}^{k}\int_0^t \sigma_l^i(y)dB_l^H(y), \quad k \in \mathbb{N}.$$

Assume that

$$\sum_{l=1}^{\infty}\int_0^T\int_0^T \|\sigma_l^i(y)\|_{L_{Q_i}^0}\|\phi_l(z)\|_{L_{Q_i}^0}|y-z|^{2H-2}dydz < \infty,$$

for all $t \geq 0$, then M^k is convergent in $L^2(\Omega, \mathcal{F}_t, \mathbb{P})$

Lemma 7.2 ([6]). *For any $p \geq 1$ and for an arbitrary L_0^2-valued predictable process $g^i(.)$ for each $i = 1,2$,*

$$\mathbb{E}\sup_{s\in[0,t]}\left\|\int_0^s g^i(u)dW(u)\right\|_X^{2p} \leq (p(2p-1))^p\left(\int_0^t (E\|g^i(s)\|_{L_0^2}^{2p})^{1/p}ds\right)^p.$$

Theorem 7.1. *Let $\mathcal{X}$ be a generalized Banach space, and*

$$P^1, P^2 : \mathcal{X} \to \mathcal{X},$$

be two operators satisfying:

1. *P^1 is a contraction, and*
2. *P^2 is completely continuous.*

Then either:

1. *The operator equation*

$$x = P^1(x) + P^2(x),$$

   has a solution, or

2. *The set*

$$\mathcal{M} = \left\{ x \in \mathcal{X} : \lambda P^1\left(\frac{(x)}{\lambda}\right) + \lambda P^2(x) = x, \text{ for some } \lambda \in (0,1) \right\},$$

   is unbounded.

We suppose that $0 \in \rho(A_i)$ (the resolvent set of A_i, for each $i = 1, 2$). That is, the semigroup $S_i(t)$ is uniformly bounded; that is to say,

$$\|S_i(t)\| \le \bar{M}_1,$$

for some constant $\bar{M}_1 \ge 1$ and for every $t \ge 0$. For $0 < \alpha \le 1$, it is possible to define the fractional power operator $(-A_i)^\alpha$ as a closed linear operator on its domain $D((-A_i)^\alpha)$ with inverse $(-A_i)^{-\alpha}$. Furthermore, the subspace $D((-A_i)^\alpha)$ is dense in X. We denote by X_α the Banach space $D((-A_i)^\alpha)$ endowed with the norm

$$\|x\|_\alpha = \|(-A_i)^\alpha x\|,$$

for $x \in D((-A_i)^\alpha)$ defines a norm on $D((-A_i)^\alpha)$, which is equivalent to the graph norm of $(-A_i)^\alpha$. We represent X_α as the space $D((-A_i)^\alpha)$ with the norm $\|\cdot\|_\alpha$.

Lemma 7.3. 1. *If*

$$0 < \beta < \alpha \le 1,$$

   then $X_\alpha \subset X_\beta$ and the embedding is compact whenever the resolvent operator of A_i is compact.

2. *For each $0 < \alpha \le 1$, there exists a positive constant C_α such that*

$$\|(-A_i)^\alpha S_i(t)\| \le \frac{C_\alpha}{t^\alpha} e^{-\lambda t}, \quad t > 0,\ \lambda > 0.$$

Lemma 7.4.[1] *Let*

$$v(\cdot),\ w(\cdot) : [0, b] \longrightarrow [0, \infty),$$

1 E. Hernández, Existence results for partial neutral functional integrodifferential equations with unbounded delay, *J. Math. Anal. Appl.* **292** (2004), 194–210.

be continuous functions. If $w(\cdot)$ is nondecreasing and there are constants $\theta > 0$, $0 < \alpha < 1$ such that

$$v(t) \leq w(t) + \theta \int_0^t \frac{v(s)}{(t-s)^{1-\alpha}} ds, \quad t \in J,$$

then

$$v(t) \leq e^{\theta^n \Gamma(\alpha)^n t^{n\alpha}/\Gamma(n\alpha)} \sum_{j=0}^{n-1} \left(\frac{\theta b^\alpha}{\alpha} \right)^j w(t),$$

for every $t \in [0, b]$ and every $n \in \mathbb{N}$ such that $n\alpha > 1$, and $\Gamma(\cdot)$ is the Gamma function.

Now, we first define the concept of a mild solution to our problem. It is worth mentioning that we can use several types of solutions (weak, strong, etc.), each of which requires a different set of assumptions, like this one: (absolutely continuous, differentiable, absolutely continuous).

Definition 7.2. A X-valued stochastic process

$$u = (x, y) \in \mathcal{D}_{\mathcal{F}_b} \times \mathcal{D}_{\mathcal{F}_b},$$

is called a mild solution of the problems (7.1)–(7.3) with respect to the probability space $(\Omega, \mathcal{F}, \mathbb{P})$ if:

1. $u(t)$ is $\mathcal{F}_t$-adapted for all $t \in J_k = (t_k, t_{k+1}], k = 1, 2, \ldots, m$;
2. $u(t)$ is right continuous and has a limit on the left almost surely;
3. The function $AS(t-s)h^i(s, u_s)$, $i = 1, 2$ is integrable;
4. $u(t)$ satisfies for all $t \in (-\infty, b]$ and almost surely that

$$\begin{cases} x(t) = S_1(t)(\phi(0) - h^1(0, \phi, \bar{\phi})) + h^1(t, x_t, y_t) + \int_0^t A_1 S_1(t-s) h^1(s, x_s, y_s) ds \\ \quad + \int_0^t S_1(t-s) f^1(s, x_s, y_s) ds + \sum_{l=1}^{\infty} \int_0^t S_1(t-s) \sigma_l^1(t) dB_l^H(s) \\ \quad + \sum_{0<t_k<t} S_1(t-t_k) I_k(x(t_k), y(t_k)) + \int_0^t S_1(t-s) g^1(s) dW(s), \quad \mathbb{P}\text{-a.s.}, \ t \in J, \end{cases}$$

$$\begin{cases} y(t) = S_2(t)(\bar{\phi}(0) - h^2(0, \phi, \bar{\phi})) + h^2(t, x_t, y_t) + \int_0^t A_2 S_2(t-s) h^2(s, x_s, y_s) ds \\ \quad + \int_0^t S_2(t-s) f^2(s, x_s, y_s) ds + \sum_{l=1}^{\infty} \int_0^t S_2(t-s) \sigma_l^2(t) dB_l^H(s) \\ \quad + \sum_{0<t_k<t} S_2(t-t_k) \bar{I}_k(x(t_k), y(t_k)) + \int_0^t S_2(t-s) g^2(s) dW(s), \quad \mathbb{P}\text{-a.s.}, \ t \in J, \end{cases}$$

and the initial value condition

$$(x_0, y_0) = (\phi, \bar{\phi}) \in \mathcal{D}_{\mathcal{F}_0} \times \mathcal{D}_{\mathcal{F}_0},$$

on $(-\infty, 0]$ satisfies

$$\|\phi\|_{\mathcal{D}_{\mathcal{F}_0}}, \|\bar{\phi}\|_{\mathcal{D}_{\mathcal{F}_0}} < \infty.$$

In order to obtain our main result, we assume that the following conditions hold:

(H_{27}) A_i is the infinitesimal generator of an analytic semigroup, $S_i(t)$ of bounded linear operators on X. Further, to avoid unnecessary notations, we suppose that $0 \in \rho(A_i)$, and that (see Lemma 7.3) there exists a constant M, such that

$$\{\|S_i(t)\|^2 \leq M\},$$

for all $t \geq 0$

$$\|(-A_i)^{1-\beta} S_i(t)\| \leq C_{1-\beta} t^{\beta-1},$$

for some constants $M, c_{1-\beta}$ and all $t \in [0, b]$.

(H_{28}) a. There exist constants $0 < \beta < 1$, $a^i_{h_i}, \bar{a}^i_{h_i} \geq 0$ and h^i is X_β-valued, $(-A_i)^\beta h^i$ is continuous, and

$$\mathbb{E}|(-A_i)^\beta h^i(t,x,y)|^2 \leq a^i_{h_i}(\|x\|^2_{\mathcal{D}_{\mathcal{F}_0}} + \|y\|^2_{\mathcal{D}_{\mathcal{F}_0}}) + \bar{a}^i_{h_i}, \quad t \in J,\ x,y \in \mathcal{D}_{\mathcal{F}_0}.$$

b. There exist constants $0 < \beta < 1$, $c_{h^i}, \bar{c}_{h^i} \geq 0$, and

$$\mathbb{E}|(-A_i)^\beta h^i(t,x,y) - (-A_i)^\beta h^i(t,\bar{x},\bar{y})|^2 \leq c^i_h \|x - \bar{x}\|^2_{\mathcal{D}_{\mathcal{F}_0}} + \bar{c}^i_h \|y - \bar{y}\|^2_{\mathcal{D}_{\mathcal{F}_0}}, \quad t \in J,$$

for all $x, y, \overline{x}, \overline{y} \in \mathcal{D}_{\mathcal{F}_0}$

c. The function

$$s \longmapsto A_i S_i(t-s) h^i(s, x_s, y_s), \quad i = 1,2,$$

is measurable, and

$$M_{b,c} = \begin{pmatrix} \sqrt{\lambda_1} & \sqrt{\lambda_2} \\ \sqrt{\bar{\lambda}_1} & \sqrt{\bar{\lambda}_2} \end{pmatrix},$$

where

$$\lambda_1 = 4c^1_h \Big(\|(-A_1)^{-\beta}\|^2 + \frac{(C_{1-\beta} b^\beta)^2}{2\beta - 1} \Big) \widetilde{K}^2,$$

$$\lambda_2 = 4\bar{c}^1_h \Big(\|(-A_1)^{-\beta}\|^2 + \frac{(C_{1-\beta} b^\beta)^2}{2\beta - 1} \Big) \widetilde{K}^2,$$

and

$$\bar{\lambda}_1 = 4c^2_h \Big(\|(-A_2)^{-\beta}\|^2 + \frac{(C_{1-\beta} b^\beta)^2}{2\beta - 1} \Big) \widetilde{K}^2,$$

$$\bar{\lambda}_2 = 4\bar{c}^2_h \Big(\|(-A_2)^{-\beta}\|^2 + \frac{(C_{1-\beta} b^\beta)^2}{2\beta - 1} \Big) \widetilde{K}^2,$$

where $\lambda_i, \bar{\lambda}_i \geq 0$ for each $i = 1, 2$.

(H_{29}) There exist constants $d_k, \bar{d}_k > 0, k = 1, \ldots, m, \ldots$ for each

$$|I_k(x, y)| \leq d_k,$$

and

$$|\bar{I}_k(x, y)| \leq \bar{d}_k,$$

for all $x, y \in X$, and $\sum_{k=1}^{m} d_k < \infty$, $\sum_{k=1}^{m} \bar{d}_k < \infty$.

(H_{30}) f^i is a L^2-Carathéodory map and for every $t \in [0, b]$ for each $i = 1, 2$ the function $t \longmapsto f^i(t, x_t, y_t)$ and $t \longmapsto f^i(t, x_t, y_t)$, $x_t, y_t \in \mathcal{D}_{\mathcal{F}_0}$ are measurable.

(H_{31}) The function $\sigma^i : J \longrightarrow L_{Q^i}(Y, X)$ satisfies

$$\sum_{l=1}^{\infty} \int_0^T \|\sigma_l^i(s)\|_{L^0_{Q^i}}^2 \, ds < \infty.$$

(H_{32}) There exists a continuous nondecreasing function $\psi : [0, \infty) \to [0, \infty)$ and $m_i \in L^1(J, \mathbb{R}_+)$, $i = 1, 2$, such that

$$\mathbb{E}|f^i(t, x, y)|^2 \leq m_i(t)\psi(\|x\|_{\mathcal{D}_{\mathcal{F}_0}}^2 + \|y\|_{\mathcal{D}_{\mathcal{F}_0}}^2),$$

and

$$\int_0^{\infty} \frac{ds}{\psi(s)} = \infty,$$

for all $t \in J$, $x, y \in \mathcal{D}_{\mathcal{F}_0}$

(H_{33}) The function

$$g^i : J \longrightarrow L^0(Y, X),$$

satisfies

$$\int_0^b \|g^i(s)\|_{L^0}^2 ds = C_1 < \infty, \quad t \in J.$$

The main result is the following theorem.

Theorem 7.2. *Assume that* (H_{27})–(H_{33}) *holds. If* $M_{b,c}$ *converges to zero, then problems* (7.1)–(7.3) *possess a unique mild solution on* $(-\infty, b]$.

Proof. Transform the problems (7.1)–(7.3) into a fixed-point problem. Consider the operator

$$N = (N_1, N_2) : \mathcal{D}_{\mathcal{F}_b} \times \mathcal{D}_{\mathcal{F}_b} \to \mathcal{D}_{\mathcal{F}_b} \times \mathcal{D}_{\mathcal{F}_b},$$

defined by

$$N_1(x,y)(t) = \begin{cases} \phi(t), \quad \text{if } t \in (-\infty, 0], \\ S_1(t)[\phi(0) - h^1(0,\phi,\bar{\phi})] + h^1(t,x_t,y_t) + \int_0^t A_1 S_1(t-s)h^1(s,x_s,y_s)ds \\ \quad + \int_0^t S_1(t-s)f^1(s,x_s,y_s)ds + \sum_{l=1}^{\infty}\int_0^t S_1(t-s)\sigma_l^1(s)dB_l^H(s) \\ \quad + \int_0^t S_1(t-s)g^1(s)dW(s) + \sum_{0<t_k<t} S_1(t-t_k)I_k(x(t_k^-),y(t_k^-)), \quad \text{if } t \in [0,b], \end{cases}$$

and

$$N_2(x,y)(t) = \begin{cases} \bar{\phi}(t), \text{ if } t \in (-\infty, 0], \\ S_2(t)[\bar{\phi}(0) - h^2(0,\phi,\bar{\phi})] + h^2(t,x_t,y_t) + \int_0^t A_2 S_2(t-s)h^2(s,x_s,y_s)ds \\ + \int_0^t S_2(t-s)f^2(s,x_s,y_s)ds + \sum_{l=1}^{\infty}\int_0^t S_2(t-s)\sigma_l^2(s)dB_l^H(s) \\ + \int_0^t S_2(t-s)g^2(s)dW(s) + \sum_{0<t_k<t} S_2(t-t_k)\bar{I}_k(x(t_k^-),y(t_k^-)), \text{ if } t \in [0,b]. \end{cases}$$

For all

$$(\phi,\bar{\phi}) \in \mathcal{D}_{\mathcal{F}_0} \times \mathcal{D}_{\mathcal{F}_0},$$

put

$$\theta(t) = \begin{cases} \phi(t), & \text{if } t \in (-\infty, 0], \\ S_1(t)\phi(0), & \text{if } t \in [0,b], \end{cases}$$

and

$$\bar{\theta}(t) = \begin{cases} \bar{\phi}(t), & \text{if } t \in (-\infty, 0], \\ S_2(t)\bar{\phi}(0), & \text{if } t \in [0,b]. \end{cases}$$

It is clear that

$$(\theta,\bar{\theta}) \in \mathcal{D}_{\mathcal{F}_b} \times \mathcal{D}_{\mathcal{F}_b}.$$

Set

$$\big(x(t),y(t)\big) = \big(z(t) + \theta(t), \bar{z}(t) + \bar{\theta}(t)\big),$$

for each $t \in (-\infty, b]$, where

$$z(t) = \begin{cases} z_0, \quad \text{if } t \in (-\infty, 0], \\ -S_1(t)h^1(0,\phi,\bar{\phi}) + h^1(t, z_t + \theta_t, \bar{z}_t + \bar{\theta}_t) + \int_0^t A_1 S_1(t-s)h^1(s, z_s + \theta_s, \bar{z}_s + \bar{\theta}_s)ds \\ \quad + \int_0^t S_1(t-s)f^1(s, z_s + \theta_s, \bar{z}_s + \bar{\theta}_s)ds + \sum_{l=1}^{\infty} \int_0^t S_1(t-s)\sigma_l^1(s)dB_l^H(s) \\ \quad + \int_0^t S_1(t-s)g^1(s)dW(s) \\ \quad + \sum_{0<t_k<t} S_1(t-t_k)I_k(z(t_k^-) + \theta(t_k^-), \bar{z}(t_k^-) + \bar{\theta}(t_k^-)), \quad \text{if } t \in [0, b], \end{cases}$$

and

$$\bar{z}(t) = \begin{cases} \bar{z}_0, \quad \text{if } t \in (-\infty, 0], \\ -S_2(t)h^2(0,\phi,\bar{\phi}) + h^2(t, z_t + \theta_t, \bar{z}_t + \bar{\theta}_t) + \int_0^t A_2 S_2(t-s)h^2(s, z_s + \theta_s, \bar{z}_s + \bar{\theta}_s)ds \\ \quad + \int_0^t S_2(t-s)f^2(s, z_s + \theta_s, \bar{z}_s + \bar{\theta}_s)ds + \sum_{l=1}^{\infty} \int_0^t S_2(t-s)\sigma_l^2(s)dB_l^H(s) \\ \quad + \int_0^t S_2(t-s)g^2(s)dW(s) \\ \quad + \sum_{0<t_k<t} S_2(t-t_k)\bar{I}_k(z(t_k^-) + \theta(t_k^-), \bar{z}(t_k^-) + \bar{\theta}(t_k^-)), \quad \text{if } t \in [0, b]. \end{cases}$$

Set

$$\widehat{\mathcal{D}}_{\mathcal{F}_b} = \{z, \bar{z} \in \mathcal{D}_{\mathcal{F}_b}, \text{ such that } (z_0, \bar{z}_0) = (0,0)\},$$

for any $z, \bar{z} \in \mathcal{D}_{\mathcal{F}_0}$ we have

$$\|z\|_{\widehat{\mathcal{D}}_{\mathcal{F}_b}} = \|z_0\|_{\mathcal{D}_{\mathcal{F}_0}} + \sup_{t \in J} \sqrt{\mathbb{E}\|z(t)\|^2} = \sup_{t \in J} \sqrt{\mathbb{E}\|z(t)\|^2}.$$

It is not difficult to check that $(\widehat{\mathcal{D}}_{\mathcal{F}_b}, \|\cdot\|_{\widehat{\mathcal{D}}_{\mathcal{F}_b}})$ is a Banach space. The set

$$\mathcal{B}_q = \{z, \bar{z} \in \widehat{\mathcal{D}}_{\mathcal{F}_b}, \ \|z\|^2_{\widehat{\mathcal{D}}_{\mathcal{F}_b}} \le q \text{ and } \|\bar{z}\|^2_{\widehat{\mathcal{D}}_{\mathcal{F}_b}} \le q, \ q \ge 0\},$$

is clearly a bounded, closed, convex set in $\widehat{\mathcal{D}}_{\mathcal{F}_b}$ for each $q \ge 0$ and for each $x \in \mathcal{B}_q$. We have

$$\begin{aligned} \|z_t + \theta_t\|^2_{\mathcal{D}_{\mathcal{F}_0}} + \|\bar{z}_t + \bar{\theta}_t\|^2_{\mathcal{D}_{\mathcal{F}_0}} &\le 2(\|z_t\|^2_{\mathcal{D}_{\mathcal{F}_0}} + \|\theta_t\|^2_{\mathcal{D}_{\mathcal{F}_0}}) + 2(\|\bar{z}_t\|^2_{\mathcal{D}_{\mathcal{F}_0}} + \|\bar{\theta}_t\|^2_{\mathcal{D}_{\mathcal{F}_0}}) \\ &\le 4(\widetilde{N}^2(\|\phi\|^2_{\mathcal{D}_{\mathcal{F}_0}} + \|\bar{\phi}\|^2_{\mathcal{D}_{\mathcal{F}_0}}) + \widetilde{K}^2(2q + M(E|\phi(0)|^2 + E|\bar{\phi}(0)|^2)) \\ &= C_{std}. \end{aligned}$$

Consider the operator

$$\widehat{N} : \widehat{\mathcal{D}}_{\mathcal{F}_b} \times \widehat{\mathcal{D}}_{\mathcal{F}_b} \to \widehat{\mathcal{D}}_{\mathcal{F}_b} \times \widehat{\mathcal{D}}_{\mathcal{F}_b},$$

defined by

$$\widehat{N}(z, \bar{z}) = (\widehat{N}_1(z, \bar{z}), \widehat{N}_2(z, \bar{z}), \quad (z, \bar{z}) \in \widehat{\mathcal{D}}_{\mathcal{F}_b} \times \widehat{\mathcal{D}}_{\mathcal{F}_b},$$

where

$$\widehat{N}_1(z,\bar{z}) = \begin{cases} 0, \quad \text{if } t \in (-\infty,0], \\ -S_1(t)h^1(0,\phi,\bar{\phi}) + h^1(t,z_t+\theta_t,\bar{z}_s+\bar{\theta}_s) \\ \quad + \int_0^t A_1S_1(t-s)h^1(s,z_s+\theta_s,\bar{z}_s+\bar{\theta}_s)ds \\ \quad + \int_0^t S_1(t-s)f^1(s,z_s+\theta_s,\bar{z}_s+\bar{\theta}_s)ds + \sum_{l=1}^{\infty}\int_0^t S_1(t-s)\sigma_l^1(s)dB_l^H(s) \\ \quad + \int_0^t S_1(t-s)g^1(s)dW(s) \\ \quad + \sum_{0<t_k<t} S_1(t-t_k)I_k(z(t_k^-)+\theta(t_k^-),\bar{z}(t_k^-)+\bar{\theta}(t_k^-)), \quad \text{if } t \in [0,b], \end{cases}$$

and

$$\widehat{N}_2(z,\bar{z}) = \begin{cases} 0, \quad \text{if } t \in (-\infty,0], \\ -S_2(t)h^2(0,\phi,\bar{\phi}) + h^2(t,z_t+\theta_t,\bar{z}_t+\bar{\theta}_t) \\ \quad + \int_0^t A_2S_2(t-s)h^2(s,z_s+\theta_s,\bar{z}_s+\bar{\theta}_s)ds \\ \quad + \int_0^t S_2(t-s)f^2(s,z_s+\theta_s,\bar{z}_s+\bar{\theta}_s)ds + \sum_{l=1}^{\infty}\int_0^t S_2(t-s)\sigma_l^2(s)dB_l^H(s) \\ \quad + \int_0^t S_2(t-s)g^2(s)dW(s) \\ \quad + \sum_{0<t_k<t} S_2(t-t_k)\bar{I}_k(z(t_k^-)+\theta(t_k^-),\bar{z}(t_k^-)+\bar{\theta}(t_k^-)), \quad \text{if } t \in [0,b]. \end{cases}$$

Now, consider the four operators $\widehat{N}_{11}, \widehat{N}_{12}$ and $\widehat{N}_{21}, \widehat{N}_{22}$,

$$\widehat{N}_{11}(z,\bar{z}) = \begin{cases} 0, \quad \text{if } t \in (-\infty,0], \\ -S_1(t)h^1(0,\phi,\bar{\phi}) + h^1(t,z_t+\theta_t,\bar{z}_s+\bar{\theta}_s) \\ \quad + \int_0^t A_1S_1(t-s)h^1(s,z_s+\theta_s,\bar{z}_s+\bar{\theta}_s)ds \\ \quad + \int_0^t S_1(t-s)g^1(s)dW(s) + \sum_{l=1}^{\infty}\int_0^t S_1(t-s)\sigma_l^1(s)dB_l^H(s), \quad \text{if } t \in [0,b], \end{cases}$$

and

$$\widehat{N}_{12}(z,\bar{z}) = \begin{cases} 0, \quad \text{if } t \in (-\infty,0], \\ \int_0^t S_1(t-s)f^1(s,z_s+\theta_s,\bar{z}_s+\bar{\theta}_s)ds \\ \quad + \sum_{0<t_k<t} S_1(t-t_k)I_k(z(t_k^-)+\theta(t_k^-),\bar{z}(t_k^-)+\bar{\theta}(t_k^-)), \quad \text{if } t \in [0,b]. \end{cases}$$

Put

$$\widehat{N}_1 = \widehat{N}_{11} + \widehat{N}_{12},$$

with

$$\widehat{N}_{21}(z,\bar{z}) = \begin{cases} 0, \quad \text{if } t \in (-\infty,0], \\ -S_2(t)h^2(0,\phi,\bar{\phi}) + h^2(t,z_t+\theta_t,\bar{z}_s+\bar{\theta}_s) \\ \quad + \int_0^t A_2S_2(t-s)h^2(s,z_s+\theta_s,\bar{z}_s+\bar{\theta}_s)ds \\ \quad + \int_0^t S_2(t-s)g^2(s)dW(s) + \sum_{l=1}^{\infty}\int_0^t S_2(t-s)\sigma_l^2(s)dB_l^H(s), \quad \text{if } t \in [0,b], \end{cases}$$

and

$$\widehat{N}_{22}(z,\bar{z}) = \begin{cases} 0, & \text{if } t \in (-\infty,0],\\ \int_0^t S_2(t-s)f^2(s,z_s+\theta_s,\bar{z}_s+\bar{\theta}_s)ds \\ \quad + \sum_{0<t_k<t} S_2(t-t_k)\bar{I}_k(z(t_k^-)+\theta(t_k^-),\bar{z}(t_k^-)+\bar{\theta}(t_k^-)), & \text{if } t \in [0,b]. \end{cases}$$

It is clear that

$$\widehat{N}_{21}+\widehat{N}_{22}=\widehat{N}_2.$$

Then the problem of finding the solution of (7.1)–(7.3) is reduced to finding the solution of the operator

$$(z(t),\bar{z}(t)) = (\widehat{N}_1(z,\bar{z})(t),\widehat{N}_2(z,\bar{z})(t)), \quad t \in (-\infty,b].$$

We shall show that the operators $\widehat{N}_{11}, \widehat{N}_{21}$ and $\widehat{N}_{12}, \widehat{N}_{22}$ satisfy all the conditions of Theorem 7.1. The proof will be given in several steps.

Part 1:
$\widehat{N}_{11}, \widehat{N}_{21}$ is a contraction.

Let

$$(v,u),(\bar{v},\bar{u}) \in \widehat{\mathcal{D}}_{\mathcal{F}_b} \times \widehat{\mathcal{D}}_{\mathcal{F}_b}.$$

Then, for $t \in [0,b]$, we have

$$\begin{aligned}
&\mathbb{E}\big|\widehat{N}_{11}(u,v)(t)-\widehat{N}_{11}(\bar{u},\bar{v})(t)\big|^2\\
&\quad \le 2\mathbb{E}\big|h^1(t,u_t+\theta_t,v_t+\bar{\theta}_t)-h^1(t,\bar{u}_t+\theta_t,\bar{v}_t+\bar{\theta}_t)\big|^2\\
&\qquad + 2\mathbb{E}\left|\int_0^t A_1S_1(t-s)(h^1(s,u_s+\theta_s,v_s+\bar{\theta}_s)-h^1(s,\bar{u}_s+\theta_s,\bar{v}_s+\bar{\theta}_s))ds\right|^2\\
&\quad \le 2\|(-A_1)^{-\beta}\|^2(c_h^1\|u-\bar{u}\|^2_{\mathcal{D}_{\mathcal{F}_0}}+\bar{c}_h^1\|v-\bar{v}\|^2_{\mathcal{D}_{\mathcal{F}_0}})\\
&\qquad + 2b\int_0^t \frac{C_{1-\beta}^2}{(t-s)^{2(1-\beta)}}(c_h^1\|u(s)-\bar{u}(s)\|^2_{\mathcal{D}_{\mathcal{F}_0}}+\bar{c}_h^1\|v(s)-\bar{v}(s)\|^2_{\mathcal{D}_{\mathcal{F}_0}})ds,\\
&\quad \le 2c_h^1\left(\|(-A_1)^{-\beta}\|^2+\frac{(C_{1-\beta}b^\beta)^2}{2\beta-1}\right)\times\left[2\widetilde{K}^2\sup_{0\le s\le b}\mathbb{E}\big|u(s)-\bar{u}(s)\big|^2+2\widetilde{N}\|u_0\|^2_{\mathcal{D}_{\mathcal{F}_0}}+2\widetilde{N}\|\bar{u}_0\|^2_{\mathcal{D}_{\mathcal{F}_0}}\right]\\
&\qquad + 2\bar{c}_h^1\left(\|(-A_1)^{-\beta}\|^2+\frac{(C_{1-\beta}b^\beta)^2}{2\beta-1}\right)\times\left[2\widetilde{K}^2\sup_{0\le s\le b}\mathbb{E}\big|v(s)-\bar{v}(s)\big|^2+2\widetilde{N}\|v_0\|^2_{\mathcal{D}_{\mathcal{F}_0}}+2\widetilde{N}\|\bar{v}_0\|^2_{\mathcal{D}_{\mathcal{F}_0}}\right],\\
&\quad \le \lambda_1\sup_{0\le s\le b}\mathbb{E}\big|u(s)-\bar{u}(s)\big|^2+\lambda_2\sup_{0\le s\le b}\mathbb{E}\big|v(s)-\bar{v}(s)\big|^2.
\end{aligned}$$

Therefore, we arrive at

$$\mathbb{E}\left|\widehat{N}_{11}(u,v)(t)-\widehat{N}_{11}(\bar{u},\bar{v})(t)\right|^2 \le \lambda_1 \sup_{0\le s\le b}\mathbb{E}\left|u(s)-\bar{u}(s)\right|^2+\lambda_2 \sup_{0\le s\le b}\mathbb{E}\left|v(s)-\bar{v}(s)\right|^2.$$

Since

$$\begin{aligned}\|u_0\|^2_{\mathcal{D}_{\mathcal{F}_0}} &= \|\bar{u}_0\|^2_{\mathcal{D}_{\mathcal{F}_0}} = 0,\\ \|v_0\|^2_{\mathcal{D}_{\mathcal{F}_0}} &= \|\bar{v}_0\|^2_{\mathcal{D}_{\mathcal{F}_0}} = 0.\end{aligned}$$

Then we deduce

$$\|\widehat{N}_{11}(u,v)(t)-\widehat{N}_{11}(\bar{u},\bar{v})(t)\|^2_{\widehat{\mathcal{D}}_{\mathcal{F}_b}} \le \lambda_1\|u(s)-\bar{u}(s)\|^2_{\widehat{\mathcal{D}}_{\mathcal{F}_b}}+\lambda_2\|v(s)-\bar{v}(s)\|^2_{\widehat{\mathcal{D}}_{\mathcal{F}_b}},$$

where

$$\lambda_1 = 4c_h^1\left(\|(-A_1)^{-\beta}\|^2+\frac{(C_{1-\beta}b^\beta)^2}{2\beta-1}\right)\widetilde{K}^2,$$

and

$$\lambda_2 = 4\bar{c}_h^1\left(\|(-A_1)^{-\beta}\|^2+\frac{(C_{1-\beta}b^\beta)^2}{2\beta-1}\right)\widetilde{K}^2.$$

Similarly,

$$\|\widehat{N}_{21}(u,v)(t)-\widehat{N}_{21}(\bar{u},\bar{v})(t)\|^2_{\widehat{\mathcal{D}}_{\mathcal{F}_b}} \le \bar{\lambda}_1\|u(s)-\bar{u}(s)\|^2_{\widehat{\mathcal{D}}_{\mathcal{F}_b}}+\bar{\lambda}_2\|v(s)-\bar{v}(s)\|^2_{\widehat{\mathcal{D}}_{\mathcal{F}_b}},$$

where

$$\bar{\lambda}_1 = 4c_h^2\left(\|(-A_2)^{-\beta}\|^2+\frac{(C_{1-\beta}b^\beta)^2}{2\beta-1}\right)\widetilde{K}^2,$$

and

$$\bar{\lambda}_2 = 4\bar{c}_h^2\left(\|(-A_2)^{-\beta}\|^2+\frac{(C_{1-\beta}b^\beta)^2}{2\beta-1}\right)\widetilde{K}^2.$$

Using (5.3), we conclude that

$$\|\widehat{N}_{11}(u,v)(t)-\widehat{N}_{11}(\bar{u},\bar{v})(t)\|_{\widehat{\mathcal{D}}_{\mathcal{F}_b}} \le \sqrt{\lambda_1}\|u(s)-\bar{u}(s)\|_{\widehat{\mathcal{D}}_{\mathcal{F}_b}}+\sqrt{\lambda_2}\|v(s)-\bar{v}(s)\|_{\widehat{\mathcal{D}}_{\mathcal{F}_b}}.$$

Similar computations for N_{21} yield

$$\|\widehat{N}_{21}(u,v)(t)-\widehat{N}_{21}(\bar{u},\bar{v})(t)\|_{\widehat{\mathcal{D}}_{\mathcal{F}_b}} \le \sqrt{\bar{\lambda}_1}\|u(s)-\bar{u}(s)\|_{\widehat{\mathcal{D}}_{\mathcal{F}_b}}+\sqrt{\bar{\lambda}_2}\|v(s)-\bar{v}(s)\|_{\widehat{\mathcal{D}}_{\mathcal{F}_b}}.$$

Thus,

$$\begin{pmatrix} \|\widehat{N}_{11}((u,v)-\widehat{N}_{11}(\bar{u},\bar{v})\|_{\widehat{\mathcal{D}}_{\mathcal{F}_b}} \\ \|\widehat{N}_{21}(u,v)-\widehat{N}_{21}(\bar{u},\bar{v})\|_{\widehat{\mathcal{D}}_{\mathcal{F}_b}} \end{pmatrix} \leq \begin{pmatrix} \sqrt{\lambda_1} & \sqrt{\lambda_2} \\ \sqrt{\bar{\lambda}_1} & \sqrt{\bar{\lambda}_2} \end{pmatrix} \begin{pmatrix} \|u-\bar{u}\|_{\widehat{\mathcal{D}}_{\mathcal{F}_b}} \\ \|v-\bar{v}\|_{\widehat{\mathcal{D}}_{\mathcal{F}_b}} \end{pmatrix}.$$

It is clear that the radius spectral $\rho(M_{b,c}) < 1$. By Lemma 3.5, it converges to zero. From Theorem 7.1, thus $\widehat{N}_{11}, \widehat{N}_{21}$ is a contraction.

Part 2:

We aim to prove that the operator $\widehat{N}_{12}, \widehat{N}_{22}$ is completely continuous.

Step 1. $\widehat{N}_{12}, \widehat{N}_{22}$ is continuous.

Let $(z^n, \bar{z}^n)$ be a sequence such that

$$(z^n, \bar{z}^n) \to (z, \bar{z}),$$

in $\widehat{\mathcal{D}}_{\mathcal{F}_b}$. Then, for $t \in J$, and thanks to (H_{27}), (H_{29}), and (H_{32}), I_k, $\bar{I}_k$, $k = 1, 2, \ldots, m$ is continuous, we have by the dominated convergence theorem that

$$\begin{aligned}
&\mathbb{E}\left|\widehat{N}_{12}(z^n, \bar{z}^n)(t) - \widehat{N}_{12}(z, \bar{z})(t)\right|^2 \\
&\leq 2\mathbb{E}\left|\int_0^t S_1(t-s)(f^1(s, z_s^n + \theta_s, \bar{z}_s^n + \bar{\theta}_s) - f^1(s, z_s + \theta_s, \bar{z}_s + \bar{\theta}_s))ds\right|^2 \\
&\quad + 2\mathbb{E}\Big|\sum_{0 \leq t_k \leq t} |S_1(t-t_k)|^2 |I_k(z^n(t_k^-) + \theta(t_k^-), \bar{z}^n(t_k^-) + \bar{\theta}(t_k^-)) \\
&\quad - I_k(z(t_k^-) + \theta(t_k^-), \bar{z}(t_k^-) + \bar{\theta}(t_k^-))\Big|^2 \\
&\leq 2Mb\int_0^t \mathbb{E}\left|(f^1(s, z_s^n + \theta_s, \bar{z}_s^n + \bar{\theta}_s) - f^1(s, z_s + \theta_s, \bar{z}_s + \bar{\theta}_s))ds\right|^2 \\
&\quad + 2mM\sum_{0 < t_k < t} \mathbb{E}\big|I_k(z^n(t_k^-) + \theta(t_k^-), \bar{z}^n(t_k^-) + \bar{\theta}(t_k^-)) \\
&\quad - I_k(z(t_k^-) + \theta(t_k^-), \bar{z}(t_k^-) + \bar{\theta}(t_k^-))\big|^2 \to 0 \quad \text{as } n \to +\infty,
\end{aligned}$$

and

$$\begin{aligned}
&\mathbb{E}\left|\widehat{N}_{22}(z^n, \bar{z}^n)(t) - \widehat{N}_{22}(z, \bar{z})(t)\right|^2 \\
&\leq 2\mathbb{E}\left|\int_0^t S_2(t-s)(f^2(s, z_s^n + \theta_s, \bar{z}_s^n + \bar{\theta}_s) - f^2(s, z_s + \theta_s, \bar{z}_s + \bar{\theta}_s))ds\right|^2 \\
&\quad + 2\mathbb{E}\Big|\sum_{0 \leq t_k \leq t} |S_2(t-t_k)|^2 |\bar{I}_k(z^n(t_k^-) + \theta(t_k^-), \bar{z}^n(t_k^-) + \bar{\theta}(t_k^-)) \\
&\quad - \bar{I}_k(z(t_k^-) + \theta(t_k^-), \bar{z}(t_k^-) + \bar{\theta}(t_k^-))\Big|^2
\end{aligned}$$

$$\leq 2Mb\int_0^t \mathbb{E}\big|f^2(s,z_s^n+\theta_s,\bar z_s^n+\bar\theta_s)-f^2(s,z_s+\theta_s,\bar z_s+\bar\theta_s)ds\big|^2$$
$$+2mM\sum_{0<t_k<t}\mathbb{E}\big|\bar I_k(z^n(t_k^-)+\theta(t_k^-),\bar z^n(t_k^-)+\bar\theta(t_k^-))$$
$$-\bar I_k(z(t_k^-)+\theta(t_k^-),\bar z(t_k^-)+\bar\theta(t_k^-))\big|^2\to 0 \quad \text{as } n\to+\infty.$$

Thus, $\widehat{N}_{12}$, $\widehat{N}_{22}$ is continuous.

Step 2. $\widehat{N}_{12}$, $\widehat{N}_{22}$ maps bounded sets into bounded sets in $\widehat{\mathcal{D}}_{\mathcal{F}_b}$.

Indeed, it is enough to show that for any $q>0$, there exists a positive constant l_1, l_2 such that for each

$$z,\bar z\in\mathcal{B}_q=\{z,\bar z\in\widehat{\mathcal{D}}_{\mathcal{F}_b}:\|z\|^2_{\widehat{\mathcal{D}}_{\mathcal{F}_b}}\leq q,\ \|\bar z\|^2_{\widehat{\mathcal{D}}_{\mathcal{F}_b}}\leq q\},$$

one has

$$\|\widehat{N}_{12}(z,\bar z)\|^2_{\widehat{\mathcal{D}}_{\mathcal{F}_b}}\leq l_1,$$
$$\|\widehat{N}_{22}(z,\bar z)\|^2_{\widehat{\mathcal{D}}_{\mathcal{F}_b}}\leq l_2.$$

Let $z,\bar z\in\mathcal{B}_q$, then for each $t\in[0,b]$, we have

$$\mathbb{E}\big|\widehat{N}_{12}(z,\bar z)(t)\big|^2$$
$$\leq 2M\mathbb{E}\left|\int_0^t f^1(s,z_s+\theta_s,\bar z_s+\bar\theta_s)ds\right|^2$$
$$+2Mm\sum_{0<t_k<t}\mathbb{E}\big|I_k(z(t_k^-)+\theta(t_k^-),\bar z(t_k^-)+\bar\theta(t_k^-))\big|^2.$$
$$\leq 2bM\int_0^t m_1(s)\psi(\|z_s+\theta_s\|^2_{\widehat{\mathcal{D}}_{\mathcal{F}_b}}+\|\bar z_s+\bar\theta_s\|^2_{\widehat{\mathcal{D}}_{\mathcal{F}_b}})ds$$
$$+2Mm\sum_{0<t_k<t}\big|I_k(z(t_k^-)+\theta(t_k^-),\bar z(t_k^-)+\bar\theta(t_k^-))\big|^2$$
$$\leq 2Mb\psi(2C_{st})\int_0^b m_1(s)ds+2Mm\sum_{k=1}^m d_k^2.$$

Then we have

$$\mathbb{E}\big|\widehat{N}_{12}(z,\bar z)(t)\big|^2\leq 2Mb\psi(2C_{st})\|m_1\|_{L^1([0,b])}+2Mm\sum_{k=1}^m d_k^2=l_1<\infty.$$

Similarly, for $\widehat{N}_{22}$, we have

$$\begin{aligned}
&\mathbb{E}\big|\widehat{N}_{22}(z,\bar{z})(t)\big|^2 \\
&\leq 2bM\int_0^t m_2(s)\psi(\|z_s+\theta_s\|^2_{\widehat{\mathcal{D}}_{\mathcal{F}_b}}+\|\bar{z}_s+\bar{\theta}_s\|^2_{\widehat{\mathcal{D}}_{\mathcal{F}_b}})ds \\
&\quad + 2Mm\sum_{0<t_k<t}|\bar{I}_k(z(t_k^-)+\theta(t_k^-),\bar{z}(t_k^-)+\bar{\theta}(t_k^-))|^2 \\
&\leq 2Mb\psi(2C_{st})\int_0^b m_2(s)ds + 2Mm\sum_{k=1}^m \bar{d}_k^2.
\end{aligned}$$

Then we have

$$\mathbb{E}\big|\widehat{N}_{22}(z,\bar{z})(t)\big|^2 \leq 2Mb\psi(2C_{st})\|m_2\|_{L^1([0,b])} + 2Mm\sum_{k=1}^m \bar{d}_k^2 = l_2 < \infty.$$

Step 3. $\widehat{N}_{12}, \widehat{N}_{22}$ maps bounded sets into equicontinuous sets of $\widehat{\mathcal{D}}_{\mathcal{F}_b}$. Let $\tau_1,\tau_2 \in J$ such that $0<\tau_1<\tau_2\leq b$ and $\mathcal{B}_q$ be a bounded set of $\widehat{\mathcal{D}}_{\mathcal{F}_b}$ as in Step 2 is an equi-continuous set of $\widehat{\mathcal{D}}_{\mathcal{F}_b}\times\widehat{\mathcal{D}}_{\mathcal{F}_b}$. Let $z,\bar{z}\in\mathcal{B}_q$. Then

$$\begin{aligned}
&\mathbb{E}|(\widehat{N}_{12}(z,\bar{z})(\tau_2)-(\widehat{N}_{12}(z,\bar{z}))(\tau_1)|^2 \\
&\leq 4b\int_0^{\tau_1}\|S_1(\tau_2-\tau_1)-I\|^2\|S_1(\tau_1-s)\|^2\mathbb{E}|f^1(s,z_s+\theta_s,\bar{z}_s+\bar{\theta}_s)|^2 ds \\
&\quad + 4b\int_{\tau_1}^{\tau_2}\|S_1(\tau_2-s)\|^2\mathbb{E}|f^1(s,z_s+\theta_s,\bar{z}_s+\bar{\theta}_s)|^2 ds \\
&\quad + 4\sum_{0<t_k<\tau_1}\|S_1(\tau_2-\tau_1)-I\|^2\|S_1(\tau_1-t_k)\|^2\mathbb{E}|I_k(z^n(t_k^-)+\theta(t_k^-),\bar{z}(t_k^-)+\bar{\theta}(t_k^-))|^2 \\
&\quad + 4\sum_{\tau_1<t_k<\tau_2}\|S_1(\tau_2-t_k)\|^2\mathbb{E}|I_k(z(t_k^-)+\theta(t_k^-),\bar{z}(t_k^-)+\bar{\theta}(t_k^-))|^2.
\end{aligned}$$

Therefore, by the assumptions, for each $t\in J$, we have

$$\begin{aligned}
&\mathbb{E}|(\widehat{N}_{12}(z,\bar{z})(\tau_2)-(\widehat{N}_{12}(z,\bar{z}))(\tau_1)|^2 \\
&\leq 4\psi(2C_{st})M\|S_1(\tau_2-\tau_1)-I\|^2\int_0^{\tau_1}m_1(s)ds + 4\psi(2C_{st})M\int_{\tau_1}^{\tau_2}m_1(s)ds \\
&\quad + 4Mm\sum_{0<t_k<\tau_1}\|S_1(\tau_2-\tau_1)-I\|^2 d_k^2 + 4Mm\sum_{\tau_1<t_k<\tau_2}\|S_1(\tau_1-t_k)\|^2 d_k^2.
\end{aligned}$$

Similar computations for $\widehat{N}_{22}$ yield

$$\begin{aligned}
&\mathbb{E}|(\widehat{N}_{22}(z,\bar{z})(\tau_2) - (\widehat{N}_{22}(z,\bar{z}))(\tau_1)|^2 \\
&\quad \leq 4\psi(2C_{st})M\|S_2(\tau_2-\tau_1)-I\|^2 \int_0^{\tau_1} m_2(s)ds + 4\psi(2C_{st})M \int_{\tau_1}^{\tau_2} m_2(s)ds \\
&\quad + 4Mm \sum_{0<t_k<\tau_1} \|S_2(\tau_2-\tau_1)-I\|^2 \bar{d}_k^2 + 4Mm \sum_{\tau_1<t_k<\tau_2} \|S_2(\tau_1-t_k)\|^2 \bar{d}_k^2.
\end{aligned}$$

Since

$$\{S_i(t),\ i=1,2\}_{t>0},$$

is equicontinuous for $t > 0$, it implies continuity in the uniform operator topology; thus,

$$\|S_i(h) - I\| \to 0,$$

as $h \to 0^+$. Therefore, the right-hand side tends to zero as $\tau_2 \to \tau_1$. This proves equicontinuity for the case where $t \neq t_k$, $k = 1, \dots, m$.

Now, we prove equicontinuity at $t = t_j^-$. Let $\xi_1 > 0$, such that

$$\{t_k : k \neq j\} \cap [t_j - \xi_1, t_j + \xi_1] \neq \emptyset.$$

For $0 < \epsilon < \xi_1$, we get

$$\begin{aligned}
&\mathbb{E}|(\widehat{N}_{12}(z,\bar{z})(t_j) - (\widehat{N}_{12}(z,\bar{z}))(t_j-\epsilon)|^2 \\
&\quad \leq 3\psi(2C_{st}) \int_0^{t_j} \|S_1(t_j-s) - S_1(t_j-\epsilon-s)\|^2 m_1(s)ds \\
&\quad + 3\psi(2C_{st}) \int_{t_j-\epsilon}^{t_j} \|S_1(t_j-\epsilon-s)\|^2 m_1(s)ds \\
&\quad + 3m \sum_{k=1}^{j} \|S_1(t_j-t_k) - S_1(t_j-\epsilon-t_k)\|^2 d_k^2.
\end{aligned}$$

The right-hand side tends to zero as $\epsilon \to 0$. Next, we prove equicontinuity at $t = t_j^+$. Fix $\xi_2 > 0$, such that

$$\{t_k : k \neq j\} \cap [t_j - \xi_2, t_j + \xi_2] \neq \emptyset.$$

For $0 < \epsilon < \xi_2$, we have

$$\begin{aligned}
&\mathbb{E}|(\widehat{N}_{12}(z,\bar{z})(t_j+\epsilon)-(\widehat{N}_{12}(z,\bar{z}))(t_j)|^2\\
&\leq 3\psi(2C_{st})\int_0^{t_j}\|S_1(t_j+\epsilon-s)-S_1(t_j-s)\|^2 m_1(s)ds\\
&+3\psi(2C_{st})\int_{t_j}^{t_j+\epsilon}\|S_1(t_j-s)\|^2 m_1(s)ds\\
&+3m\sum_{k=1}^{j}\|S_1(t_j+\epsilon-t_k)-S_1(t_j-t_k)\|^2 d_k^2.
\end{aligned}$$

The right-hand side tends to zero as $\epsilon \to 0$. By a similar method, we can prove the equicontinuity for

$$\widehat{N}_{12}(\mathcal{B}_q\times\mathcal{B}_q),$$

and

$$\widehat{N}_{22}(\mathcal{B}_q\times\mathcal{B}_q).$$

Step 4. $(\widehat{N}_{12}\mathcal{B}_q)(t)$, $(\widehat{N}_{22}\mathcal{B}_q)(t)$ is precompact in X As a consequence of Steps 1 to 3, together with the Arzelá–Ascoli theorem, it suffices to show that $\widehat{N}_{12}$, $\widehat{N}_{22}$ maps $\mathcal{B}_q$ into a precompact set in X. Let $0<t<b$ be fixed and let ϵ be a real number satisfying $0<\epsilon<t$. For $z,\bar{z}\in\mathcal{B}_q$, define

$$\begin{aligned}
\widehat{N}_{12}^{\epsilon}(z,\bar{z})(t) &= \int_0^{t-\epsilon} S_1(t-s)f^1(s,z_s+\theta_s,\bar{z}_s+\bar{\theta}_s)ds\\
&\quad+\sum_{0<t_k<t-\epsilon} S_1(t-t_k)I_k(z(t_k^-)+\theta(t_k^-),\bar{z}(t_k^-)+\bar{\theta}(t_k^-))\\
&= S_1(\epsilon)\int_0^{t-\epsilon} S_1(t-s-\epsilon)f^1(s,z_s+\theta_s,\bar{z}_s+\bar{\theta}_s)ds\\
&\quad+S_1(\epsilon)\sum_{0<t_k<t-\epsilon} S_1(t-t_k-\epsilon)I_k(z(t_k^-)+\theta(t_k^-),\bar{z}(t_k^-)+\bar{\theta}(t_k^-)),
\end{aligned}$$

and

$$\begin{aligned}
\widehat{N}_{22}^{\epsilon}(z,\bar{z})(t) &= S_2(\epsilon)\int_0^{t-\epsilon} S_2(t-s-\epsilon)f^2(s,z_s+\theta_s,\bar{z}_s+\bar{\theta}_s)ds\\
&\quad+S_2(\epsilon)\sum_{0<t_k<t-\epsilon} S_2(t-t_k-\epsilon)\bar{I}_k(z(t_k^-)+\theta(t_k^-),\bar{z}(t_k^-)+\bar{\theta}(t_k^-)).
\end{aligned}$$

Since $S_i(t)$ is a compact operator, the set

$$H_\epsilon(t) = \{\widehat{N}_{12}^{\epsilon}(z,\bar{z})(t), \widehat{N}_{22}^{\epsilon}(z,\bar{z})(t) : z, \bar{z} \in \mathcal{B}_q\},$$

is precompact in X for every ϵ, $0 < \epsilon < t$. Moreover, for every $z, \bar{z} \in \mathcal{B}_q$ we have

$$\mathbb{E}\left|\widehat{N}_{12}(z,\bar{z})(t) - (\widehat{N}_{12}^{\epsilon}(z,\bar{z})(t)\right|^2 \leq 2M\psi(2C_{st}) \int_{t-\epsilon}^{t} m_1(s)ds + 2Mm \sum_{t-\epsilon<t_k<t} d_k^2.$$

Similarly,

$$\mathbb{E}|\widehat{N}_{22}(z,\bar{z})(t) - (\widehat{N}_{22}^{\epsilon}(z,\bar{z})(t)|^2 \leq 2M\psi(2C_{st}) \int_{t-\epsilon}^{t} m_2(s)ds + 2Mm \sum_{t-\epsilon<t_k<t} \bar{d}_k^2.$$

The right-hand term tends to 0 uniformly in t as $\epsilon \to 0$. This implies that the set

$$\widehat{N}_{12}^{\epsilon}(\mathcal{B}_q \times \mathcal{B}_q),$$

is relatively compact for $t \in [0, b]$.

By a similar way as above, we prove that

$$\widehat{N}_{12}^{\epsilon}(\mathcal{B}_q \times \mathcal{B}_q),$$

is also relatively compact. This implies that

$$\widehat{N}_{12}, \widehat{N}_{22} : \widehat{\mathcal{D}}_{\mathcal{F}_b} \times \widehat{\mathcal{D}}_{\mathcal{F}_b} \to \widehat{\mathcal{D}}_{\mathcal{F}_b},$$

is completely continuous.

Step 5. Now it remains to show that the set

$$\mathcal{M} = \left\{z, \bar{z} \in \widehat{\mathcal{D}}_{\mathcal{F}_b} : \begin{pmatrix} z \\ \bar{z} \end{pmatrix} = \begin{pmatrix} \lambda\widehat{N}_{11}(z,\bar{z}) + \lambda\widehat{N}_{12}(\frac{z}{\lambda}, \frac{\bar{z}}{\lambda}) \\ \lambda\widehat{N}_{21}(z,\bar{z}) + \lambda\widehat{N}_{12}(\frac{z}{\lambda}, \frac{\bar{z}}{\lambda}), \end{pmatrix}, \text{ for some } 0 < \lambda < 1\right\},$$

is bounded.

$$\begin{aligned} z(t) &= -S_1(t)h^1(0,\phi,\bar{\phi}) + h^1(t, z_t + \theta_t, \bar{z}_t + \bar{\theta}_t) + \int_0^t A_1 S_1(t-s)h^1(s, z_s + \theta_s, \bar{z}_s + \bar{\theta}_s)ds \\ &+ \int_0^t S_1(t-s)f^1(s, z_s + \theta_s, \bar{z}_s + \bar{\theta}_s)ds + \sum_{l=1}^{\infty} \int_0^t S_1(t-s)\sigma_l^1(s)dB_l^H(s) \\ &+ \int_0^t S_1(t-s)g^1(s)dW(s) + \sum_{0<t_k<t} S_1(t-t_k)I_k(z(t_k^-) + \theta(t_k^-), \bar{z}(t_k^-) + \bar{\theta}(t_k^-)), \end{aligned}$$

and

$$\bar{z}(t) = -S_2(t)h^2(0,\phi,\bar{\phi}) + h^2(t,z_t+\theta_t,\bar{z}_t+\bar{\theta}_t) + \int_0^t A_2S_2(t-s)h^2(s,z_s+\theta_s,\bar{z}_s+\bar{\theta}_s)ds$$
$$+ \int_0^t S_2(t-s)f^2(s,z_s+\theta_s,\bar{z}_s+\bar{\theta}_s)ds + \sum_{l=1}^{\infty}\int_0^t S_2(t-s)\sigma_l^2(s)dB_l^H(s)$$
$$+ \int_0^t S_2(t-s)g^2(s)dW(s) + \sum_{0<t_k<t} S_2(t-t_k)\bar{I}_k(z(t_k^-)+\theta(t_k^-),\bar{z}(t_k^-)+\bar{\theta}(t_k^-)).$$

Thus, for $t \in [0,b]$, namely

$$\mathbb{E}|z(t)|^2$$
$$\le 7\mathbb{E}|S_1(t)h^1(0,\phi,\bar{\phi})|^2 + 7\mathbb{E}|h^1(t,z_t+\theta_t,\bar{z}_s+\bar{\theta}_s)|^2$$
$$+ 7\mathbb{E}\left|\int_0^t A_1S_1(t-s)h^1(s,z_s+\theta_s,\bar{z}_s+\bar{\theta}_s)ds\right|^2$$
$$+ 7\mathbb{E}\left|\int_0^t S_1(t-s)f^1(s,z_s+\theta_s,\bar{z}_s+\bar{\theta}_s)ds\right|^2 + 7\mathbb{E}\left|\sum_{l=1}^{\infty}\int_0^t S_1(t-s)\sigma_l^1(s)dB_l^H(s)\right|^2$$
$$+ 7\mathbb{E}\left|\int_0^t S_1(t-s)g^1(s)dW(s)\right|^2$$
$$+ 7\mathbb{E}\left|\sum_{0<t_k<t} S_1(t-t_k)I_k(z(t_k^-)+\theta(t_k^-),\bar{z}(t_k^-)+\bar{\theta}(t_k^-))\right|^2.$$

From assumption (H_{27})–(H_{33}) and Lemma 7.2 and Lemma 7.1 yield that

$$\mathbb{E}|z(t)|^2$$
$$\le 7\|(-A_1)^{-\beta}\|^2 M\alpha_{h_1}^1(\|\phi\|_{\mathcal{D}_{\mathcal{F}_0}}^2 + \|\bar{\phi}\|_{\mathcal{D}_{\mathcal{F}_0}}^2) + 7\|(-A_1)^{-\beta}\|^2\alpha_{h_1}^1(\|z_t+\theta_t\|_{\mathcal{D}_{\mathcal{F}_0}}^2 + \|\bar{z}_t+\bar{\theta}_t\|_{\mathcal{D}_{\mathcal{F}_0}}^2)$$
$$+ 7b\int_0^t C_{1-\beta}^2(t-s)^{-2(1-\beta)}(\alpha_{h_1}^1(\|z_s+\theta_s\|_{\mathcal{D}_{\mathcal{F}_0}}^2 + \|\bar{z}_s+\bar{\theta}_s\|_{\mathcal{D}_{\mathcal{F}_0}}^2) + \alpha_{h_1}^1)ds$$
$$+ 7Mb\int_0^t m_1(s)\psi(\|z_s+\theta_s\|_{\mathcal{D}_{\mathcal{F}_0}}^2 + \|\bar{z}_s+\bar{\theta}_s\|_{\mathcal{D}_{\mathcal{F}_0}}^2)ds + 7Mc_H H(2H-1)t^{2H-1}\int_0^t \|\sigma^1(s)\|^2 ds$$
$$+ \int_0^t \|g^1(s)\|_{L^0}^2 ds + 7Mm\sum_{k=1}^m d_k^2 + 7\|(-A_1)^{-\beta}\|^2\bar{\alpha}_{h_1}^1(M+1).$$

Therefore,

$$\begin{aligned}\mathbb{E}|z(t)|^2 \leq D_1 &+ 7\|(-A_1)^{-\beta}\|^2 \alpha_{h_1}^1 (\|z_t + \theta_t\|^2_{\mathcal{D}_{\mathcal{F}_0}} + \|\bar{z}_t + \bar{\theta}_t\|^2_{\mathcal{D}_{\mathcal{F}_0}}) \\ &+ 7b\alpha_{h_1}^1 \int_0^t C_{1-\beta}^2 (t-s)^{-2(1-\beta)} (\|z_s + \theta_s\|^2_{\mathcal{D}_{\mathcal{F}_0}} + \|\bar{z}_s + \bar{\theta}_s\|^2_{\mathcal{D}_{\mathcal{F}_0}}) ds \\ &+ 7Mb \int_0^t m_1(s) \psi(\|z_s + \theta_s\|^2_{\mathcal{D}_{\mathcal{F}_0}} + \|\bar{z}_s + \bar{\theta}_s\|^2_{\mathcal{D}_{\mathcal{F}_0}}) ds.\end{aligned}$$

Similarly,

$$\begin{aligned}\mathbb{E}|\bar{z}(t)|^2 \leq D_2 &+ 7\|(-A_2)^{-\beta}\|^2 \alpha_{h_2}^2 (\|z_t + \theta_t\|^2_{\mathcal{D}_{\mathcal{F}_0}} + \|\bar{z}_t + \bar{\theta}_t\|^2_{\mathcal{D}_{\mathcal{F}_0}}) \\ &+ 7b\alpha_{h_2}^2 \int_0^t C_{1-\beta}^2 (t-s)^{-2(1-\beta)} (\|z_s + \theta_s\|^2_{\mathcal{D}_{\mathcal{F}_0}} + \|\bar{z}_s + \bar{\theta}_s\|^2_{\mathcal{D}_{\mathcal{F}_0}}) ds \\ &+ 7Mb \int_0^t m_2(s) \psi(\|z_s + \theta_s\|^2_{\mathcal{D}_{\mathcal{F}_0}} + \|\bar{z}_s + \bar{\theta}_s\|^2_{\mathcal{D}_{\mathcal{F}_0}}) ds,\end{aligned}$$

where

$$D_1 = \Lambda_1 + \Lambda_2, \quad \Lambda_1 = 7C_1 + 7\|(-A_1)^{-\beta}\|^2 M \alpha_{h_1}^1 (\|\phi\|^2_{\mathcal{D}_{\mathcal{F}_0}} + \|\bar{\phi}\|^2_{\mathcal{D}_{\mathcal{F}_0}}) + 7\|(-A)^{-\beta}\|^2 \bar{\alpha}_{h_1}^1 (M+1),$$

and

$$\Lambda_2 = 7Mm \sum_{k=1}^m d_k^2 + 7Mc_H H(2H-1) t^{2H-1} \int_0^t \|\sigma^1(s)\|^2 ds + 7b\bar{\alpha}_{h_1}^1 \int_0^t C_{1-\beta}^2 (t-s)^{-2(1-\beta)} ds,$$

and

$$D_2 = \Lambda_3 + \Lambda_4, \quad \Lambda_3 = 7C_1 + 7\|(-A_2)^{-\beta}\|^2 M \alpha_{h_2}^2 (\|\phi\|^2_{\mathcal{D}_{\mathcal{F}_0}} + \|\bar{\phi}\|^2_{\mathcal{D}_{\mathcal{F}_0}}) + 7\|(-A)^{-\beta}\|^2 \bar{\alpha}_{h_2}^2 (M+1),$$

and

$$\Lambda_4 = 7Mm \sum_{k=1}^m \bar{d}_k^2 + 7Mc_H H(2H-1) t^{2H-1} \int_0^t \|\sigma^2(s)\|^2 ds + 7b\bar{\alpha}_{h_2}^2 \int_0^t C_{1-\beta}^2 (t-s)^{-2(1-\beta)} ds.$$

By the two above inequalities, we get

$$\begin{aligned}\mathbb{E}|z(t)|^2 + \mathbb{E}|\bar{z}(t)|^2 \leq B_* &+ 7\|(-A)^{-\beta}\|^2 \alpha_h (\|z_t + \theta_t\|^2_{\mathcal{D}_{\mathcal{F}_0}} + \|\bar{z}_t + \bar{\theta}_t\|^2_{\mathcal{D}_{\mathcal{F}_0}}) \\ &+ 7b\alpha_h \int_0^t C_{1-\beta}^2 (t-s)^{-2(1-\beta)} (\|z_s + \theta_s\|^2_{\mathcal{D}_{\mathcal{F}_0}} + \|\bar{z}_s + \bar{\theta}_s\|^2_{\mathcal{D}_{\mathcal{F}_0}}) ds\end{aligned}$$

$$+ 7Mb \int_0^t m(s)\psi(\|z_s + \theta_s\|^2_{\mathcal{D}_{\mathcal{F}_0}} + \|\bar{z}_s + \bar{\theta}_s\|^2_{\mathcal{D}_{\mathcal{F}_0}})ds,$$

where

$$B_* = D_1 + D_2, \quad \alpha_h = \sup\{\alpha^1_{h_1}, \alpha^2_{h_2}\}, \quad m(t) = \sup\{m_1(t), m_2(t)\},$$

and

$$\|(-A)^{-\beta}\|^2 = \sup\{\|(-A_1)^{-\beta}\|^2, \|(-A_2)^{-\beta}\|^2\}.$$

That is to say,

$$\begin{aligned}
&\|z_t + \theta_t\|^2_{\mathcal{D}_{\mathcal{F}_0}} + \|\bar{z}_t + \bar{\theta}_t\|^2_{\mathcal{D}_{\mathcal{F}_0}} \\
&\quad \le 4\widetilde{K}^2 \sup_{s\in[0,b]} (E|z(s)|^2 + \mathbb{E}|\bar{z}(s)|^2) + 4\widetilde{K}^2 M(\mathbb{E}|\phi(0)|^2 + E|\bar{\phi}(0)|^2) + 4\widetilde{N}^2(\|\phi\|^2_{\mathcal{D}_{\mathcal{F}_0}} + \|\bar{\phi}\|^2_{\mathcal{D}_{\mathcal{F}_0}}).
\end{aligned}$$

If we set $v(t)$ the right-hand side of the above inequality, we have that

$$\|z_t + \theta_t\|^2_{\mathcal{D}_{\mathcal{F}_0}} + \|\bar{z}_t + \bar{\theta}_t\|^2_{\mathcal{D}_{\mathcal{F}_0}} \le v(t). \tag{7.6}$$

Consequently,

$$\begin{aligned}
&\mathbb{E}|z(t)|^2 + \mathbb{E}|\bar{z}(t)|^2 \\
&\quad \le B_* + 7\|(-A)^{-\beta}\|^2 \alpha_h(\|z_t + \theta_t\|^2_{\mathcal{D}_{\mathcal{F}_0}} + \|\bar{z}_t + \bar{\theta}_t\|^2_{\mathcal{D}_{\mathcal{F}_0}}) \\
&\quad + 7b\alpha_h \int_0^t C^2_{1-\beta}(t-s)^{-2(1-\beta)}(\|z_s + \theta_s\|^2_{\mathcal{D}_{\mathcal{F}_0}} + \|\bar{z}_s + \bar{\theta}_s\|^2_{\mathcal{D}_{\mathcal{F}_0}})ds \\
&\quad + 7Mb \int_0^t m(s)\psi(\|z_s + \theta_s\|^2_{\mathcal{D}_{\mathcal{F}_0}} + \|\bar{z}_s + \bar{\theta}_s\|^2_{\mathcal{D}_{\mathcal{F}_0}})ds.
\end{aligned}$$

This implies, for each $t \in J$,

$$\mathbb{E}|z(t)|^2 + \mathbb{E}|\bar{z}(t)|^2 \le \Upsilon(t), \tag{7.7}$$

where

$$\Upsilon(t) = B_* + 7\|(-A)^{-\beta}\|^2 \alpha_h v(t) + 7b\alpha_h \int_0^t C^2_{1-\beta}(t-s)^{-2(1-\beta)} v(s)ds + 7Mb \int_0^t m(s)\psi(v(s))ds.$$

Using (7.6) in the definition of v, we have that

$$v(t) \le 4\widetilde{K}^2(B_* + 7\|(-A)^{-\beta}\|^2 \alpha_h v(t) + 7b\alpha_h \int_0^t C_{1-\beta}^2 (t-s)^{-2(1-\beta)} v(s)ds$$
$$+ 7Mb \int_0^t m(s)\psi(v(s)ds) + 4\widetilde{K}^2 M(\mathbb{E}|\phi(0)|^2 + \mathbb{E}|\bar{\phi}(0)|^2) + 4\widetilde{N}^2(\|\phi\|^2_{\mathcal{D}_{\mathcal{F}_0}} + \|\bar{\phi}\|^2_{\mathcal{D}_{\mathcal{F}_0}})$$
$$\le B_1 + B_2 \int_0^t v(s)(t-s)^{-2(1-\beta)} ds + B_3 \int_0^t m_*(s)\psi(v(s))ds, \tag{7.8}$$

where

$$B_1 = \frac{4\widetilde{K}^2 M(E|\phi(0)|^2 + E|\bar{\phi}(0)|^2) + 4\widetilde{N}^2(\|\phi\|^2_{\mathcal{D}_{\mathcal{F}_0}} + \|\bar{\phi}\|^2_{\mathcal{D}_{\mathcal{F}_0}}) + 4\widetilde{K}^2 B_*}{1 - 28\widetilde{K}^2\|(-A)^{-\beta}\|^2 \bar{p}_*},$$
$$B_2 = \frac{28b\widetilde{K}^2 C_{1-\beta}^2 \bar{p}_*}{1 - 28\widetilde{K}^2(-A)^{-\beta}\alpha_h},$$

and

$$B_3 = \frac{28b\widetilde{K}^2 M}{1 - 28\widetilde{K}^2(-A)^{-\beta}\alpha_h}.$$

Now from Lemma 7.4, we have

$$v(t) \le L_1 + L_2 \int_0^t m(s)\psi(v(s))ds, \tag{7.9}$$

where

$$L_1 = B_1 \exp(B_2^n(\Gamma(2\beta-1))^n b^{n(2\beta-1)}/\Gamma(n(2\beta-1))) \sum_{j=0}^{n-1}\left(\frac{B_2 b^{2\beta-1}}{2\beta-1}\right),$$

and

$$L_2 = B_3 \exp(B_2^n(\Gamma(2\beta-1))^n b^{n(2\beta-1)}/\Gamma(n(2\beta-1))) \sum_{j=0}^{n-1}\left(\frac{B_2 b^{2\beta-1}}{2\beta-1}\right).$$

Let us denote the right-hand side of the inequality (7.9) by $v(t)$. Then we have

$$w(0) = L_1, \; v(t) \le w(t), \quad t \in J,$$

and

$$w'(t) = L_2 m(t)\psi(v(t)), \quad t \in J.$$

Using the increasing property of ψ, we obtain

$$w'(t) \leq L_2 m(t)\psi(w(t)), \quad \text{for } a.e.\ t \in J.$$

Applying the Bihari lemma, we obtain

$$\int_{w(0)}^{w(t)} \frac{ds}{\psi(s)} \leq L_2 \int_0^b m(s)ds \quad \Longrightarrow \quad H(w(t)) \leq L_2 \|m\|_{L^1},$$

where H is a nondecreasing function defined by

$$H(u) = \int_{L_1}^{w(2)} \frac{du}{\psi(u)}.$$

It follows from this inequality that $w(t)$ is a bounded function. Thus, we obtain that there is a constant K such that

$$w(t) \leq H^{-1}(L_2\|m\|_{L^1}) = K \quad \Longrightarrow \quad v(t) \leq w(t) \leq K, \quad t \in J.$$

Finally, from (7.7) there exists a constant $R > 0$, such that

$$\begin{aligned} &\mathbb{E}|z(t)|^2 + \mathbb{E}|\bar{z}(t)|^2 \\ &\quad \leq B_* + 7\|(-A)^{-\beta}\|^2 a_h K + 7b \int_0^t C_{1-\beta}^2 (t-s)^{-2(1-\beta)} a_h K ds + 7Mb \int_0^t m(s)\psi(K) ds = R. \end{aligned}$$

Thus, we get

$$\|z\|^2_{\widehat{\mathcal{D}}_{\mathcal{F}_b}} \leq R \quad \text{and} \quad \|\bar{z}\|^2_{\widehat{\mathcal{D}}_{\mathcal{F}_b}} \leq R.$$

As a consequence of Theorem 7.1, we deduce that $\widehat{N}$ has a fixed point since

$$(x(t), y(t)) = (z(t) + \theta(t), \bar{z}(t) + \bar{\theta}(t)), \quad t \in (-\infty, b].$$

Then (x, y) is a fixed point of the operator $N = (N_1, N_2)$, which is a mild solution for the problems (7.1)–(7.3). The proof is complete. □

7.2 Example

We consider infinite fractional Brownian motion and the Wiener process.

Example 7.2. Consider the impulsive stochastic partial functional neutral differential equations of the form

$$
\begin{cases}
d[z(t,\xi) - \int_{-\infty}^{t} \vartheta_1(t)\vartheta_2(s-t)(z(s,\xi)+\bar{z}(s,\xi))ds] = \frac{\partial^2}{\xi^2} z(t,\xi) \\
\quad + \int_{-\infty}^{t} v_1(s)v_1(s-t)z(s,\xi)dsdt + \sum_{l=1}^{\infty} \sigma_l^1(t)dB_l^H(t) \\
\quad + \int_{-\infty}^{t} \bar{w}_1(t)\bar{w}_1(s-t)(z(s,\xi)+\bar{z}(s,\xi))dsdW(t), \quad t \in [0,b],\ t \neq t_k,\ 0 \le \xi \le \pi, \\
d[\bar{z}(t,\xi) - \frac{2}{3}\int_{-\infty}^{t} \vartheta_1(t)\vartheta_2(s-t)(z(s,\xi)+\bar{z}(s,\xi))ds] = \frac{\partial^2}{\xi^2} \bar{z}(t,\xi) \\
\quad + \int_{-\infty}^{t} v_1(s)v_1(s-t)\bar{z}(s,\xi)dsdt + \sum_{l=1}^{\infty} \sigma_l^2(t)dB_l^H(t) \\
\quad + \int_{-\infty}^{t} \bar{w}_1(t)\bar{w}_1(s-t)(z(s,\xi)-\bar{z}(s,\xi))dsdW(t), \quad t \in [0,b],\ t \neq t_k,\ 0 \le \xi \le \pi, \\
z(t_k^+,\xi) - z(t_k^-,\xi) = a_k z(t_k^-,\xi), \quad k = 1,\dots,m, \\
\bar{z}(t_k^+,\xi) - \bar{z}(t_k^-,\xi) = \bar{a}_k \bar{z}(t_k^-,\xi), \quad k = 1,\dots,m, \\
z(t,0) = z(t,\pi) = 0, \quad t \ge 0, \\
\bar{z}(t,0) = \bar{z}(t,\pi) = 0, \quad t \ge 0, \\
z(\tau,\xi) = \phi_1(\tau,\xi), \quad -\infty < \tau \le 0,\ 0 \le \xi \le \pi, \\
0\bar{z}(\tau,\xi) = \phi_2(\tau,\xi), \quad -\infty < \tau \le 0,\ 0 \le \xi \le \pi,
\end{cases}
\tag{7.10}
$$

where ϕ_1, ϕ_2 is continuous. $W(t)$ denotes a one-dimensional standard Wiener process in X defined on a stochastic space $(\Omega, \mathcal{F}, \mathbb{P})$, and we take

$$X = Y = L^2([0,\pi]),$$

with the norm $\|\cdot\|$ and define the operators $A : X \to X$ by $Au = u''$ with the domain

$$D(I) = \{u \in X, u'' \in X \text{ and } u(0) = u(\pi) = 0\}.$$

Then it is well known that

$$Au = -\sum_{n=1}^{\infty} n^2 \langle u, z_n\rangle z_n, \quad u \in D(I),$$

where

$$z_n(\xi) = \sqrt{\frac{2}{\pi}} \sin(nx), \quad n = 1,2,,3,\dots,$$

is an orthogonal set of eigenvectors of $-A$. It is well known that A generates a strongly continuous semigroup that

$$\{S(t)\}_{t\ge 0},$$

is analytic and compact; there exists a constant $M > 0$ such that

$$\|S(t)\|^2 \le M,$$

and it satisfies (H_{27}), which is given by

$$S(t)u = \sum_{n=1}^{\infty} e^{-n^2 t}\langle u, z_n\rangle z_n, \quad u \in X.$$

The bounded linear operator $(-A)^{\frac{3}{4}}$ is given by

$$(-A)^{\frac{3}{4}}u = \sum_{n=1}^{\infty} n^{\frac{3}{2}} \langle u, z_n \rangle z_n,$$

on the space

$$D\big((-A)^{\frac{3}{4}}\big) = \left\{ u(\cdot) \in X : \sum_{n=1}^{\infty} n^{\frac{3}{2}} \langle u, z_n \rangle z_n \in X \right\},$$

and

$$(-A)^{\frac{3}{4}}u = \sum_{n=1}^{\infty} \frac{1}{n^{\frac{3}{2}}} \langle u, z_n \rangle z_n,$$

for every $u \in X$ and $\|(-A)^{\frac{3}{4}}\|$ is bounded.

Let $r \geq 0, 1 \leq p < +\infty$, and

$$\widetilde{h} : (-\infty, -r] \to \mathbb{R},$$

be nonnegative measurable functions that satisfy the conditions $(h - 5)$ and $(h - 6)$ in the terminology in [10]. Briefly, this means that $\widetilde{h}$ is locally integrable, and there exists a nonnegative, locally bounded function y on $(-\infty, 0]$ such that

$$\widetilde{h}(\delta + \tau) \leq y(\delta)\widetilde{h}(\tau),$$

holds for all $\delta \leq 0$ and $y \in (-\infty, -r)\ N_\xi$, where $N_\xi \subset (-\infty, -r)$ is a set whose Lebesgue measure is zero. We denote by

$$C_r \times L^p(\widetilde{h}, X),$$

the set that consists of all classes of functions

$$\phi_i : (-\infty, 0] \to X,$$

such that

$$\phi_i|_{[-r,0]} \in C_r = C\big([-r, 0], X\big),$$

ϕ_i is Lebesgue measurable on $(-\infty, -r)$, and $\widetilde{h}\|\phi_i\|^p$ is Lebesgue integrable on $(-\infty, -r)$.

The seminorm is given by

$$\|\phi_i\|_{\mathcal{D}_{\mathcal{F}_0}} = \sup_{\tau\in[-r,0]} \|\phi_i(\tau)\| + \left(\int_{-\infty}^{-\tau} \widetilde{h}(\tau)\|\phi_i\|^p d\tau \right)^{1/p}.$$

The space

$$\mathcal{D}_{\mathcal{F}_0} = C_r \times L^p(\widetilde{h}, X),$$

satisfies the axioms (1)–(3). Moreover, when $r = 0$ and $p = 2$, we can take

$$L = 1, \quad N(t) = y(-t)^{1/2},$$

and

$$K(t) = 1 + \int_{-r}^{0} \widetilde{h}(\tau)d\tau)^{1/2},$$

for $t \geq 0$.

Additionally, assume that the following conditions hold:

1. The functions

$$\bar{w}_2, v_2, \vartheta_2 : \mathbb{R} \to \mathbb{R},$$

are continuous with

$$L_{\vartheta_2} = \int_{-\infty}^{0} \frac{(\vartheta_2(s))^2}{\widetilde{h}(s)} ds < \infty,$$

$$L_{v_2} = \int_{-\infty}^{0} \frac{(v_2(s))^2}{\widetilde{h}(s)} ds < \infty,$$

$$L_{\bar{w}_2} = \int_{-\infty}^{0} \frac{(\bar{w}_2(s))^2}{\widetilde{h}(s)} ds < \infty.$$

2. The functions

$$\eta_i : \mathbb{R} \to \mathbb{R}, \quad j = 1, 2, \ldots, N,$$

are continuous with

$$\widetilde{L}_i = \int_{-\infty}^{0} \frac{(\eta_j(s))^2}{\widetilde{h}(s)} ds < \infty,$$

for every $j = 1, 2, \ldots, N$.

Take

$$\phi_i \in \mathcal{D}_{\mathcal{F}_0} = C_0 \times L^2(\widetilde{h}, X),$$

with

$$\phi_i(\theta)(\xi) = \phi_i(\theta, \xi), \quad (\theta, \xi) \in (-\infty, 0] \times \mathcal{D}_{\mathcal{F}_0}.$$

Let

$$\big(z(s)(\xi), \bar{z}(s)(\xi)\big) = \big(z(s, \xi), \bar{z}(s, \xi)\big).$$

Let

$$h^i : [0, b] \times \mathcal{D}_{\mathcal{F}_0} \times \mathcal{D}_{\mathcal{F}_0} \to X,$$

$$f^i : [0, b] \times \mathcal{D}_{\mathcal{F}_0} \times \mathcal{D}_{\mathcal{F}_0} \to X,$$

the operators defined by

$$(-A)^{\frac{3}{4}} h^1(t, \phi_1, \phi_2)(\xi) = \int_{-\infty}^{0} \vartheta_1(t)\vartheta_2(\theta)\big(\phi_1(\theta)(\xi) + \phi_2(\theta)(\xi)\big)d\theta,$$

$$(-A)^{\frac{3}{4}}h^2(t,\phi_1,\phi_2)(\xi)=\frac{2}{3}\int_{-\infty}^{0}\vartheta_1(t)\vartheta_2(\theta)\big(\phi_1(\theta)(\xi)+\phi_2(\theta)(\xi)\big)d\theta,$$

$$f^1(t,\phi_1,\phi_2)(\xi)=\int_{-\infty}^{0}v_1(\theta)v_1(\theta)\big(\phi_1(\theta)(\xi)+\phi_2(\theta)(\xi)\big)d\theta,$$

$$f^2(t,\phi_1,\phi_2)(\xi)=\int_{-\infty}^{0}\frac{v_1(\theta)v_1(\theta)}{3}\big(\phi_1(\theta)(\xi)+\phi_2(\theta)(\xi)\big)d\theta,$$

$$g^1(t,\phi_1,\phi_2)=\int_{-\infty}^{0}\bar{w}_1(t)\bar{w}_2(\theta)\big(\phi_1(\theta)(\xi)+\phi_2(\theta)(\xi)\big)d\theta,$$

$$g^2(t,\phi_1,\phi_2)=\int_{-\infty}^{0}\bar{w}_1(t)\bar{w}_2(\theta)\big(\phi_1(\theta)(\xi)-\phi_2(\theta)(\xi)\big)d\theta.$$

Put

$$\sigma^1(t)=\sum_{l=1}^{\infty}\sigma_l^1(t),$$

$$\sigma^2(t)=\sum_{l=1}^{\infty}\sigma_l^2(t),$$

where

$$\sigma_l^1(t)=a_{2l+1}\int_0^t\big(\sin(l^2)\big)e(s)ds+a_{2l}\int_0^t\big(\cos(l^2)\big)e(s)ds,\quad e\in L^2([0,b]),$$

and

$$\sigma_l^2(t)=b_{2l+1}\int_0^t\big(\sin(l^2)\big)e(s)ds+b_{2l}\int_0^t\big(\cos(l^2)\big)e(s)ds,\quad e\in L^2([0,b]),$$

where

$$\sum_{l=1}^{\infty}a_l<\infty,$$

$$\sum_{l=1}^{\infty}b_l<\infty.$$

Under the above assumptions, problem (7.10) can be written as (7.1)–(7.3). Moreover, for any $t\in[0,b]$, $x,y\in\mathcal{D}_{\mathcal{F}_0}$, we have

$$\begin{aligned}&\mathbb{E}\big\|(-A)^{\frac{3}{4}}h^1(t,x,y)\big\|^2\\&=\mathbb{E}\left[\int_0^{\pi}\left(\int_{-\infty}^{0}\vartheta_1(t)\vartheta_2(s)\big(x(s)(\xi)+y(s)(\xi)\big)ds\right)^2d\xi\right]\\&\le\left(\int_{-\infty}^{0}\vartheta_1(t)\vartheta_2(s)\big\|x(s)+y(s)\big\|ds\right)^2\end{aligned}$$

$$\leq \|\vartheta_1\|_\infty^2 \left(\int\limits_{\infty}^{0} \frac{(\vartheta_2(s))^2}{\widetilde{h}(s)} ds \right) \left(\int\limits_{-\infty}^{0} \widetilde{h}(s) \left(\|x(s)\|^2 + \|y(s)\|^2 \right) ds \right)$$
$$\leq \bar{L}_{h_1} \left(\|x\|^2_{\mathcal{D}\mathcal{F}_0} + \|y\|^2_{\mathcal{D}\mathcal{F}_0} \right).$$

Similarly,

$$\mathbb{E}\left\| (-A)^{\frac{3}{4}} h^2(t,x,y) \right\|^2 \leq \bar{L}_{h_2} \left(\|x\|^2_{\mathcal{D}\mathcal{F}_0} + \|y\|^2_{\mathcal{D}\mathcal{F}_0} \right),$$

where

$$\bar{L}_{h_1} = 2\|\vartheta_1\|_\infty^2 L_{\vartheta_2},$$
$$\bar{L}_{h_2} = \frac{8}{9}\|\vartheta_1\|_\infty^2 L_{\vartheta_2}.$$

For any

$$t \in [0,b], \quad x, \bar{x}, y, \bar{y} \in \mathcal{D}_{\mathcal{F}_0}.$$

We have

$$\mathbb{E}\left\| (-A)^{\frac{3}{4}} \left(h^1(t,x,y) - h^1(t,\bar{x},\bar{y}) \right) \right\|^2$$
$$= \mathbb{E}\left[\int\limits_0^\pi \left(\int\limits_{-\infty}^{0} \vartheta_1(t)\vartheta_2(s) \left[\left(x(s)(\xi) - \bar{x}(s)(\xi) \right) + \left(y(s)(\xi) - \bar{y}(s)(\xi) \right) \right] ds \right)^2 d\xi \right]$$
$$\leq \left(\int\limits_{-\infty}^{0} \vartheta_1(t)\vartheta_2(s) \left(\|x(s) - \bar{x}(s)\| + \|y(s) - \bar{y}(s)\| \right) ds \right)^2$$
$$\leq \|\vartheta_1\|^2 \left(\int\limits_{\infty}^{0} \frac{(\vartheta_2(s))^2}{\widetilde{h}(s)} ds \right) \left(\int\limits_{-\infty}^{0} \widetilde{h}(s) \|x(s) - \bar{x}(s)\|^2 ds \right)$$
$$\leq L_{h_1} \|x - \bar{x}\|^2_{\mathcal{D}\mathcal{F}_0} + L_{h_1} \|y - \bar{y}\|^2_{\mathcal{D}\mathcal{F}_0}.$$

Similarly,

$$\mathbb{E}\left\| (-A)^{\frac{3}{4}} \left(h^2(t,x,y) - h^2(t,\bar{x},\bar{y}) \right) \right\|^2 \leq L_{h_2} \|x - \bar{x}\|^2_{\mathcal{D}\mathcal{F}_0} + L_{h_2} \|y - \bar{y}\|^2_{\mathcal{D}\mathcal{F}_0},$$

where

$$L_{h_1} = 2\|\vartheta_1\|_\infty^2 L_{\vartheta_2},$$
$$L_{h_2} = \frac{8}{9}\|\vartheta_1\|_\infty^2 L_{\vartheta_2}.$$

For any $t \in [0,b], x, \bar{x}, y, \bar{y} \in \mathcal{D}_{\mathcal{F}_0}$.

$$\mathbb{E}\left\| f^1(t,x,y) \right\|^2 = \mathbb{E}\left[\int\limits_0^\pi \left(\int\limits_{-\infty}^{t} v_1(t)v_2(s) \left(x(s)(\xi) + y(s)(\xi) \right) ds \right)^2 d\xi \right]$$
$$\leq \left(\int\limits_{-\infty}^{0} v_1(t)v_2(s) \|x(s) + y(s)\| ds \right)^2$$

$$\leq \|v_1\|_\infty^2 \left(\int_\infty^0 \frac{(v_2(s))^2}{\widetilde{h}(s)} ds \right) \left(\int_{-\infty}^0 \widetilde{h}(s) \left(\|x(s)\|^2 + \|y(s)\|^2 \right) ds \right)$$
$$\leq L_{f_1} \left(\|x\|^2_{\mathcal{D}_{\mathcal{F}_0}} + \|y\|^2_{\mathcal{D}_{\mathcal{F}_0}} \right).$$

Similarly,

$$\mathbb{E}\left\| f^2(t,x,y) \right\|^2 \leq L_{f_2} \left(\|x\|^2_{\mathcal{D}_{\mathcal{F}_0}} + \|y\|^2_{\mathcal{D}_{\mathcal{F}_0}} \right),$$

and

$$\mathbb{E}\|f^1(t,x,y) - f^1(t,\bar{x},\bar{y}))\|^2$$
$$= \mathbb{E}\left[\int_0^\pi \left(\int_{-\infty}^0 v_1(t)v_2(s) \left[\left(x(s)(\xi) - \bar{x}(s)(\xi) \right) + \left(y(s)(\xi) - \bar{y}(s)(\xi) \right) \right] ds \right)^2 d\xi \right]$$
$$\leq \left(\int_{-\infty}^0 v_1(t)v_2(s) \left(\|x(s) - \bar{x}(s)\| + \|y(s) - \bar{y}(s)\| \right) ds \right)^2$$
$$\leq \|v_1\|^2 \left(\int_\infty^0 \frac{(v_2(s))^2}{\widetilde{h}(s)} ds \right) \left(\int_{-\infty}^0 \widetilde{h}(s) \left(\|x(s) - \bar{x}(s)\|^2 + \|y(s) - \bar{y}(s)\|^2 \right) ds \right)$$
$$\leq L_{f_1} \|x - \bar{x}\|^2_{\mathcal{D}_{\mathcal{F}_0}} + L_{f_1} \|y - \bar{y}\|^2_{\mathcal{D}_{\mathcal{F}_0}}.$$

Similarly,

$$\mathbb{E}\left\| f^2(t,x,y) - f^2(t,\bar{x},\bar{y}) \right\|^2 \leq L_{f_2} \|x - \bar{x}\|^2_{\mathcal{D}_{\mathcal{F}_0}} + L_{f_2} \|y - \bar{y}\|^2_{\mathcal{D}_{\mathcal{F}_0}},$$

where

$$L_{f_1} = 2\|v_1\|_\infty^2 L_{v_2},$$
$$L_{f_2} = \frac{2}{9}\|v_1\|_\infty^2 L_{v_2},$$
$$\mathbb{E}\left\| g^i(t,x,y) \right\|^2 \leq L_{g_i} \left(\|x\|^2_{\mathcal{D}_{\mathcal{F}_0}} + \|y\|^2_{\mathcal{D}_{\mathcal{F}_0}} \right) < \infty,$$

where

$$L_{g_i} = 2\|\bar{w}_1\|_\infty^2 L_{\bar{w}_2}.$$

In order to define the operator

$$Q_i : X \longrightarrow X, \quad i = 1,2,$$

we choose a sequence

$$\{\sigma_{n,i}\}_{n\geq 1} \subset \mathbb{R}^+,$$

set $Q_i e_n = \sigma_{n,i} e_n$ and assume that

$$\operatorname{tr}(Q_i) = \sum_{n=1}^{\infty} \sqrt{\sigma_{n,i}} < \infty.$$

Define the process $B^H_{Q_i}(s)$ by

$$B^H_{Q_i} = \sum_{n=1}^{\infty} \sqrt{\sigma_{n,i}} \gamma^H_n(t) e_n,$$

where $H \in (1/2, 1)$, and $\{\gamma^H_n\}_{n\in\mathbb{N}}$ is a sequence of two-sided, one-dimensional fractional Brownian motions that are mutually independent.

Clearly, for every $k \in \mathbb{N}$, σ^1_l, σ^2_l are linear continuous operators. Let $(e_i)_{i\in\mathbb{N}}$ be an orthonormal basis of X,

$$\begin{aligned}\|\sigma^1_l(t,x,y)\|^2_{L^0_{Q_1}} &= \sum_{n=1}^{\infty} \int_0^{\pi} \left(\left| \sqrt{\sigma_n} a_{2l+1} \int_0^t [\sin l^2] e_n(s) ds + \sqrt{\sigma_n} a_{2l} \int_0^t [\cos l^2] e_n(s) ds \right|^2 \right) dt \\ &\le 2\pi^2 \left(a^2_{2l+1} + a^2_{2l} \right) \sum_{n=1}^{\infty} \sigma_{n,1}.\end{aligned}$$

Therefore,

$$\|\sigma^1_l(t)\|_{L^0_{Q_1}} \le \pi \sqrt{2\left(a^2_{2l+1} + a^2_{2l}\right)} \sum_{n=1}^{\infty} \sqrt{\sigma_{n,1}} < \infty.$$

Moreover, we have

$$\|\sigma^1_l(t)\|^2_{L^0_{Q_1}} \le L_* \left(a^2_{2l+1} + a^2_{2l}\right), \quad L_* = 2\pi^2 \sum_{n=1}^{\infty} \sqrt{\sigma_{n,1}}.$$

We deduce

$$\begin{aligned}\|\sigma^1(t)\|^2 &= \sum_{l=1}^{\infty} \|\sigma^1_l\|_{L^0_{Q_1}} \\ &\le L_* \sum_{l=1}^{\infty} \left(a^2_{2l+1} + a^2_{2l}\right) \\ &\le L_* \sum_{l=1}^{\infty} a^2_l < \infty,\end{aligned}$$

and

$$\begin{aligned}\|\sigma^2(t)\|^2 &= \sum_{l=1}^{\infty} \|\sigma^2_l\|^2_{L^0_{Q_2}} \\ &\le L_* \sum_{l=1}^{\infty} b^2_l < \infty,\end{aligned}$$

and

$$M_{tice} = 3 \begin{pmatrix} \sqrt{a_1} & \sqrt{a_2} \\ \sqrt{\bar{a}_1} & \sqrt{\bar{a}_2} \end{pmatrix},$$

where

$$a_1 = 4L_{h_1} \left(\|(-A)^{-\frac{3}{4}}\|^2 + \frac{(C_{1-\beta} b^{\frac{3}{4}})^2}{2(\frac{3}{4}) - 1} \right) \widetilde{K}^2,$$

$$a_2 = 4L_{h_1}\left(\left\|(-A)^{-\frac{3}{4}}\right\|^2 + \frac{(C_{1-\beta}b^{\frac{3}{4}})^2}{2(\frac{3}{4})-1}\right)\widetilde{K}^2,$$

and

$$\bar{a}_1 = 4L_{h_1}\left(\left\|(-A)^{-\frac{3}{4}}\right\|^2 + \frac{(C_{1-\frac{3}{4}}b^{\frac{3}{4}})^2}{2(\frac{3}{4})-1}\right)\widetilde{K}^2,$$

$$\bar{a}_2 = 4\bar{L}_{h_1}\left(\left\|(-A)^{-\beta}\right\|^2 + \frac{(C_{1-\frac{3}{4}}b^{\frac{3}{4}})^2}{2(\frac{3}{4})-1}\right)\widetilde{K}^2,$$

In the case of $t \in (-\infty, b]$, assume now that:

1. There exists a positive number $d_k, \bar{d}_k$, $k \in \{1,\dots,m\}$, such that

$$\left|I_k(\xi)\right| \le d_k, \quad \left|\bar{I}_k(\xi)\right| \le \bar{d}_k,$$

for any $\xi \in \mathbb{R}$.

2. The function $h^i : [0,b] \times X \times X \longrightarrow X$ defined by $h^i(t,x,y)(.)$ is continuous, and we impose suitable conditions to verify (H_{28}).

3. Assume that there exists an integrable function

$$m_i : [0,b] \longrightarrow \mathbb{R}^+,$$

such that

$$\mathbb{E}\left|f^i(t,x,y)\right|^2 \le m_i(t)\psi\left(\|x\|^2_{\mathcal{D}_{\mathcal{F}_0}} + \|y\|^2_{\mathcal{D}_{\mathcal{F}_0}}\right),$$

for any $t \in [0,b]$ and

$$m_1(t) = L_{f_1}, \quad m_2(t) = L_{f_2},$$

where

$$\psi : (0,\infty) \longrightarrow (0,\infty),$$

is continuous and nondecreasing with

$$\int_0^\infty \frac{ds}{\psi(s)} = +\infty.$$

Obviously, $\psi(s) = s$.

4. The functions

$$\sigma_i^j : [0,b] \longrightarrow L_{Q_i}(Y,X),$$

and

$$g^j : J \rightarrow L^0(Y,X),$$

are bounded for each $i = 1,2$, that is, there exists a positive constant L_1, c_1, such that

$$\sum_{l=1}^{\infty}\int_0^b \left\|\sigma_l^j(s)\right\|_{L_{Q_l}^0}^2 < L_1, \quad \forall b > 0.$$

and

$$\int_0^b \left\|g^j(s)\right\|_{L^0}^2 ds = C_1 < \infty, \quad t \in J.$$

If the associated conditions M_{tice} also converge to zero, then they satisfy all the assumptions given in Theorem 7.2, and we can conclude that problem (7.10) has at least one mild solution on $[0, b]$.

8 Stochastic functional differential equations with measure of noncompactness

We consider the following impulsive neutral stochastic functional differential equations driven by fBm with a noncompact semigroup of the form:

$$\begin{cases} d[x(t) - g_1(t, x_t, y_t)] = A_1 x(t) + f_1(t, x_t, y_t)dt + h_1(t, x_t, y_t)dW_1(t) + \sigma_1(t)\frac{dB_1^H(t)}{dt}, \\ \quad t \in [0, T],\ t \neq t_k, \\ d[y(t) - g_2(t, x_t, y_t)] = A_2 x(t) + f_2(t, x_t, y_t)dt + h_2(t, x_t, y_t)dW_2(t) + \sigma_1(t)\frac{dB_2^H(t)}{dt}, \\ \quad t \in [0, T],\ t \neq t_k, \\ \Delta x(t_k) = I_k^1(x_{t_k}, y_{t_k}), \quad k = 1, 2, \dots, m, \\ \Delta y(t_k) = I_k^2(x_{t_k}, y_{t_k}), \quad k = 1, 2, \dots, m, \\ x_0 = \phi_1(t), \quad t \in (-\infty, 0], \\ y_0 = \phi_2(t), \quad t \in (-\infty, 0]. \end{cases} \tag{8.1}$$

Here, $x(\cdot), y(\cdot)$ takes the value in the separable Hilbert space X with inner product $\langle \cdot, \cdot \rangle$ induced by the norm $\| \cdot \|$;

$$A_i : D(A_i) \subset X \longrightarrow X,$$

is the infinitesimal generator of a strongly continuous semigroup of bounded linear operators $(S_i(t))_{t\geq 0}$ in X for each $i = 1, 2$. B_i^H is fractional Brownian motion on a real and separable Hilbert space Y, with the Hurst parameter $H \in (\frac{1}{2}, 1)$. Let

$$\{W_i(t),\ t \in \mathbb{R},\ i = 1, 2\},$$

be a standard cylindrical Wiener process with values in Y and with respect to a complete probability space $(\Omega, \mathcal{F}, \mathcal{F}_t, \mathbb{P})$ furnished with a family of right continuous and increasing σ-algebras

$$\{\mathcal{F}_t, t \in J\},$$

satisfying $\mathcal{F}_t \subset \mathcal{F}$. Here, $\mathcal{D}_{\mathcal{F}_0}$ is a linear space of a family of $\mathcal{F}_0$-measurable functions from $(-\infty, 0]$ into X, which will also be defined in the next section. The impulse times t_k satisfy

$$0 < t_1 < t_2 < \cdots < t_m < b.$$

As for x_t, we mean the segment solution, which is defined in the usual way, that is, if

$$x(\cdot, \cdot) : (-\infty, b] \times \Omega \to X,$$

https://doi.org/10.1515/9783112251898-008

then for any $t \geq 0$,

$$x_t(\cdot,\cdot) : (-\infty, 0] \times \Omega \to X,$$

is given by

$$x_t(\theta, \omega) = x(t + \theta, \omega), \quad \text{for } \theta \in (-\infty, 0],\ \omega \in \Omega,$$

and belong to some abstract phase space $\mathcal{D}_{\mathcal{F}_0}$, that is, a phase space defined axiomatically. Thus $\mathcal{D}_{\mathcal{F}_0}$ is a linear space of functions mapping $[0, b]$ into X endowed with a seminorm $\|\cdot\|_{\mathcal{D}_{\mathcal{F}_{b_1}}}$. Consider the following space of piecewise continuous functions:

$$\begin{aligned}\mathcal{D}_{\mathcal{F}_{b_1}} = \Big\{x\colon [0, b] \times \Omega \to X \text{ is random processes, } x|_{J_k}(\cdot, w) \in C(J_k, X) \\ \text{for } k = 1, \dots m, \text{ and there exist } x(t_k^-, .) \text{ and } x(t_k^+, .) \text{ with } x(t_k, w) = x(t_k^-, w), \\ k = 1, \dots, m, \text{ and } \sup_{t \in [0,b]} \mathbb{E}(|x(t)|^2) < \infty\Big\},\end{aligned}$$

where x_k denotes the restriction of x to $J_k = (t_{k-1}, t_k]$, $k = 1, 2, \dots, m$, and $J_0 = (-\infty, 0]$, $i = 1, 2$. We endow $\mathcal{D}_{\mathcal{F}_{b_1}}$ with a norm $\|\cdot\|_{\mathcal{D}_{\mathcal{F}_{b_1}}}$ on $\mathcal{D}_{\mathcal{F}_{b_1}}$ defined by

$$\|x\|_{\mathcal{D}_{\mathcal{F}_{b_1}}} = \sup_{0 \leq s \leq b} \Big(\sqrt{\mathbb{E}\|x(s)\|^2}\Big).$$

We will use an axiomatic definition of the phase space $\mathcal{D}_{\mathcal{F}_0}$.

Remark 8.1. Let $\phi \in \mathcal{D}_{\mathcal{F}_0}$ and $t \leq 0$. The notation ϕ_t represents the function defined by $\phi_t = \phi(t + \theta)$. Consequently, if the function $x(.)$ in axiom (i) is such that $x_0 = \phi$, then $x_t = \phi_t$. We observe that ϕ_t is well-defined for $t < 0$ since the domain of ϕ is $(-\infty, 0]$.

Remark 8.2. In retarded functional differential equations without impulses, the axioms of the abstract phase space $\mathcal{D}_{\mathcal{F}_0}$ include the continuity of the function $t \to x_t$. Due to the impulsive effect, this property is not satisfied in impulsive delay systems and, for this reason, has been eliminated in our abstract description of $\mathcal{D}_{\mathcal{F}_0}$.

Example 8.1. Let $(X, \|\cdot\|_X)$ be a Banach space and

$$\Psi \in C\big((-\infty, 0], X\big).$$

Assume that

$$h : (-\infty, 0] \to [0, +\infty),$$

is a continuous function with $l = \int_{-\infty}^{0} h(s)ds < \infty$, define

$$\|\Psi\|_X = \sup_{\theta \le 0} \sqrt{\left(\mathbb{E}|\Psi(\theta)|^2\right)} < \infty,$$

$$\mathcal{C}_{\mathcal{F}_0} = \Bigg\{ x\colon (-\infty, 0] \times \Omega \to X, \text{ for any, } a > 0 \quad \sqrt{\mathbb{E}\left(|\Psi(\theta)|^2\right)}$$

$$\text{is bounded and measurable function on } [-a, 0],$$

$$\text{and } \int_{-\infty}^{0} h(s) \sup_{s \in [0,\theta]} \sqrt{\mathbb{E}\left(|\Psi(\theta)|^2\right)} < \infty. \Bigg\}.$$

If $\mathcal{C}_{\mathcal{F}_0}$ is endowed with the norm

$$\|\phi\|_{\mathcal{D}_{\mathcal{F}_0}} = \int_{-\infty}^{0} h(s) \sup_{s \in [0,\theta]} \sqrt{\mathbb{E}\left(|\Psi(\theta)|^2\right)},$$

then

$$\left(\mathcal{C}_{\mathcal{F}_0}, \|\cdot\|_{\mathcal{D}_{\mathcal{F}_0}}\right),$$

is a Banach space that satisfies the axioms in Definition 7.1 with $\hat{K} = l, \hat{N} = 1$.

Now, for a given $b > 0$, we define

$$\mathcal{D}_{\mathcal{F}_T} = \{x\colon (-\infty, b] \times \Omega \to X,\ x \in \mathcal{D}_{\mathcal{F}_0} \cap \mathcal{D}_{\mathcal{F}_{b_1}}\}.$$

Endowed with the norm

$$\|x\|_{\mathcal{D}_{\mathcal{F}_T}} = \|x_0\|_{\mathcal{D}_{\mathcal{F}_0}} + \sup_{0 \le s \le b}\left(\sqrt{\mathbb{E}\|x(s)\|^2}\right), \quad x \in \mathcal{D}_{\mathcal{F}_b}.$$

Then we will consider our initial data

$$(x_0, y_0) = \phi, \bar{\phi} \in \mathcal{D}_{\mathcal{F}_0}.$$

Let Y be another real separable Hilbert space, and suppose that $B_Q^H = B^H$ is a Y-valued fractional Brownian motion with increment covariance given by a non-negative trace class operator Q_i (see the next section for more details). We denote by $\mathcal{L}(Y, X)$ the space of all bounded, continuous, and linear operators from Y into X.

Then we assume that

$$h^i : J \times \mathcal{D}_{\mathcal{F}_0} \times \mathcal{D}_{\mathcal{F}_0} \to X,$$
$$f^i : J \times \mathcal{D}_{\mathcal{F}_0} \times \mathcal{D}_{\mathcal{F}_0} \to X,$$

and

$$\sigma^i : J \to L^0_{Q_i}(Y, X),$$

and

$$g^i : J \to L^0(Y,X).$$

Here, $L^0_{Q_i}(Y,X)$ denotes the space of all Q_i-Hilbert–Schmidt operators from Y into X, and

$$L^0(Y,X) = L_2(Q^{1/2}Y,X),$$

is the space of all Hilbert–Schmidt operators from $Q^{1/2}Y$ to X with the inner product

$$\langle \psi, \varphi \rangle_{L^0} = \mathrm{tr}(\psi Q \varphi^*),$$

and

$$L^0_{Q_i} \neq L^0.$$

As for the impulse functions, we will assume that

$$\begin{aligned} I_k, \bar{I}_k &\in C(\mathcal{D}_{\mathcal{F}_T} \times \mathcal{D}_{\mathcal{F}_T}, X) \quad (k = 1, \dots, m), \\ \Delta x|_{t=t_k} &= x(t_k^+) - x(t_k^-), \\ x(t_k^+) &= \lim_{h \to 0^+} x(t_k + h), \end{aligned}$$

and

$$x(t_k^-) = \lim_{h \to 0^+} x(t_k - h).$$

8.1 Global existence result

Recall that $(\Omega, \mathcal{F}, \mathcal{F}_t, P)$ is a complete probability space furnished with a family of right continuous increasing σ-algebras

$$\{\mathcal{F}_t,\ t \in J\},$$

satisfying $\mathcal{F}_t \subset \mathcal{F}$.

Definition 8.1. Given $H \in (0,1)$, a continuous centered Gaussian process

$$\beta^H = \{\beta^H(t),\ t \in \mathbb{R}\},$$

with the covariance function

$$R_H(t,s) = E[\beta^H(t)\beta^H(s)] = \frac{1}{2}(|t|^{2H} + |s|^{2H} - |t-s|^{2H}), \quad t, s \in \mathbb{R},$$

is called a two-sided one-dimensional fractional Brownian motion, and H is the Hurst parameter.

Notice that if

$$\sum_{n=1}^{\infty} \|\varphi Q^{1/2} e_n\|_{L^{1/H}([0,T];X)} < \infty, \tag{8.2}$$

Lemma 8.1. *If*

$$\varphi : [0,T] \longrightarrow L_Q^0(Y,X),$$

satisfies

$$\int_0^T \|\varphi(s)\|^2_{L_Q^0(Y,X)} ds < \infty,$$

then the series in (8.2) is well-defined as a X-valued random variable, and we have

$$E\left\| \int_0^t \varphi(s) dB_Q^H(s) \right\|^2 \leq 2Ht^{2H-1} \int_0^t \|\varphi(s)\|^2_{L_Q^0(Y,X)} ds.$$

Lemma 8.2. 1. *If*

$$0 < \beta < \alpha \leq 1,$$

then $\mathcal{H}_\alpha \subset \mathcal{H}_\beta$ and the embedding is compact whenever the resolvent operator of A is compact.

2. *For each $0 < \alpha \leq 1$, there exists a positive constant C_α such that*

$$\|(-A)^\alpha S(t)\| \leq \frac{C_\alpha}{t^\alpha} e^{-\lambda t}, \quad t > 0,\ \lambda > 0.$$

Next, we introduce the Hausdorff measure of non compactness $\chi(\cdot)$ defined on each bounded subset C of Banach space X,

$$\chi(C) = \{\epsilon \in \mathbb{R}_+^n : \text{there exists } n \in \mathbb{N} \text{ such that } C \text{ has finite } \epsilon\text{-net}\}.$$

Some basic properties of $\chi(\cdot)$ are listed in the following lemma.

Lemma 8.3. *Let X be a real Banach space and C_0, c_1 bounded; then the following properties are satisfied:*

1. *Monotone if*

$$C_0, C_1 \in \mathcal{P}(X), \quad C_0 \subset C_1,$$

implies

$$\beta(C_0) \le \beta(C_1).$$

2. *Nonsingular if*

$$\chi(\{a\} \cup C) = \chi(C_0),$$

for every $a \in X$, $C \in \mathcal{P}(X)$.
3. *Invariant with respect to the union with compact sets if*

$$\beta(K \cup C) = \beta(C),$$

for every relatively compact set $K \subset X$, *and* $C \in \mathcal{P}(X)$.
4. *Semiadditive if*

$$\chi(C_0 \cup C_1) = \max(\chi(C_0), \chi(C_1)),$$

for every $C_0, c_1 \in \mathcal{P}(X)$.
5. *Lower additive if* χ *is real and*

$$\chi(C_0 + C_1) \le \chi(C_0) + \chi(C_1),$$

for every $C_0, c_1 \in \mathcal{P}(X)$.
6. *Regular if the condition* $\chi(C) = 0$ *is equivalent to the relative compactness of* C.
7. *If* $f \in C([0,T])$ *is bounded, then*

$$\chi(f(t)) \le \chi(f) \quad \textit{for all } t \in [0,T],$$

where

$$f(t) = \{u(t) : u \in f \subset X\}.$$

Further, if f *is equicontinuous on* $[0,T]$, *then* $t \to f(t)$ *is continuous on* $[0,T]$, *and*

$$\chi(f) = \sup\{f(t) : t \in [0,T]\}.$$

8. *If* $f \in C([0,T], X)$ *is bounded and equicontinuous, then* $t \to \chi(W(t))$ *is continuous on* $[0,T]$ *and*

$$\chi\left(\int_0^t f(s)ds\right) \le \int_0^t \chi(f(s))ds \quad \textit{for all } t \in [0,T],$$

where

$$\int_0^t f(s)ds = \left\{\int u(s)ds : u \in f\right\}.$$

9. *Let $\{u_n\}_{n=1}^{\infty}$ be a sequence of Bochner integrable functions from J to X with $\|u_n\| \le \widehat{m}(t)$ for almost all $t \in J$ and every $n \ge 1$, where $\widehat{m}(t) \in L(J, R+)$. Then the function*

$$\psi(t) = \chi(\{u_n\}_{n=1}^{\infty}) \in L(J, R^+),$$

and satisfies

$$\chi\left\{\int_0^t u_n(s)ds : n \ge 1\right\} \le 2\int_0^t \psi(s)ds.$$

Here, we denote by χ the Hausdorff measure of noncompactness of $C([0,T],X)$ and by $\chi_{\mathcal{D}_{\mathcal{F}_T}}$ the Hausdorff measure of noncompactness of $\mathcal{D}_{\mathcal{F}_T}$. To discuss the existence, we need the following results.

Lemma 8.4 ([2]). *If $f_i \in C([0,T],X)$ is bounded, then*

$$\chi(f_i(t)) \le \chi_C(f_i),$$

for any $t \in [0,T]$, where

$$f_i(t) = \{u_i(t) : u_i \in f_i \subseteq X\}.$$

Furthermore, if f_i is piecewise equicontinuous on $[0,T]$, then $f(t)$ is continuous for $t \in [0,T]$ and

$$\chi_C(f_i(t)) = \sup\{f_i(t) \ : \ t \in [0,T]\}.$$

Lemma 8.5 ([2]). *If $f_i \in \mathcal{D}_{\mathcal{F}_{b_1}}$ is bounded, then*

$$\chi(f_i(t)) \le \chi_{\mathcal{D}_{\mathcal{F}_T}}(f_i),$$

for any $t \in [0,T]$, where

$$f_i(t) = \{u_i(t) : u_i \in f_i \subseteq X\}.$$

Furthermore, if f_i is piecewise equicontinuous on $[0,T]$, then $f_i(t)$ is continuous for $t \in [0,T]$ and

$$\chi_{\mathcal{D}_{\mathcal{F}_T}}(f_i(t)) = \sup\{f_i(t) : t \in [0,T]\}.$$

Definition 8.2. Let X, Y be two generalized normed spaces and

$$F : X \to \mathcal{P}(Y),$$

be a multivalued map. F is called M_{trice}-contraction (with respect to χ) if there exists $M_{trice} \in \mathcal{M}(\mathbb{R}_+)$ converging to zero such that, for every $D \in \mathcal{P}(X)$, we have

$$\chi(F(D)) \le M_{trice}\chi(D).$$

Lemma 8.6 ([2]). *If*

$$f_i \in C([0,T], L^0_{Q_i}(Y,X)),$$

W_i is a standard Wiener process, then

$$\chi\left(\int_0^t f_i(s)dW_i(s)\right) = \sqrt{T}\chi(f_i(s)),$$

where

$$\int_0^t f_i(s)dW_i(s) = \left\{\int_0^t u(s)dW_i(s) : \mathit{for\ all}\ u_i \in f_i,\ t \in [0,T]\right\}.$$

Lemma 8.7 ([2]). *Suppose that D is a closed convex subset of X, $0 \in D$. If the map $N : D \to X$ is continuous and of Mönch type, that is, N satisfies the property*

$$M \subset D, M \text{ is countable},\ M \subset \overline{co}(\{0\} \cup N(M)) \quad \Longrightarrow \quad \overline{M} \text{ is compact},$$

then N has a fixed point in D.

Remark 8.3. The function σ_i is independent of

$$x(t), y(t), \quad t \in [0,T].$$

From the functional point of view, we know that

$$\chi\left(\int_0^t S(t-s)\sigma_i(s)dB_1^H(s)\right) = 0.$$

8.2 Existence results

In this section, we present and prove the existence of mild solutions of system (8.1). To develop our results, we first need to introduce the following concept of a mild solution.

Definition 8.3. A X-valued stochastic process

$$u = (x, y) \in \mathcal{D}_{\mathcal{F}_T} \times \mathcal{D}_{\mathcal{F}_T},$$

is called a mild solution of the problem (8.1) with respect to the probability space $(\Omega, \mathcal{F}, \mathbb{P})$, if:

1. $u(t)$ is $\mathcal{F}_t$-adapted for all $t \in J_k = (t_k, t_{k+1}]$, $k = 1, 2, \ldots, m$.
2. $u(t)$ is right continuous and has a limit on the left almost surely.
3. The function

$$A_i S_i(t-s) g_i(s, u_s), \quad i = 1, 2,$$

   is integrable.
4. $u(t)$ satisfies for all $t \in (-\infty, T]$ and almost surely that

$$\begin{cases} x(t) = S_1(t)(\phi_1(0) - g_1(0, \phi_1(0), \phi_2(0))) + g_1(t, x_t, y_t) \\ \quad + \int_0^t A_1 S_1(t-s) g_1(s, x_s, y_s) ds \\ \quad + \int_0^t S_1(t-s) f_1(s, x_s, y_s) ds + \int_0^t S_1(t-s) G_1(s, x_s, y_s) dW_1(s) \\ \quad + \sum_{0<t_k<t} S_1(t-t_k) I_k^1(x_{t_k}, y_{t_k}) + \int_0^t S(t-s)\sigma_1(s) dB_1^H(s), \quad \mathbb{P}\text{-a.s.}, \ t \in J, \end{cases}$$

   and

$$\begin{cases} y(t) = S(t)(\phi_2(0) - g_2(0, \phi_1(0), \phi_2(0))) + g_2(t, x_t, y_t) \\ \quad + \int_0^t A_1 S_2(t-s) g_1(s, x_s, y_s) ds \\ \quad + \int_0^t S_2(t-s) f_2(s, x_s, y_s) ds + \int_0^t S_2(t-s) G_2(s, x_s, y_s) dW_2(s) \\ \quad + \sum_{0<t_k<t} S_2(t-t_k) I_k^2(x_{t_k}, y_{t_k}) + \int_0^t S_2(t-s)\sigma_2(s) dB_2^H(s), \quad \mathbb{P}\text{-a.s.}, \ t \in J. \end{cases}$$

   $(x_0(\cdot), y_0(\cdot)) = \phi_1, \phi_2 \in \mathcal{D}$ on J_0 satisfies $\|\phi_i\|_{\mathcal{D}} < \infty$.

We are now in a position to state and prove our existence result for the problem (8.1). First, we will list the following hypotheses that will be imposed in our main theorem.

(H_{34}) A_i is the infinitesimal generator of an analytic semigroup of bounded linear operators $S_i(t)$, $t \geq 0$, for each $i = 1, 2$ and there exists a constant M such that

$$\{\|S(t)\|^2 \leq M\}, \quad t \geq 0,$$

and

$$\|(-A_i)^{1-\beta} S_i(t)\| \leq \frac{M_{1-\beta}}{t^{1-\beta}}, \quad t > 0.$$

(H_{35}) There exist positive constants β such that g_i is X_β-valued, $(-A)^\beta g_i$ is continuous, and

a. There exist constants $\widetilde{L_{g_i}} \in \mathbb{R}^+$ for each $i = 1, 2$ such that

$$\mathbb{E}\left|(-A_i)^\beta g_i(t, x_1, y_1)\right|^2 \le \tilde{L}_{g_i}(1 + \|x_1\|^2_{\mathcal{D}_{\mathcal{F}_0}} + \|y_1\|^2_{\mathcal{D}_{\mathcal{F}_0}}), \quad t \in J,\ x_1, y_1 \in \mathcal{D}_{\mathcal{F}_0},$$
$$\mathbb{E}\left|(-A_i)^\beta g_i(t, x_1, y_1) - (-A_i)^\beta g_i(t, x_2, y_2)\right|^2 \le L_{g_i}\|x_1 - x_2\|^2_{\mathcal{D}_{\mathcal{F}_0}} + \bar{L}_{g_i}\|y_1 - y_2\|^2_{\mathcal{D}_{\mathcal{F}_0}},$$

for all x_1, y_1 and $x_2, y_2 \in \mathcal{D}_{\mathcal{F}_0}$ and $t \in J$

(H_{36}) There exists a positive function

$$l_{g_i}, \bar{l}_{g_i} \in L^1(J, \mathbb{R}^+),$$

for any bounded subsets $D_1, d_2 \subset \mathcal{D}_{\mathcal{F}_0}$ satisfies

$$\chi((-A_i)^\beta g_i(t, d_1, d_2)) \le l_{g_i} \sup_{\theta \in (-\infty, 0]} \chi(D_1(\theta)) + \bar{l}_{g_i} \sup_{\theta \in (-\infty, 0]} \chi(D_2(\theta)).$$

(H_{37}) The function

$$f_i : J \times \mathcal{D}_{\mathcal{F}_0} \times \mathcal{D}_{\mathcal{F}_0} \to X,$$

satisfies

a. For the initial value $\phi_i \in \mathcal{D}_{\mathcal{F}_0}$, there exists a continuous nondecreasing function

$$\psi_{f_i} : [0, \infty) \to [0, \infty),$$

and

$$p_{f_i} \in L^1(J, \mathbb{R}_+),$$

such that

$$\mathbb{E}\left|f_i(t, x, y)\right|^2 \le p_{f_i}(t)\psi_{f_i}(\|x\|^2_{\mathcal{D}_{\mathcal{F}_0}} + \|y\|^2_{\mathcal{D}_{\mathcal{F}_0}}),$$

for a. e. $t \in J$ and $x, y \in \mathcal{D}_{\mathcal{F}_0}$.

b. There exists a positive function

$$l_{f_i}, \bar{l}_{f_i} \in L^1(J, \mathbb{R}^+),$$

for any bounded subsets

$$D_1, d_2 \subset \mathcal{D}_{\mathcal{F}_0},$$

that satisfies

$$\chi(f_i(t, d_1, d_2)) \le l_{f_i} \sup_{\theta \in (-\infty, 0]} \chi(D_1(\theta)) + \bar{l}_{f_i} \sup_{\theta \in (-\infty, 0]} \chi(D_2(\theta)).$$

c. f_i is a L^2-Caratheodory map and for every $t \in [0,T]$ the function

$$t \to f(t,x,y), \quad x,y \in \mathcal{D}_{\mathcal{F}_0},$$

is measurable.

(H_{38}) The function

$$h_i : J \times \mathcal{D}_{\mathcal{F}_0} \times \mathcal{D}_{\mathcal{F}_0} \to L_Q^0(Y,X),$$

satisfies

a. For the initial value

$$\phi_i \in \mathcal{D}_{\mathcal{F}_0},$$

there exists a continuous nondecreasing function

$$\psi_{h_i} : [0,\infty) \to [0,\infty),$$

and

$$p_{h_i} \in L^1(J,\mathbb{R}_+),$$

such that

$$\mathbb{E}|h_i(t,x,y)|^2 \le p_{h_i}(t)\psi_{h_i}(\|x\|^2_{\mathcal{D}_{\mathcal{F}_0}} + \|y\|^2_{\mathcal{D}_{\mathcal{F}_0}}),$$

for a. e. $t \in J$ and $x,y \in \mathcal{D}_{\mathcal{F}_0}$.

(H_{39}) There exists a positive function

$$l_{h_i}, \bar{l}_{h_i} \in L^1(J,\mathbb{R}^+),$$

for any bounded subsets

$$D_1, d_2 \subset \mathcal{D}_{\mathcal{F}_0},$$

that satisfies

$$\chi(h_i(t,d_1,d_2)) \le l_{h_i} \sup_{\theta\in(-\infty,0]} \chi(D_1(\theta)) + \bar{l}_{h_i} \sup_{\theta\in(-\infty,0]} \chi(D_2(\theta)).$$

(H_{40}) The continuous function

$$I_k^i : \mathcal{D}_{\mathcal{F}_T} \times \mathcal{D}_{\mathcal{F}_T} \to X,$$

satisfies

a. There exist $c_k^i, \bar{c}_k^i > 0$, $k = 1, 2, \ldots, m$ such that for any $x, y \in \mathcal{D}_{\mathcal{F}_T}$ and

$$\mathbb{E}\|I_k^1(x_1, y_1) - I_k^1(x_2, y_2)\| \leq c_k^i \|x_1 - x_2\|^2_{\mathcal{D}_{\mathcal{F}_T}} + \bar{c}_k^i \|x_1 - x_2\|^2_{\mathcal{D}_{\mathcal{F}_T}}.$$

b. There exists a positive function $l_{k_i}, \bar{l}_{k_i} > 0$, for any bounded subsets $D_1, d_2 \subset \mathcal{D}_{\mathcal{F}_0}$ that satisfies

$$\chi(I_k^i(t, d_1, d_2)) \leq l_{k_i} \sup_{\theta \in (-\infty, 0]} \chi(D_1(\theta)) + \bar{l}_{h_i} \sup_{\theta \in (-\infty, 0]} \chi(D_2(\theta)).$$

c.
$$28\left(M^2 \|(-A_i)^{-\beta}\|^2 \widetilde{L}_{g_i} + \frac{M_{1-\beta}^2 T^{2\beta}}{2\beta - 1} \widetilde{L}_{g_i} + \frac{1}{2} M^2 m \sum_{k=1}^{m} (c_k^i + \bar{c}_k^i) \right) \widetilde{K}^2$$
$$+ 7M^2 \left(T \int_0^t p_{f_i}(s) \limsup_{\widetilde{K} \to \infty} \frac{\psi_{f_i}(2\widetilde{K})}{\widetilde{K}} ds + 7M^2 \int_0^t p_{h_i}(s) \limsup_{\widetilde{K} \to \infty} \frac{\psi_{h_i}(2\widetilde{K})}{\widetilde{K}} ds \right) < 1,$$

and

$$M_{b,c} = \begin{pmatrix} \alpha_1 & \alpha_2 \\ \alpha_3 & \alpha_4 \end{pmatrix}, \quad \alpha_i \geq 0, \ i = 1, \ldots, 4.$$

Theorem 8.1. *Assume that hypotheses* $(H_{34}) - (H_{40})$ *hold. If* $M_{b,c}$ *converges to zero. Then the problem* (8.1) *has at least one integral solution on* $(-\infty, T]$.

Proof. We first transform the problem (8.1) into a fixed-point one. Consider the operator

$$N = (N_1, N_2) : \mathcal{D}_{\mathcal{F}_T} \times \mathcal{D}_{\mathcal{F}_T} \to \mathcal{D}_{\mathcal{F}_T} \times \mathcal{D}_{\mathcal{F}_T},$$

defined by

$$N_1(x, y)(t) = \begin{cases} \phi_1(t), \quad t \in (-\infty, 0], \\ S_1(t)(\phi_1(0) - g_1(0, \phi_1(0), \phi_2(0))) + g_1(t, x_t, y_t) + \int_0^t A_1 S_1(t - s) g_1(s, x_s, y_s) ds \\ \quad + \int_0^t S_1(t - s) f_1(s, x_s, y_s) ds + \int_0^t S_1(t - s) h_1(s, x_s, y_s) dW_1(s) \\ \quad + \sum_{0 < t_k < t} S_1(t - t_k) I_k^1(x_{t_k}, y_{t_k}) + \int_0^t S_1(t - s) \sigma_1(s) dB_1^H(s), \quad \mathbb{P}\text{-a.s.}, \ t \in J, \end{cases}$$

and

$$N_2(x, y)(t) = \begin{cases} \phi_2(t), \quad t \in (-\infty, 0], \\ S_2(t)(\phi_2(0) - g_2(0, \phi_1(0), \phi_2(0))) + g_2(t, x_t, y_t) \\ \quad + \int_0^t A_2 S_2(t - s) g_2(s, x_s, y_s) ds \\ \quad + \int_0^t S_2(t - s) f_2(s, x_s, y_s) ds + \int_0^t S_2(t - s) h_2(s, x_s, y_s) dW_2(s) \\ \quad + \sum_{0 < t_k < t} S_2(t - t_k) I_k^2(x_{t_k}, y_{t_k}) \\ \quad + \int_0^t S_2(t - s) \sigma_2(s) dB_2^H(s), \quad \mathbb{P}\text{-a.s.}, \ t \in J. \end{cases}$$

For $\phi_1, \phi_2 \in \mathcal{D}_{\mathcal{F}_0}$, we define $\widehat{\phi}$ by

$$\widehat{\phi}^i(t) = \begin{cases} \phi_i(t), & t \in (-\infty, 0], \\ S_i(t)\phi_i(0), & t \in [0, T]; \end{cases}$$

then $\widehat{\phi}_i \in \mathcal{D}_{\mathcal{F}_T}$.

Let

$$(x(t), y(t)) = (z^1(t) + \widehat{\phi}^1(t), z^2(t) + \widehat{\phi}^2(t)), \quad -\infty < t \leq T.$$

It is evident that z^1, z^2 satisfies

$$z_0^1 = z_0^2 = 0,$$

for all $t \in (-\infty, 0]$,

$$z^i(t) = \begin{cases} S_i(t)(-g_i(0, \phi_1(0), \phi_2(0))) + g_i(t, z_t^1 + \widehat{\phi}_t^1, z_t^2 + \widehat{\phi}_t^2) \\ \quad + \int_0^t A_i S_i(t-s) g_i(s, z_s^1 + \widehat{\phi}_s^1, z_s^2 + \widehat{\phi}_s^2) ds \\ \quad + \int_0^t S_i(t-s) f_i(s, z_s^1 + \widehat{\phi}_s^1, z_s^2 + \widehat{\phi}_s^2) ds \\ \quad + \int_0^t S_i(t-s) h_i(s, z_s^1 + \widehat{\phi}_s^1, z_s^2 + \widehat{\phi}_s^2) dW_i(s) \\ \quad + \sum_{0<t_k<t} S_i(t-t_k) I_k^i(z_{t_k}^1 + \widehat{\phi}_{t_k}^1, z_{t_k}^2 + \widehat{\phi}_{t_k}^2) \\ \quad + \int_0^t S_i(t-s)\sigma_i(s) dB_i^H(s), \quad \mathbb{P}\text{-a. s., } t \in J. \end{cases}$$

Set

$$\mathcal{D}'_{\mathcal{F}_T} = \{z^i \in \mathcal{D}_{\mathcal{F}_T}, \text{ such that } z_0^i = 0 \in \mathcal{D}_{\mathcal{F}_0}\},$$

and for any $z^j \in \mathcal{D}_{\mathcal{F}_0}$, we have

$$\|z^i\|_{\mathcal{F}_T} = \|z_0^i\|_{\mathcal{D}_{\mathcal{F}_0}} + \sup_{t \in J} (\mathbb{E}\|z^i(t)\|^2)^{\frac{1}{2}} = \sup_{t \in J} (\mathbb{E}\|z^i(t)\|^2)^{\frac{1}{2}}. \tag{8.3}$$

It is not difficult to check that

$$(\widehat{\mathcal{D}}_{\mathcal{F}_b}, \|\cdot\|_{\widehat{\mathcal{D}}_{\mathcal{F}_b}})$$

is a Banach space.

Let us first define

$$\mathcal{B}_q = \left\{z \in \mathcal{D}'_{\mathcal{F}_T}, \|z^i\|^2_{\mathcal{D}'_{\mathcal{F}_T}} \leq q\right\},$$

where $q > 0$ is a given number. The set $\mathcal{B}_q$ is clearly a bounded, closed, convex set in $\mathcal{D}'_{\mathcal{F}_T}$. We then have

$$
\begin{aligned}
\|z_t^i + \widehat{\phi}_t^i\|^2_{\mathcal{D}_{\mathcal{F}_0}} &\le 2(\|z_t^i\|^2_{\mathcal{D}_{\mathcal{F}_0}} + \|\widehat{\phi}_t^i\|^2_{\mathcal{D}_{\mathcal{F}_0}}) \\
&\le 4\Big(\widetilde{K}^2 \sup_{s\in[0,T]} \mathbb{E}|z^i(s)|^2 + \widetilde{K}^2 M^2 E\|\phi_i(0)\|^2 + \widetilde{N}^2 \|\phi_i\|^2_{\mathcal{D}_{\mathcal{F}_0}}\Big) \\
&\le 4(\widetilde{K}^2 q + (\widetilde{K}^2 M^2 \mathcal{L}^2 + \widetilde{N}^2)\|\phi_i\|^2_{\mathcal{D}_{\mathcal{F}_0}}) \\
&= q'.
\end{aligned} \tag{8.4}
$$

Let the operator

$$\widehat{N} : \mathcal{D}'_{\mathcal{F}_T} \times \mathcal{D}'_{\mathcal{F}_T} \to \mathcal{D}'_{\mathcal{F}_T} \times \mathcal{D}'_{\mathcal{F}_T},$$

be defined by

$$(\widehat{N}_1(z^1, z^2)(t), \widehat{N}_2(z^1, z^2)(t)) = (0, 0), \quad t \in (-\infty, 0],$$

and

$$
\widehat{N}_i(z^1, z^2)(t) = \begin{cases}
S_i(t)(-g_i(0, \phi_1(0), \phi_2(0))) + g_i(t, z_t^1 + \widehat{\phi}_t^1, z_t^2 + \widehat{\phi}_t^2) \\
\quad + \int_0^t A_i S_i(t-s) g_i(s, z_s^1 + \widehat{\phi}_s^1, z_s^2 + \widehat{\phi}_s^2) ds \\
\quad + \int_0^t S_i(t-s) f_i(s, z_s^1 + \widehat{\phi}_s^1, z_s^2 + \widehat{\phi}_s^2) ds \\
\quad + \int_0^t S_i(t-s) h_i(s, z_s^1 + \widehat{\phi}_s^1, z_s^2 + \widehat{\phi}_s^2) dW_i(s) \\
\quad + \sum_{0<t_k<t} S_i(t-t_k) I_k^i(z_{t_k}^1 + \widehat{\phi}_{t_k}^1, z_{t_k}^2 \widehat{\phi}_{t_k}^2) \\
\quad + \int_0^t S_i(t-s)\sigma_i(s) dB_i^H(s), \quad \mathbb{P}\text{-a. s.}, \ t \in J.
\end{cases}
$$

Obviously, the operator N has a fixed point is equivalent to proving that $\widehat{N}$ has a fixed point. For better readability, we break the proof into a sequence of steps.

Step 1: There exists a positive number q such that

$$\widehat{N}(\mathcal{B}_q \times \mathcal{B}_q) \subset \mathcal{B}_q \times \mathcal{B}_q.$$

Let us assume that

$$\widehat{N}(\mathcal{B}_q \times \mathcal{B}_q) \not\subset \mathcal{B}_q \times \mathcal{B}_q,$$

then for each positive number q, there exists a function $z^{1,q} \in \mathcal{B}_q$ but

$$\widehat{N}(\mathcal{B}_q \times \mathcal{B}_q) \notin \mathcal{B}_q \times \mathcal{B}_q,$$

that implies that for some $t = t(q) \in J$ and

$$\mathbb{E}\|\widehat{N}_i(z^1, z^2)(t)\|^2 > q.$$

In fact, we have

$$
\begin{aligned}
q &\le \mathbb{E}\big|\widehat{N}_i(z^1, z^2)(t)\big|^2 \\
&\le 7\mathbb{E}\big\|S_i(t)(-g_i(0, \phi_1(0), \phi_2(0)))\big\|^2 + \mathbb{E}\big\|g_i(t, z_t^1 + \widehat{\phi}_t^1, z_t^2 + \widehat{\phi}_t^2)\big\|^2 \\
&\quad + 7\mathbb{E}\left\|\int_0^t A_i S_i(t-s) g_i(s, z_s^1 + \widehat{\phi}_s^1, z_s^2 + \widehat{\phi}_s^2) ds\right\|^2 \\
&\quad + 7\mathbb{E}\left\|\int_0^t S_i(t-s) f_i(s, z_s^1 + \widehat{\phi}_s^1, z_s^2 + \widehat{\phi}_s^2) ds\right\|^2 \\
&\quad + 7\mathbb{E}\left\|\int_0^t S_i(t-s) h_i(s, z_s^1 + \widehat{\phi}_s^1, z_s^2 + \widehat{\phi}_s^2) dW_i(s)\right\|^2 \\
&\quad + 7\mathbb{E}\left\|\sum_{0<t_k<t} S_i(t-t_k) I_k^j(z_{t_k}^1 + \widehat{\phi}_{t_k}^1, z_{t_k}^2 + \widehat{\phi}_{t_k}^2)\right\|^2 \\
&\quad + 7\mathbb{E}\left\|\int_0^t S_i(t-s)\sigma_i(s) dB_i^H(s)\right\|^2 \\
&= 7\sum_{j=1}^{7} I_{j,i}.
\end{aligned}
$$

From (H_{34}) and (H_{35}), we have

$$
\begin{aligned}
I_{1,i} &= \mathbb{E}\big\|S_i(t)(-g_i(0, \phi_1(0), \phi_2(0)))\big\|^2 \\
&\le M^2 \big\|(-A_i)^{-\beta}\big\|^2 \widetilde{L}_{g_i}(1 + \|\phi_1\|^2_{\mathcal{D}_{\mathcal{F}_0}} + \|\phi_2\|^2_{\mathcal{D}_{\mathcal{F}_0}}) \qquad (8.5) \\
I_{2,i} &= \mathbb{E}\big\|g_i(t, z_t^1 + \widehat{\phi}_t^1, z_t^2 + \widehat{\phi}_t^2)\big\|^2 \\
&\le M^2 \big\|(-A_i)^{-\beta}\big\|^2 \widetilde{L}_{g_i}\left(1 + \sum_{i=1}^{2} \|z_t^i + \widehat{\phi}_t^i\|^2_{\mathcal{D}_{\mathcal{F}_0}}\right) \\
&\le M^2 \big\|(-A_i)^{-\beta}\big\|^2 \widetilde{L}_{g_i}(1 + 2q), \qquad (8.6) \\
I_{3,i} &= \mathbb{E}\left\|\int_0^t A_i S_i(t-s) g_i(s, z_s^1 + \widehat{\phi}_s^1, z_s^2 + \widehat{\phi}_s^2) ds\right\|^2 \\
&= \mathbb{E}\left\|\int_0^t (-A_i)^{1-\beta} S_i(t-s)(-A_i)^{\beta} g_i(s, z_s^1 + \widehat{\phi}_s^1, z_s^2 + \widehat{\phi}_s^2) ds\right\|^2 \\
&\le M_{1-\beta}^2 T \int_0^t (t-s)^{-2(1-\beta)} \widetilde{L}_{g_i}\left(1 + \sum_{j=1}^{2} \|z_s^i + \widehat{\phi}_s^i\|^2_{\mathcal{D}_{\mathcal{F}_0}}\right) ds \\
&\le \frac{M_{1-\beta}^2 T^{2\beta}}{2\beta - 1} \widetilde{L}_{g_i}(1 + 2q). \qquad (8.7)
\end{aligned}
$$

By Lemma 8.1 and the conditions $(H_{36})-(H_{40})$, we have

$$\begin{aligned}
I_{i,4} &= \mathbb{E}\left\|\int_0^t S_i(t-s)f_i(s,z_s^1+\widehat{\phi}_s^1,z_s^2+\widehat{\phi}_s^2)ds\right\|^2\\
&\le M^2T\int_0^t p_{f_i}(s)\psi_{f_i}\left(\sum_{j=1}^2\|z_s^j+\widehat{\phi}_s^j\|^2_{\mathcal{D}_{\mathcal{F}_0}}\right)\\
&\le M^2T\int_0^t p_{f_i}(s)\psi_{f_i}(2q')ds, \qquad (8.8)\\
I_{i,5} &= \mathbb{E}\left\|\int_0^t S_i(t-s)h_i(s,z_s^1+\widehat{\phi}_s^1,z_s^2+\widehat{\phi}_s^2)dW(s)\right\|^2\\
&\le M^2T\int_0^t p_{h_i}(s)\psi_{h_i}\left(\sum_{i=1}^2\|z_s^i+\widehat{\phi}_s^i\|^2_{\mathcal{D}_{\mathcal{F}_0}}\right)\\
&\le M^2T\int_0^t p_{h_i}(s)\psi_{h_i}(2q')ds, \qquad (8.9)\\
I_{i,6} &= \mathbb{E}\left\|\sum_{0<t_k<t} S_i(t-t_k)I_k^i(z_{t_k}^1+\widehat{\phi}_{t_k}^1,z_{t_k}^2+\widehat{\phi}_{t_k}^2)\right\|^2\\
&\le M^2m\sum_{k=1}^m r(c_k^i+\bar{c}_k^i), \qquad (8.10)\\
I_{i,7} &= \mathbb{E}\left\|\int_0^t S_i(t-s)\sigma_i(s)dB_i^H(s)\right\|^2\\
&\le M^2 2HT^{2H-1}\sup_{s\in[0,T]}\|\sigma_i(s)\|^2_{L_Q^0(\mathcal{K},\mathcal{H})}. \qquad (8.11)
\end{aligned}$$

Thus, it follows from (8.5)–(8.11)

$$\begin{aligned}
q \le \mathbb{E}|\widehat{N}_i(z^1,z^2)(t)|^2 &\le 7M^2\|(-A_i)^{-\beta}\|^2\widetilde{L}_{g_i}(1+\|\phi_1\|^2_{\mathcal{D}_{\mathcal{F}_0}}+\|\phi_2\|^2_{\mathcal{D}_{\mathcal{F}_0}})\\
&\quad+7M^2\|(-A_i)^{-\beta}\|^2\widetilde{L}_{g_i}(1+2q')+7\frac{M_{1-\beta}^2T^{2\beta}}{2\beta-1}\widetilde{L}_{g_1}(1+2q')\\
&\quad+7M^2T\int_0^t p_{f_i}(s)\psi_{f_i}(2q')ds+7M^2T\int_0^t p_{h_j}(s)\psi_{h_i}(2q')ds\\
&\quad+7M^2m\sum_{k=1}^m(c_k^iq'+\bar{c}_k^iq')+7M^2 2HT^{2H-1}\sup_{s\in[0,T]}\|\sigma_i(s)\|^2_{L_Q^0(\mathcal{K},\mathcal{H})}.
\end{aligned}$$

Therefore, we arrive at

$$\begin{aligned} q \leq 14\Bigg(M^2\|(-A_i)^{-\beta}\|^2\widetilde{L}_{g_i} + \frac{M_{1-\beta}^2 T^{2\beta}}{2\beta-1}\widetilde{L}_{g_i} + \frac{1}{2}M^2 m\sum_{k=1}^{m}(c_k^i + \bar{c}_k^i)\Bigg)q' \\ + 7M^2\|(-A_i)^{-\beta}\|^2\widetilde{L}_{g_i}(1 + \|\phi_1\|^2_{\mathcal{D}_{\mathcal{F}_0}} + \|\phi_2\|^2_{\mathcal{D}_{\mathcal{F}_0}}) \\ + 7M^2\|(-A_i)^{-\beta}\|^2\widetilde{L}_{g_i} + 7\frac{M_{1-\beta}^2 T^{2\beta}}{2\beta-1}\widetilde{L}_{g_i} + 7M^2 2HT^{2H-1}\sup_{s\in[0,T]}\|\sigma_i(s)\|^2_{L_Q^0(\mathcal{K},\mathcal{H})} \\ + 7M^2\Bigg(T\int_0^t p_{f_i}(s)\psi_{f_i}(2q')ds + 7M^2\int_0^t p_{h_i}(s)\psi_{h_i}(2q')ds\Bigg). \end{aligned} \tag{8.12}$$

Dividing both sides of (8.12) by q and letting $q \to \infty$, we have

$$\lim_{q\to\infty}\frac{q'}{q} = \lim_{r\to\infty}\frac{4(\widetilde{K}^2 q + (\widetilde{K}^2 M^2\mathcal{L}^2 + \widetilde{N}^2)\|\phi_i\|^2_{\mathcal{D}_{\mathcal{F}_0}})}{q} = 4\widetilde{K}^2.$$

Combining

$$\begin{aligned} 1 \leq 28\Bigg(M^2\|(-A_i)^{-\beta}\|^2\widetilde{L}_{g_i} + \frac{M_{1-\beta}^2 T^{2\beta}}{2\beta-1}\widetilde{L}_{g_i} + \frac{1}{2}M^2 m\sum_{k=1}^{m}(c_k^i + \bar{c}_k^i)\Bigg)\widetilde{K}^2 \\ + 7M^2\Bigg(T\int_0^t p_{f_i}(s)\limsup_{\widetilde{K}\to\infty}\frac{\psi_{f_i}(2\widetilde{K})}{\widetilde{K}}ds + 7M^2\int_0^t p_{h_i}(s)\limsup_{\widetilde{K}\to\infty}\frac{\psi_{h_i}(2\widetilde{K})}{\widetilde{K}}ds\Bigg). \end{aligned}$$

This contradicts (H_{44}). Therefore, there exists some positive q' such that

$$\widehat{N}(\mathcal{B}_q \times \mathcal{B}_q) \subset \mathcal{B}_q \times \mathcal{B}_q.$$

Step 2: $\widehat{\Phi}_2$ **is continuous.**

Let

$$\{(z^{n,1}, z^{n,2})\}_{n=1}^{\infty} \subseteq \mathcal{D}'_{\mathcal{F}_T} \times \mathcal{D}'_{\mathcal{F}_T},$$

be a sequence such that

$$(z^{n,1}, z^{n,2}) \to (z^1, z^2), \quad (n \to \infty),$$

in

$$\mathcal{D}'_{\mathcal{F}_T} \times \mathcal{D}'_{\mathcal{F}_T}.$$

Then there is a number $q > 0$ such that $\|z^{i,n}\|^2 \leq q$ for all n and a. e. $t \in J$, so $z^{1,n}, z^{2,n} \in \mathcal{B}_q$ and $z^{,n}, z^{,n} \in \mathcal{B}_q$. Again, similar to (8.4), we have

$$\|z_t^i + \widehat{\phi}_t^i\|^2_{\mathcal{D}_{\mathcal{F}_0}} \leq q'.$$

Furthermore, from axiom (1), we know that

$$\begin{aligned}\|z_t^{i,n} - z_t^i\|_{\mathcal{D}_{\mathcal{F}_0}} &\leq K(t)\sup\{\|z^{i,n}(s) - z^i(s)\| : 0 \leq s \leq t\} + N(t)\|\widehat{\phi}_t^{i,n} - \widehat{\phi}_t^i\|_{\mathcal{D}_{\mathcal{F}_0}} \\ &\leq K(t)\sup\{\|z^{i,n}(s) - z^i(s)\| : 0 \leq s \leq t\} \\ &\leq \widehat{K}\|z^{i,n} - z^i\| \to 0 \quad (n \to \infty), t \in [0,T].\end{aligned}$$

Thus, by (H_{36}) and (H_{37}), we see that, for $t \in [0,b]$,

$$f_i(t, z_t^{1,n} + \widehat{\phi}_t^1, z_t^{2,n} + \widehat{\phi}_t^2) \to f_i(t, z_t^1 + \widehat{\phi}_t^1, z_t^2 + \widehat{\phi}_t^2) \quad \text{as } n \to \infty,$$

and

$$h_i(t, z_t^{1,n} + \widehat{\phi}_t^1, z_t^{2,n} + \widehat{\phi}_t^2) \to f_i(t, z_t^1 + \widehat{\phi}_t^1, z_t^2 + \widehat{\phi}_t^2) \quad \text{as } n \to \infty,$$

for each $t \in [0,T]$, and since

$$\mathbb{E}|h_i(t, z_t^{1,n} + \widehat{\phi}_t^1, z_t^{2,n} + \widehat{\phi}_t^2) - h_i(t, z_t^1 + \widehat{\phi}_t^1, z_t^2 + \widehat{\phi}_t^2)|^2 \leq 2p_{h_i}(t)\psi_{h_i}(2q'),$$

and

$$\mathbb{E}|f_i(t, z_t^{1,n} + \widehat{\phi}_t^1, z_t^{2,n} + \widehat{\phi}_t^2) - f_i(t, z_t^1 + \widehat{\phi}_t^1, z_t^2 + \widehat{\phi}_t^2)|^2 \leq 2p_{f_i}(t)\psi_{f_i}(2q').$$

We have, by the dominated convergence theorem, that for any $t \in J$,

$$\begin{aligned}&\mathbb{E}\|\widehat{N}_i(z^{n,1}, z^{n,2})(t) - \widehat{N}_i(z^1, z^2)(t)\|^2 \\ &\leq 5\mathbb{E}|g_i(t, z_t^{n,1} + \widehat{\phi}_t^1, z_t^{n,2} + \widehat{\phi}_t^2) - g_i(t, z_t^1 + \widehat{\phi}_t^1, z_t^2 + \widehat{\phi}_t^2)|^2 \\ &\quad + \frac{5M_{1-\beta}^2 T^{2\beta}}{2\beta - 1}(L_{g_i}\|z_n^1 - z^1\|^2_{\mathcal{D}_{\mathcal{F}_0}} + \bar{L}_{g_j}\|z_t^{n,2} - z_t^{n,2}\|^2_{\mathcal{D}_{\mathcal{F}_0}}) \\ &\quad + 5TM^2 \int_0^t \mathbb{E}\|f_i(s, z_s^{n,1} + \widehat{\phi}_s^1, z_s^{n,2} + \widehat{\phi}_s^2) - f_i(s, z_s^1 + \widehat{\phi}_s^1, z_s^2 + \widehat{\phi}_s^2)\|^2 ds \\ &\quad + 5M^2 \int_0^t \mathbb{E}\|h_i(s, z_s^{n,1} + \widehat{\phi}_s^1, z_s^{n,2} + \widehat{\phi}_s^2) - h_i(s, z_s^1 + \widehat{\phi}_s^1, z_s^2 + \widehat{\phi}_s^2)\|^2 ds \\ &\quad + 5M^2 m \sum_{k=1}^m \mathbb{E}\|I_k^i(z_{t_k}^{n,1} + \widehat{\phi}_{t_k}^1, z_{t_k}^{n,2} + \widehat{\phi}_{t_k}^2) - I_k^i(z_{t_k}^1 + \widehat{\phi}_{t_k}^1, z_{t_k}^2 + \widehat{\phi}_{t_k}^2)\|^2.\end{aligned}$$

This implies that $\widehat{N}_j$ is continuous in $\mathcal{B}_q \times \mathcal{B}_q$.

Step 3: The operator $\widehat{N}(\mathcal{B}_q \times \mathcal{B}_q)$ is equicontinuous on J.

It is easy to see that the function

$$\{\widehat{N}_i(z^1,z^2) : z^1,z^2 \in \mathcal{B}_q\}, \quad t = 0,$$

is equicontinuous. For $0 < \tau_1 < \tau_2 < T$ and $z^1, z^2 \in \mathcal{B}_q$ from the above assumptions, we have

$$\begin{aligned}
&\mathbb{E}|\widehat{N}_i(z^1,z^2)(\tau_2) - \widehat{N}_i(z^1,z^2)(\tau_1)|^2 \\
&\quad \le 14\|S_i(\tau_2) - S(\tau_1)\|^2 \mathbb{E}\|g_i(0,\phi_1(0),\phi_2(0))\|^2 \\
&\quad + 14\mathbb{E}\|g_i(\tau_2, z^1_{\tau_2} + \widehat{\phi}^1_{\tau_2}, z^2_{\tau_2} + \widehat{\phi}^2_{\tau_2}) - g_i(\tau_1, z^1_{\tau_1} + \widehat{\phi}^1_{\tau_1}, z^2_{\tau_1} + \widehat{\phi}^2_{\tau_1})\|^2 \\
&\quad + 14\mathbb{E}\left\|\int_0^{\tau_1} (-A_i)^{1-\beta}(S_i(\tau_2 - s) - S_i(\tau_1 - s))(-A_i)^{-\beta} g_i(s, z^1_s + \widehat{\phi}^1_s, z^2_s + \widehat{\phi}^2_s)ds\right\|^2 \\
&\quad + 14\mathbb{E}\left\|\int_{\tau_1}^{\tau_2} A_i S_i(\tau_2 - s) g_i(s, z^1_s + \widehat{\phi}^1_s, z^2_s + \widehat{\phi}^2_s)ds\right\|^2 \\
&\quad + 14\mathbb{E}\left\|\int_0^{\tau_1} (S_i(\tau_2 - s) - S_i(\tau_1 - s)) f_i(s, z^1_s + \widehat{\phi}^1_s, z^2_s + \widehat{\phi}^2_s)ds\right\|^2 \\
&\quad + 14\mathbb{E}\left\|\int_{\tau_1}^{\tau_2} S_i(\tau_2 - s) f_i(s, z^1_s + \widehat{\phi}^1_s, z^2_s + \widehat{\phi}^2_s)ds\right\|^2 \\
&\quad + 14\mathbb{E}\left\|\int_0^{\tau_1} (S_i(\tau_2 - s) - S_i(\tau_1 - s)) h_i(s, z^1_s + \widehat{\phi}^1_s, z^2_s + \widehat{\phi}^2_s)dW(s)\right\|^2 \\
&\quad + 14\mathbb{E}\left\|\int_{\tau_1}^{\tau_2} S_i(\tau_2 - s) h_i(s, z^1_s + \widehat{\phi}^1_s, z^2_s + \widehat{\phi}^2_s)dW_i(s)\right\|^2 \\
&\quad + 14\mathbb{E}\left\|\sum_{0<t_k<\tau_1} (S_i(\tau_2 - t_k) - S_i(\tau_1 - t_k)) I^i_k(z^1_{t_k} + \widehat{\phi}^1_{t_k}, z^2_{t_k} + \widehat{\phi}^2_{t_k})\right\|^2 \\
&\quad + 14\mathbb{E}\left\|\sum_{\tau_1<t_k<\tau_2} S_i(\tau_2 - t_k) I^i_k(z^1_{t_k} + \widehat{\phi}^1_{t_k}, z^2_{t_k} + \widehat{\phi}^2_{t_k})\right\|^2 \\
&\quad + 14\mathbb{E}\left\|\int_0^{\tau_1} (S_i(\tau_2 - s) - S_i(\tau_1 - s))\sigma_i(s)dB^H_i(s)\right\|^2 \\
&\quad + 14\mathbb{E}\left\|\int_{\tau_1}^{\tau_2} (S_i(\tau_2 - s) - S_i(\tau_1 - s))\sigma_i(s)dB^H_i(s)\right\|^2 \to 0 \quad \text{as } \tau_2 \to \tau_1.
\end{aligned}$$

This implies that

$$\{\widehat{N}_i(z^1,z^2) : z^1,z^2 \in \mathcal{B}_q\},$$

is equicontinuous on J.

Step 4: We verify that Mönch condition is established.

Let V be a subset of $\mathcal{B}_q$ such that

$$V \subset \operatorname{conv}(\widehat{N}_i(V) \cup \{0\}).$$

Now, we prove that $\chi(V) = 0$. Without loss of generality, we assume that

$$V = \{(z^{1,n}, z^{2,n})\}_{n=1}^{\infty},$$

is bounded and equicontinuous; therefore, the function

$$t \to v(t) = \chi(V(t)),$$

is continuous on J. By $(H_{35})-(H_{39})$ and the properties of the measure χ, we have for each $t \in J$. From Lemma 8.6, we have

$$\chi(\widehat{N}(z^1,z^2)(t)) = \chi(N(z^1,z^2)(t)),$$

$$\widehat{N}_i(z^1,z^2)(t) = \begin{cases} [S_i(t)(-g_i(0,\phi_1(0),\phi_2(0))) + g_i(t, z_t^1 + \widehat{\phi}_t^1, z_t^2 + \widehat{\phi}_t^2) \\ \quad + \int_0^t A_i S_i(t-s) g_i(s, z_s^1 + \widehat{\phi}_s^1, z_s^2 + \widehat{\phi}_s^2) ds] \\ \quad + [\int_0^t S_i(t-s) f_i(s, z_s^1 + \widehat{\phi}_s^1, z_s^2 + \widehat{\phi}_s^2) ds \\ \quad + \int_0^t S_i(t-s) h_i(s, z_s^1 + \widehat{\phi}_s^1, z_s^2 + \widehat{\phi}_s^2) dW_i(s) \\ \quad + \int_0^t S_i(t-s)\sigma_i(s) dB_i^H(s)] \\ \quad + [\sum_{0<t_k<t} S_i(t-t_k) I_k^i(z_{t_k}^1 + \widehat{\phi}_{t_k}^1, z_{t_k}^2 \widehat{\phi}_{t_k}^2). \end{cases}$$

Let

$$\widehat{N}_i(z^1,z^2) = P_{0,i}(z^1,z^2) + P_{1,i}(z^1,z^2) + P_{2,i}(z^1,z^2) + P_{3,i}(z^1,z^2),$$

and

$$\chi(P_{0,i}(\{z^{n,1}(t)\}_{n=1}^{\infty}, \{z^{n,2}\}_{n=1}^{\infty})) = 0.$$

By (H_{34}) and (2) of (H_{35}), we have

$$\begin{aligned} &\chi(\{P_{1,1}(z^{n,1}, z^{n,2})(t)\}_{n=1}^{\infty}) \\ &\quad \le \|(-A_1)^{-\beta}\| \Big(C_{g_1} \sup_{\theta \in (-\infty,0]} \chi(\{z_t^{n,1}(\theta) + \widehat{\phi}_t^1(\theta)\}_{n=1}^{\infty}) \\ &\quad + \bar{C}_{g_1} \sup_{\theta \in (-\infty,0]} \chi(\{z_t^{n,1}(\theta) + \widehat{\phi}_t^1(\theta)\}_{n=1}^{\infty}) \Big) \end{aligned}$$

$$
\begin{aligned}
&+2M_{1-\beta}\int_0^t (t-s)^{-(1-\beta)}\Big(l_{g_1}(s)\sup_{\theta\in(-\infty,0]}\chi(\{z_s^{n,1}(\theta)+\widehat{\phi}_s^1(\theta)\}_{n=1}^{\infty})\\
&+\bar{l}_{g_1}(s)\sup_{\theta\in(-\infty,0]}\chi(\{z_s^{n,2}(\theta)+\widehat{\phi}_s^2(\theta)\}_{n=1}^{\infty})\Big)ds\\
&\le\left(\|(-A_1)^{-\beta}\|C_{g_1}+2\frac{M_{1-\beta}T^\beta}{\beta}\int_0^T l_{g_1}(s)ds\right)\sup_{\theta\in(-\infty,0]}\chi(\{z_t^{n,1}(\theta)+\widehat{\phi}_t^1(\theta)\}_{n=1}^{\infty})\\
&+\left(\|(-A_1)^{-\beta}\|\bar{C}_{g_1}+2\frac{M_{1-\beta}T^\beta}{\beta}\int_0^T \bar{l}_{g_1}(s)ds\right)\sup_{\theta\in(-\infty,0]}\chi(\{z_t^{n,2}(\theta)+\widehat{\phi}_t^2(\theta)\}_{n=1}^{\infty})\\
&=\left(\|(-A_1)^{-\beta}\|C_{g_1}+2\frac{M_{1-\beta}T^\beta}{\beta}\int_0^T l_{g_1}(s)ds\right)\sup_{t\in[0,T]}\chi(\{z^{n,1}(t)\}_{n=1}^{\infty})\\
&+\left(\|(-A_1)^{-\beta}\|\bar{C}_{g_1}+2\frac{M_{1-\beta}T^\beta}{\beta}\int_0^T \bar{l}_{g_1}(s)ds\right)\sup_{t\in[0,T]}\chi(\{z^{n,2}(t)\}_{n=1}^{\infty}),
\end{aligned}
$$

and hence

$$
\begin{aligned}
&\chi\left(\left\{\int_0^t S_1(t-s)h_1(s,z_s^{n,1}+\widehat{\phi}_s^1,z_s^{n,2}+\widehat{\phi}_s^2)dW_1(s)\right\}_{n=1}^{\infty}\right),\\
&\qquad\le\sqrt{M}\left(\int_0^t \chi(\{h_1(s,z_s^{n,1}+\widehat{\phi}_s^1,z_s^{n,2}+\widehat{\phi}_s^2)\}_{n=1}^{\infty})^2ds\right)^{\frac{1}{2}},
\end{aligned}
$$

and

$$
\begin{aligned}
&\chi(\{P_{2,1}(z^{n,1},z^{n,2})(t)\}_{n=1}^{\infty})\\
&\quad\le\chi\left(\left\{\int_0^t S_1(t-s)f_1(s,z_s^{n,1}+\widehat{\phi}_s^1,z_s^{n,2}+\widehat{\phi}_s^2)ds\right\}_{n=1}^{\infty}\right)\\
&\quad+\chi\left(\left\{\int_0^t S_1(t-s)h_1(s,z_s^{n,1}+\widehat{\phi}_s^1,z_s^{n,2}+\widehat{\phi}_s^2)dW(s)\right\}_{n=1}^{\infty}\right).
\end{aligned}
$$

By Lemma 7.2 and Lemma 8.6 and (2) of (H_{36}), (3) of (H_{37}), we have

$$
\begin{aligned}
&\chi(\{P_{2,1}(z^{n,1},z^{n,2})(t)\}_{n=1}^{\infty})\\
&\quad\le\sqrt{M}\int_0^t \chi(\{f_1(s,z_s^{n,1}+\widehat{\phi}_s^1,z_s^{n,2}+\widehat{\phi}_s^2)\}_{n=1}^{\infty})ds
\end{aligned}
$$

$$+\sqrt{M}\left(\int_0^t \chi(\{h_1(s,z_s^{n,1}+\widehat{\phi}_s^1, z_s^{n,2}+\widehat{\phi}_s^2)\}_{n=1}^{\infty})^2 ds\right)^{\frac{1}{2}},$$

then

$$\begin{aligned}
&\chi(\{P_{2,1}(z^{n,1},z^{n,2})(t)\}_{n=1}^{\infty})\\
&\quad\le 2\sqrt{M}T\int_0^t\Big(l_{f_1}(s)\sup_{\theta\in(-\infty,0]}\chi(\{z_s^{n,1}(\theta)+\widehat{\phi}_s^1(\theta)\}_{n=1}^{\infty})ds\\
&\qquad+2\sqrt{M}T\int_0^t \bar{l}_{f_1}(s)\sup_{\theta\in(-\infty,0]}\chi(\{z_s^{n,2}(\theta)+\widehat{\phi}_s^2(\theta)\}_{n=1}^{\infty})\Big)ds\\
&\qquad+2\sqrt{M}\left(\int_0^t\Big(l_{h_1}(s)\sup_{\theta\in(-\infty,0]}\chi(\{z_s^{n,1}(\theta)+\widehat{\phi}_s^1(\theta)\}_{n=1}^{\infty})\Big)^2 ds\right)^{\frac{1}{2}}\\
&\qquad+2\sqrt{M}\left(\int_0^t\Big(\bar{l}_{h_1}(s)\sup_{\theta\in(-\infty,0]}\chi(\{z_s^{n,2}(\theta)+\widehat{\phi}_s^2(\theta)\}_{n=1}^{\infty})\Big)^2 ds\right)^{\frac{1}{2}},
\end{aligned}$$

then

$$\begin{aligned}
&\chi(\{P_{2,1}(z^{n,1},z^{n,2})(t)\}_{n=1}^{\infty})\\
&\quad\le 2\sqrt{M}T\int_0^t l_{f_1}(s)ds\sup_{t\in[0,T]}\chi(\{z^{n,1}(t)\}_{n=1}^{\infty})\\
&\qquad+2\sqrt{M}T\int_0^t \bar{l}_{f_1}(s)ds\sup_{t\in[0,T]}\chi(\{z^{n,2}(t)\}_{n=1}^{\infty})\\
&\qquad+2\sqrt{M}\left(\int_0^t l_{h_1}^2(s)ds\right)^{\frac{1}{2}}\sup_{t\in[0,T]}\chi(\{z^{n,1}(t)\}_{n=1}^{\infty})\\
&\qquad+2\sqrt{M}\left(\int_0^t \bar{l}_{h_1}^2(s)ds\right)^{\frac{1}{2}}\sup_{t\in[0,T]}\chi(\{z^{n,2}(t)\}_{n=1}^{\infty})\\
&\quad\le 2\sqrt{M}\left(T\int_0^t l_{f_1}(s)ds+\left(\int_0^t l_{h_1}^2(s)ds\right)^{\frac{1}{2}}\right)\sup_{t\in[0,T]}\chi(\{z^1(t)\}_{n=1}^{\infty})\\
&\qquad+2\sqrt{M}\left(T\int_0^t \bar{l}_{f_1}(s)ds+\left(\int_0^t \bar{l}_{h_1}^2(s)ds\right)^{\frac{1}{2}}\right)\sup_{t\in[0,T]}\chi(\{z^2(t)\}_{n=1}^{\infty}),
\end{aligned}$$

and

$$\chi\left(\int_0^t S_i(t-s)\sigma_i(s)dB_i^H(s)\right)=0.$$

By (H_{34}), (2) of (H_{38}), we have

$$\begin{aligned}
&\chi(\{P_{3,1}(z^{n,1},z^{n,2})(t)\}_{n=1}^{\infty})\\
&\quad\le \sqrt{M}\left(\sum_{k=1}^{m} l_k \sup_{\theta\in(-\infty,0]}\chi(\{z_{t_k}^{n,1}(\theta)+\widehat{\phi}_{t_k}^1(\theta)\}_{n=1}^{\infty})+\sum_{k=1}^{m}\bar{l}_k \sup_{\theta\in(-\infty,0]}\chi(\{z_{t_k}^{n,2}(\theta)+\widehat{\phi}_{t_k}^2(\theta)\}_{n=1}^{\infty})\right)\\
&\quad\le \sqrt{M}\left(\sum_{k=1}^{m} l_k \sup_{t\in[0,T]}\chi(\{z^{n,1}(t)\}_{n=1}^{\infty})+\sum_{k=1}^{m}\bar{l}_k \sup_{t\in[0,T]}\chi(\{z^{n,2}(t)\}_{n=1}^{\infty})\right).
\end{aligned}$$

Thus, for all $t\in[0,T]$, we have

$$\sup_{t\in[0,T]}\chi(\{\widehat{N}_1(z^{n,1},z^{n,2})(t)\}_{n=1}^{\infty})\le \alpha_1\sup_{t\in[0,T]}\chi(\{z^{n,1}(t)\}_{n=1}^{\infty})+\alpha_2\sup_{t\in[0,T]}\chi(\{z^{n,2}(t)\}_{n=1}^{\infty}),$$

where

$$\begin{aligned}
\alpha_1&=\left(\|(-A_1)^{-\beta}\|C_{g_1}+2\frac{M_{1-\beta}T^{\beta}}{\beta}\int_0^T l_{g_1}(s)ds\right)\\
&\quad+2\sqrt{M}\left(T\int_0^t l_{f_1}(s)ds+\left(\int_0^t l_{h_1}^2(s)ds\right)^{\frac{1}{2}}+\sum_{k=1}^{m}\frac{l_k}{2}\right),\\
\alpha_2&=\left(\|(-A_1)^{-\beta}\|\bar{C}_{g_1}+2\frac{M_{1-\beta}T^{\beta}}{\beta}\int_0^T \bar{l}_{g_1}(s)ds\right)\\
&\quad+2\sqrt{M}\left(T\int_0^t \bar{l}_{f_1}(s)ds+\left(\int_0^t \bar{l}_{h_1}^2(s)ds\right)^{\frac{1}{2}}+\sum_{k=1}^{m}\frac{\bar{l}_k}{2}\right).
\end{aligned}$$

Similarly, we have

$$\chi_{\mathcal{D}_{\mathcal{F}_T}}(\{\widehat{N}_2(z^{n,1},z^{n,2})\}_{n=1}^{\infty})\le \alpha_3\sup_{t\in[0,T]}\chi(\{z^{n,1}(t)\}_{n=1}^{\infty})+\alpha_4\sup_{t\in[0,T]}\chi(\{z^{n,2}(t)\}_{n=1}^{\infty}),$$

where

$$\begin{aligned}
\alpha_3&=\left(\|(-A_2)^{-\beta}\|C_{g_2}+2\frac{M_{1-\beta}T^{\beta}}{\beta}\int_0^T l_{g_2}(s)ds\right)\\
&\quad+2\sqrt{M}\left(T\int_0^t l_{f_2}(s)ds+\left(\int_0^t l_{h_2}^2(s)ds\right)^{\frac{1}{2}}+\sum_{k=1}^{m}\frac{l_k}{2}\right),
\end{aligned}$$

$$a_4 = \left(\|(-A_2)^{-\beta}\|\bar{C}_{g_2} + 2\frac{M_{1-\beta}T^{\beta}}{\beta} \int_0^T \bar{l}_{g_2}(s)ds \right) + 2\sqrt{M}\left(T \int_0^t \bar{l}_{f_2}(s)ds + \left(\int_0^t \bar{l}_{h_2}^2(s)ds \right)^{\frac{1}{2}} + \sum_{k=1}^{m} \frac{\bar{l}_k}{2} \right).$$

So,

$$\chi_{\mathcal{D}_{\mathcal{F}_T}}(\{\widehat{N}(z^{n,1}, z^{n,2})\}_{n=1}^{\infty}) = \begin{pmatrix} \chi_{\mathcal{D}_{\mathcal{F}_T}}(\{\widehat{N}_1(z^{n,1}, z^{n,2})\}_{n=1}^{\infty}) \\ \chi_{\mathcal{D}_{\mathcal{F}_T}}(\{\widehat{N}_1(z^{n,1}, z^{n,2})\}_{n=1}^{\infty}) \end{pmatrix}.$$

From (10.1), we have

$$\begin{pmatrix} \chi_{\mathcal{D}_{\mathcal{F}_T}}(\{\widehat{N}_1(z^{n,1}, z^{n,2})\}_{n=1}^{\infty}) \\ \chi_{\mathcal{D}_{\mathcal{F}_T}}(\{\widehat{N}_1(z^{n,1}, z^{n,2})\}_{n=1}^{\infty}) \end{pmatrix} \leq M_{b,c} \begin{pmatrix} \chi_{\mathcal{D}_{\mathcal{F}_T}}(\{z^{n,1}\}_{n=1}^{\infty}) \\ \chi_{\mathcal{D}_{\mathcal{F}_T}}(\{z^{n,2}\}_{n=1}^{\infty}) \end{pmatrix},$$

where

$$M_{b,c} = \begin{pmatrix} a_1 & a_2 \\ a_3 & a_4 \end{pmatrix}.$$

Since the spectral radius $\rho(M_{b,c}) < 1$,

$$\chi_{\mathcal{D}_{\mathcal{F}_T}}(\{\widehat{N}_1(z^{n,1}, z^{n,2})(t)\}_{n=1}^{\infty}) = 0,$$

and

$$\chi_{\mathcal{D}_{\mathcal{F}_T}}(\{\widehat{N}_2(z^{n,1}, z^{n,2})(t)\}_{n=1}^{\infty}) = 0.$$

This implies

$$\chi_{\mathcal{D}_{\mathcal{F}_T}}(\{\widehat{N}(z^{n,1}, z^{n,2})\}_{n=1}^{\infty}) = 0.$$

As a consequence of $V(t)$, it is a relatively compact set. In view of the Ascoli–Arzela theorem, V is relatively compact in $\mathcal{B}_q$. As a consequence of the Mönch fixed theorem, we deduce that N has a fixed point, which is a mild solution to problem (8.1). □

9 Differential inclusions stochastic

In this chapter, we prove the existence of mild solutions for first-order impulsive semilinear stochastic functional differential inclusions driven by fractional Brownian motion. We consider the cases in which the right-hand side is convex or nonconvex valued. The results are obtained by using two different fixed-point theorems for multivalued mappings. We are interested in the existence problem of the following stochastic differential inclusions:

$$\begin{cases} dx(t) \in (Ax(t) + F^1(t,x(t),y(t)))dt + \sum_{l=1}^{\infty} \sigma_l^1(t,x(t)),y(t))dB_l^H(t), & t \in J = [0,b],\ t \neq t_k, \\ dy(t) \in (Ay(t) + F^2(t,x(t),y(t)))dt + \sum_{l=1}^{\infty} \sigma_l^2(t,x(t),y(t))dB_l^H(t), & t \in [0,b],\ t \neq t_k, \\ \Delta x(t) = I_k(x(t_k)), & t = t_k,\ k = 1,2,\dots,m \\ \Delta y(t) = \overline{I}_k(y(t_k)), \\ x(0) = x(2), \\ y(0) = y(2), \end{cases} \tag{9.1}$$

where $J = [0,b]$, X is a real separable Hilbert space with inner product $\langle\cdot,\cdot\rangle$ induced by the norm $\|\cdot\|$,

$$A : D(I) \subset X \longrightarrow X,$$

is the infinitesimal generator of a strongly continuous semigroup of bounded linear operators $(S(t))_{t\geq 0}$ in X, and

$$F^1, F^2 : [0,b] \times X \times X \longrightarrow \mathcal{P}(X),$$

are given set-valued functions, where $\mathcal{P}(X)$ denotes the family of nonempty subsets of X,

$$I_k \in C(X,X), \quad (k = 1,2,\dots,m),$$
$$\sigma_l^1, \sigma_l^2 : J \times X \times X \to L_Q^0(Y,X).$$

Here, $L_Q^0(Y,X)$ denotes the space of all Q-Hilbert–Schmidt operators from Y into X, which will also be defined in the next section. Moreover, the fixed times t_k satisfy

$$0 < t_1 < t_2 < \cdots < t_m < T,$$

and $y(t_k^-)$ and $y(t_k^+)$ denote the left and right limits of $y(t)$ at $t = t_k$,

$$\begin{cases} \sigma(.,x) = (\sigma_1(.,x), \sigma_2(.,x), \dots), \\ \|\sigma(.,x)\|^2 = \sum_{l=1}^{\infty} \|\sigma_l(.,x)\|_{L_Q^0}^2 < \infty \end{cases}$$

with $\sigma(.,x) \in \ell^2$ for all $x \in X$, where

$$\ell^2 = \left\{ \phi = (\phi_l)_{l\geq 1} : X \times X \to L_Q^0(Y,X) : \|\phi(x)\|^2 = \sum_{l=1}^{\infty} \|\phi_l(x)\|_{L_Q^0}^2 < \infty \right\}.$$

https://doi.org/10.1515/9783112251898-009

9.1 Existence results

In this section, we prove the existence of a mild solution to the problem (9.1). Our approach is based on multivalued versions of Schaefer's fixed-point theorem.

9.1.1 The convex case

In this section, we will show some results concerning the existence of mild solutions for the convex case of system (9.1). Let

$$J_k = (t_k, t_{k+1}], \quad k = 1, 2, \dots, m.$$

To define a solution for problem (9.1), consider the following space of piecewise continuous functions:

$$\begin{aligned} PC = \Big\{ x : \ & \Omega \times [0, b] \longrightarrow X,\ x \in C(J_k, X)),\ k = 1, \dots, m \text{ such that} \\ & x(t_k^+, .) \text{ and } x(t_k^-, .) \text{ exist with } x(t_k^-, .) = x(t_k, .) \quad \text{and} \\ & \sup_{t\in[0,b]} E|x(t, .)|_X^2 < \infty \text{ almost surely} \Big\}. \end{aligned}$$

Endowed with the norm

$$\|x\|_{PC} = \Big(\sup_{s\in[0,b]} E|x(s, .)|_X^2 \Big)^{\frac{1}{2}},$$

it is not difficult to check that PC is a Banach space with norm $\|\cdot\|_{PC}$.

$AC^i(J, X)$ is the space of functions $y : J \to X$ i-differentiable in which the ith derivative, $y^{(i)}$, is absolutely continuous.

Let A be the infinitesimal generator of a C_0-semigroup

$$\{S(t)\}_{t\geq 0},$$

such that $1 \in \Lambda(S(2))$, and let $f : J \to X$ be a continuous function.

Lemma 9.1. *If $x, y \in PC$ is a mild solution to the problem*

$$\begin{cases} dx(t) = (f^1(t) + Ax(t))dt + \sum_{l=1}^{\infty} \sigma_l^1(t, x(t), y(t))dB_l^H(t), & t \in J,\ t \neq t_k, \\ x(t_k^+) - x(t_k) = I_k(x(t_k)), & t = t_k,\ k = 1, 2, \dots, m, \\ y(t_k^+) - y(t_k) = \overline{I}_k(y(t_k)), & \\ x(0) = x(2), & \end{cases} \tag{9.2}$$

then it is given by

$$x(t) = S(t)(I - S(2))^{-1}\left(\sum_{k=1}^{m}(S(b - t_k))I_k(x(t_k)) + \int_0^b S(b - s)f^1(s)ds\right.$$
$$\left.+ \sum_{l=1}^{\infty}\int_0^b S(b - s)\sigma_l^1(t, x(s), y(s))dB_l^H(s)\right) + \int_0^t S(t - s)f^1(s)ds$$
$$+ \sum_{l=1}^{\infty}\int_0^t S(t - s)\sigma_l^1(t, x(s), y(s))dB_l^H(s) + \sum_{0<t_k<t} S(t - t_k)I_k(x(t_k)), \quad \text{for } t \in J,$$

and

$$y(t) = S(t)(I - S(2))^{-1}\left(\sum_{k=1}^{m}(S(b - t_k))\bar{I}_k(y(t_k)) + \int_0^b S(b - s)f^2(s)ds\right.$$
$$\left.+ \sum_{l=1}^{\infty}\int_0^b S(b - s)\sigma_l^2(t, x(s), y(s))dB_l^H(s)\right) + \int_0^t S(t - s)f^2(s)ds$$
$$+ \sum_{l=1}^{\infty}\int_0^t S(t - s)\sigma_l^2(t, x(s), y(s))dB_l^H(s) + \sum_{0<t_k<t} S(t - t_k)\bar{I}_k(y(t_k)), \quad \text{for } t \in J.$$

Proof. Let (x, y) be a solution of problem (9.2) and

$$L_1(s) = S(t - s)x(s),$$

and

$$L_2(s) = S(t - s)y(s),$$

for fixed $t \in J$. We have

$$\begin{aligned} L_1'(s) &= -S'(t - s)x(s) + S(t - s)x'(s) \\ &= -AS(t - s)x(s) + S(t - s)x'(s) \\ &= S(t - s)(x'(s) - Ax(s)) \\ &= S(t - s)(f^1(s)ds + \sum_{l=1}^{\infty}\sigma_l^1(t, x(s), y(s))dB_l^H(s). \end{aligned}$$

Let $0 < t < t_1$. Integrating the previous equation, we deduce for $k = 1$,

$$L_1(t) - L_1(0) = \int_0^t S(t - s)f^1(s)ds + \sum_{l=1}^{\infty}\int_0^t S(t - s)\sigma_l^1(t, x(s), y(s))dB_l^H(s).$$

Hence,

$$x(t) = S(t)x(0) + \int_0^t S(t-s)f^1(s)ds + \sum_{l=1}^{\infty} \int_0^t S(t-s)\sigma_l^1(t, x(s), y(s))dB_l^H(s).$$

More generally, for $t_k < t < t_{k+1}$,

$$\begin{aligned}\int_0^{t_1} L_1'(s) + \int_{t_2}^{t_1} L_1'(s) + \cdots + \int_{t_k}^{t} L_1'(s) &= \int_0^t S(t-s)f^1(s)ds + \sum_{l=1}^{\infty} \int_0^t S(t-s)\sigma_l^1(t, x(s), y(s))dB_l^H(s) \\ &= L_1(t_1^-) - L_1(0) + L_1(t_2^-) - L_1(t_1^+) + \cdots + L_1(t) - L_1(t_k^+) \\ &= \int_0^t S(t-s)f^1(s)ds + \sum_{l=1}^{\infty} \int_0^t S(t-s)\sigma_l^1(t, x(s), y(s))dB_l^H(s).\end{aligned}$$

Therefore,

$$\begin{aligned}x(t) = S(t)x(0) &+ \sum_{0<t_k<t} (L_1(t_k^+ - L_1(t_k^-))) + \int_0^t S(t-s)f^1(s)ds \\ &+ \sum_{l=1}^{\infty} \int_0^t S(t-s)\sigma_l^1(s, x(s), y(s))dB_l^H(s).\end{aligned}$$

Since $x(0) = x(2)$ and $1 \in \rho(S(T))$, then $(I - S(2))$ is invertible. Hence, we obtain after substitution

$$\begin{aligned}x(t) &= S(t)(I - S(2))^{-1}\Bigg(\sum_{k=1}^{m}(S(b-t_k))I_k(x(t_k)) + \int_0^b S(b-s)f^1(s)ds \\ &+ \sum_{l=1}^{\infty}\int_0^b S(b-s)\sigma_l^1(s, x(s), y(s))dB_l^H(s)\Bigg) + \int_0^t S(t-s)f^1(s)ds \\ &+ \sum_{l=1}^{\infty}\int_0^t S(t-s)\sigma_l^1(s, x(s), y(s))dB_l^H(s) + \sum_{0<t_k<t} S(t-t_k)I_k(x(t_k)), \quad \text{for } t \in J,\end{aligned}$$

and we can proceed similarly with the y-component:

$$\begin{aligned}y(t) &= S(t)(I - S(2))^{-1}\Bigg(\sum_{k=1}^{m}(S(b-t_k))\bar{I}_k(y(t_k)) + \int_0^b S(b-s)f^2(s)ds \\ &+ \sum_{l=1}^{\infty}\int_0^b S(b-s)\sigma_l^2(s, x(s), y(s))dB_l^H(s)\Bigg) + \int_0^t S(t-s)f^2(s)ds\end{aligned}$$

$$+\sum_{l=1}^{\infty}\int_0^t S(t-s)\sigma_l^2(s,x(s),y(s))dB_l^H(s)+\sum_{0<t_k<t} S(t-t_k)\bar{I}_k(y(t_k)), \quad \text{for } t\in J. \ \square$$

This lemma leads to the definition of a mild solution.

Definition 9.1. A X-valued stochastic process $u=(x,y)\in PC\times PC$ is said to be a mild solution of (9.1) with respect to the probability space $(\Omega,\mathcal{F},\mathbb{P})$ if:

1) $u(t)$ is $\mathcal{F}_t$-adapted for all $t\in J_k=(t_k,t_{k+1}]$, $k=1,2,\dots,m$.
2) $u(t)$ is right continuous and has a limit on the left, and there exist selections f^i; $i=1,2$ such that $f^i(t)\in F^i(t,u(t))$ a. e. $t\in J$.
3) $u(t)$ satisfies

$$\begin{cases} x(t)=S(t)(I-S(2))^{-1}(\sum_{k=1}^m(S(b-t_k))I_k(x(t_k))+\int_0^b S(b-s)f^1(s)ds \\ \quad +\sum_{l=1}^{\infty}\int_0^b S(b-s)\sigma_l^1(t,x(s),y(s))dB_l^H(s))+\int_0^t S(t-s)f^1(s)ds \\ \quad +\sum_{l=1}^{\infty}\int_0^t S(t-s)\sigma_l^1(t,x(s),y(s))dB_l^H(s) \\ \quad +\sum_{0<t_k<t} S(t-t_k)I_k(x(t_k)), \quad t\in J, \\ y(t)=S(t)(I-S(2))^{-1}(\sum_{k=1}^m(S(b-t_k))\bar{I}_k(y(t_k))+\int_0^b S(b-s)f^2(s)ds \\ \quad +\sum_{l=1}^{\infty}\int_0^b S(b-s)\sigma_l^2(t,x(s),y(s))dB_l^H(s))+\int_0^t S(t-s)f^2(s)ds \\ \quad +\sum_{l=1}^{\infty}\int_0^t S(t-s)\sigma_l^2(t,x(s),y(s))dB_l^H(s) \\ \quad +\sum_{0<t_k<t} S(t-t_k)\bar{I}_k(y(t_k)), \quad t\in J. \end{cases}$$

Now, we assume again that $1\in\Lambda(S(2))$. We are now in a position to state and prove our existence result for the problem (9.1). First, we will list the following hypotheses that will be imposed in our main theorem.

Consider the following assumptions: In this part, we assume that $S(t)$ is compact for $t>0$ and that there exists $M>0$ such that

$$\|S(t)\|\le M, \quad \text{for every } t\in[0,b].$$

(H_{41}) The function

$$\sigma_l^i : J\times X\times X\to L_Q^0(Y,X).$$

There exist positive constants α_i and β_i and c_i for each $i=1,2$ such that

$$\|\sigma^1(t,x,y)\|^2\le\alpha_1|x|_X^2+\beta_1|y|_X^2+c_1, \quad \|\sigma^2(t,x,y)\|^2\le\alpha_2|x|_X^2+\beta_2|y|_X^2+c_2,$$

and

$$\sum_{l=1}^{\infty}\int_0^b\|\sigma_l^1(t,x,y)\|_{L_Q^0}^2 dt<\infty,$$

for all $x, y \in X$ and $t \in J$.

(H_{42}) $F^i : [0,b] \times X \times X \longrightarrow \mathcal{P}_{cv,cp}(X)$ is an integrably bounded multivalued map, that is, there exists $p_i \in L^2(J,X)$, $i = 1,2$, such that

$$|F^i(t,x,y)|_X^2 = \sup_{f^i \in F^i(t,x,y)} |f^i(t)|_X^2 \le p_i(t), \quad \forall t \in J,\ \forall (x,y) \in X \times X.$$

(H_{43}) There exist constants $d_k, \overline{d}_k \ge 0$, and $e_k, \overline{e}_k \ge 0$ for each $k = 1, \dots, m$, such that

$$\begin{aligned} |I_k(x)|_X^2 &\le d_k |x|_X^2 + e_k, \\ |\overline{I}_k(y)|_X^2 &\le \overline{d}_k |y|_X^2 + \overline{e}_k, \quad \text{for all } x, y \in X. \end{aligned}$$

Consider the following operator:

$$N(x,y) = (N_1(x,y), N_2(x,y)),\ (x,y) \in PC \times PC,$$

defined by

$$N(x,y) = \{(h, \overline{h}) \in PC \times PC\},$$

given by

$$\begin{aligned} h(t) = {} & S(t)(I - S(2))^{-1}\Bigg(\sum_{k=1}^{m}(S(b - t_k))I_k(x(t_k)) \\ & + \int_0^b S(t-s)f^1(s)ds + \sum_{l=1}^{\infty}\int_0^b S(t-s)\sigma_l^1(t,x(s),y(s))dB_l^H(s)\Bigg) \\ & + \int_0^t S(t-s)f^1(s)ds + \sum_{l=1}^{\infty}\int_0^t S(t-s)\sigma_l^1(t,x(s),y(s))dB_l^H(s) \\ & + \sum_{0<t_k<t} S(t-t_k)I_k(x(t_k)), \quad \text{if } t \in [0,b] \\ \overline{h}(t) = {} & S(t)(I - S(2))^{-1}\Bigg(\sum_{k=1}^{m}(S(b - t_k))\overline{I}_k(y(t_k)) \\ & + \int_0^b S(t-s)f^2(s)ds + \sum_{l=1}^{\infty}\int_0^b S(t-s)\sigma_l^2(t,x(s),y(s))dB_l^H(s)\Bigg) \\ & + \int_0^t S(t-s)f^2(s)ds + \sum_{l=1}^{\infty}\int_0^t S(t-s)\sigma_l^2(t,x(s),y(s))dB_l^H(s) \\ & + \sum_{0<t_k<t} S(t-t_k)\overline{I}_k(y(t_k)), \quad \text{if } t \in J, \end{aligned}$$

where

$$f^i \in S_{F^i,u} = \{f^i \in L^2(J,X) : f^i(t) \in F^i(t,x,y) \text{ for a. e. } t \in J\}.$$

Lemma 9.2. *Assume that*

$$F^i : J \times X \times X \longrightarrow \mathcal{P}_{cv,cp}(X),$$

is a Carathèodory map satisfying (H_{41})–(H_{43}). *Then the operator is completely continuous and u.s.c.*

Proof. First, we show that $N = (N_1, N_2)$ is completely continuous.

Step 1. N maps bounded sets into bounded sets in $PC \times PC$.

Indeed, it is enough to show that for any $q > 0$ there exists a positive constant $l = (l_1, l_2)$ such that for each

$$(x,y) \in B_q = \{(x,y) \in PC \times PC : E|x|_X^2 \le q,\ E|y|_X^2 \le q\},$$

one has

$$\|h\|_X^2 \le l_1, \quad \|\overline{h}\|_X^2 \le l_2.$$

Let $(h, \overline{h}) \in (N_1, N_2)$. There exists $f^i(t) \in F^i(t,x,y)$ for each $t \in J$, and we get

$$\begin{aligned} E|h(t)|_X^2 = E\Bigg| & S(t)(I - S(2))^{-1}\Bigg(\sum_{k=1}^{m}(S(b - t_k))I_k(x(t_k)) \\ & + \int_0^b S(t-s)f^1(s)ds + \sum_{l=1}^{\infty}\int_0^b S(t-s)\sigma_l^1(t,x(s),y(s))dB_l^H(s)\Bigg) \\ & + \int_0^t S(t-s)f^1(s)ds + \sum_{l=1}^{\infty}\int_0^t S(t-s)\sigma_l^1(t,x(s),y(s))dB_l^H(s) \\ & + \sum_{0<t_k<t} S(t-t_k)I_k(x(t_k))\Bigg|_X^2. \end{aligned}$$

Also,

$$\begin{aligned} E|h(t)|_X^2 \le 4E\Bigg| & S(t)(I - S(2))^{-1}\Bigg(\sum_{k=1}^{m}(S(b - t_k))I_k(x(t_k)) \\ & + \int_0^b S(t-s)f^1(s)ds + \sum_{l=1}^{\infty}\int_0^b S(t-s)\sigma_l^1(t,x(s),y(s))dB_l^H(s)\Bigg)\Bigg|_X^2 \end{aligned}$$

$$+ 4E\left|\int_0^t S(t-s)f^1(s)ds\right|_X^2 + 4E\left|\sum_{l=1}^{\infty}\int_0^t S(t-s)\sigma_l^1(t,x(s),y(s))dB_l^H(s)\right|_X^2$$
$$+ 4E\left|\sum_{0<t_k<t} S(t-t_k)I_k(x(t_k))\right|_X^2.$$

Using (H_{41})–(H_{43}) and (7.5), we have

$$\begin{aligned}
E|h(t)|_X^2 \le{}& 12M^4\|(I-S(2))^{-1}\|^2\Bigg(m\sum_{k=1}^m \sup_{z\in B(0,q)} E|I_k(z)|_X^2 + \|p_1\|_{L^1}^2 \\
&+ c_H H(2H-1)b^{2H}(\alpha_1 E|x(t)|_X^2 + \beta_1 E|y(t)|_X^2 + c_1)\Bigg) + 4M^2\|p_1\|_{L^1} \\
&+ 4M^2(c_H H(2H-1)b^{2H}(\alpha_1 E|x(t)|_X^2 + \beta_1 E|y(t)|_X^2 + c_1)) \\
&+ 4M^2 m\sum_{k=1}^m (d_k q + e_k) \\
\le{}& 4M^4\|(I-S(2))^{-1}\|^2\Bigg(\sum_{k=1}^m (d_k q + e_k) + \|p_1\|_{L^1} \\
&+ c_H H(2H-1)b^{2H}(\alpha_1 q + \beta_1 q + c_1)\Bigg) + 4M^2\|p_1\|_{L^1} \\
&+ 4M^2(c_H H(2H-1)b^{2H}(\alpha_1 q + \beta_1 q + c_1)) \\
&+ 4M^2 m\sum_{k=1}^m (d_k q + e_k) = l_1.
\end{aligned}$$

Similarly, we have

$$\begin{aligned}
E|\overline{h}(t)|_X^2 \le{}& 4M^4\|(I-S(2))^{-1}\|^2\left(\sum_{k=1}^m \overline{d}_k q + \overline{e}_k\right) + \|p_2\|_{L^1} \\
&+ c_H H(2H-1)b^{2H}(\alpha_2 q + \beta_2 q + c_2)) + 4M^2\|p_2\|_{L^1} \\
&+ 4M^2(c_H H(2H-1)b^{2H}(\alpha_2 q + \beta_2 q + c_2)) \\
&+ 4M^2 m\sum_{k=1}^m (\overline{d}_k q + \overline{e}_k)) = l_2.
\end{aligned}$$

Therefore,

$$\begin{pmatrix}\mathbb{E}|h(t)|_X^2\\ \mathbb{E}|\overline{h}(t)|_X^2\end{pmatrix} \le \begin{pmatrix} l_1\\ l_2\end{pmatrix}.$$

Step 2. N maps bounded sets into equicontinuous sets of $PC \times PC$.

Let B_q be a bounded set in $PC \times PC$ as in Step 1. Let $\tau_1, \tau_2 \in J$, $\tau_1 < \tau_2$, and $(x,y) \in B_q$. There exists

$$f^i(t) \in F^i(t,x,y), \quad i = 1,2,$$

such that

$$\begin{aligned}
&E|h(\tau_2) - h(\tau_1)|_X^2 \\
&\le 12M^2\left(\|S(\tau_2) - S(\tau_1)\|^2 \|(I - S(2))^{-1}\|^2 \left(m \sup_{z\in B(0,q)} \sum_{k=1}^{m} E|I_k(z)|_X^2 + \|p_1\|_{L^1} \right.\right. \\
&\quad \left.\left. + c_H H(2H-1) b^{2H-1}(\alpha_1 E|x(t)|_X^2 + \beta_1 E|y(t)|_X^2 + c_1)\right)\right. \\
&\quad + 12\int_0^{\tau_1} \|S(\tau_2 - s) - S(\tau_1 - s)\|^2 p_1(s)ds + 12\int_{\tau_1}^{\tau_2} \|S(\tau_2 - s)\|^2 p_1(s)ds \\
&\quad + 12c_H H(2H-1)(\tau_2)^{2H-1} \int_0^{\tau_2} \|S(\tau_2 - s) - S(\tau_1 - s)\|^2 (\alpha_1 E|x(t)|_X^2 \\
&\quad \left. + \beta_1 E|y(t)|_X^2 + c_1)ds \right) \\
&\quad + 12c_H H(2H-1)(t_2 - t_1)^{2H-1} \int_{\tau_1}^{\tau_2} \|S(\tau_1 - s)\|^2 (\alpha_1 E|x(t)|_X^2 + \beta_1 E|y(t)|_X^2 + c_1)ds \\
&\quad + 12m \sum_{0<t_k<\tau_1} \|S(\tau_2 - t_k) - S(\tau_1 - t_k)\|^2 \sup_{z\in B(0,q)} E|I_k(z)|_X^2 \\
&\quad + 12m \sum_{\tau_1<t_k<\tau_2} \|S(\tau_2 - t_k)\|^2 \sup_{z\in B(0,q)} E|I_k(z)|_X^2 .
\end{aligned}$$

The right-hand term tends to zero as $|\tau_2 - \tau_1| \to 0$ since $S(t)$ is a strongly continuous operator and the compactness of $S(t)$ for $t > 0$ implies the continuity in the uniform operator topology. This proves the equicontinuity.

Step 3. $N(B_q)(t)$ is precompact in $X \times X$. As a consequence of steps a to b, together with the Arzelá–Ascoli theorem, it suffices to show that N maps B_q into a pre-compact set in $X \times X$. Let $0 < t < b$ be fixed and let ϵ be a real number satisfying $0 < \epsilon < t$. For $(x,y) \in B_q$, we define

$$\begin{aligned}
h_\epsilon(t) &= S(\epsilon)S(t-\epsilon)(I - S(2))^{-1}\left(\sum_{k=0}^{m} S(b - t_k)I_k(x(t_k)) + \int_0^b S(b-s)f^1(s)ds \right. \\
&\quad \left. + \sum_{l=1}^{\infty} \int_0^b S(b-s)\sigma_l^1(s,(s),y(s))dB_l^H(s) \right) + S(\epsilon)\int_0^{t-\epsilon} S(t-s)f^1(s)ds
\end{aligned}$$

$$+\sum_{l=1}^{\infty} S(\epsilon)\int_0^{t-\epsilon} S(t-s)\sigma_l^1(s,x(s),y(s))dB_l^H(s)$$

$$+S(\epsilon)\sum_{0<t_k<t-\epsilon} S(t-\epsilon-t_k)I_k(x(t_k)).$$

Since $S(t)$ is a compact operator, the set

$$H_\epsilon = \{\widetilde{h}_\epsilon(t) = (h_\epsilon(t),\overline{h}_\epsilon(t)) : \widetilde{h}_\epsilon \in N_\epsilon(x,y)\ (x,y) \in B_q\}.$$

$$\begin{aligned}
E|h(t)-h_\epsilon(t)|_X^2 &\le 3E\left|\int_{t-\epsilon}^t S(t-s)f^1(s)ds\right|_X^2 \\
&\quad + 3E\left|\sum_{l=1}^{\infty}\int_{t-\epsilon}^t S(t-s)\sigma_l^1(s,x(s),y(s))dB_l^H(s)\right|_X^2 \\
&\quad + 3E\left|\sum_{t-\epsilon<t_k<t} S(t-t_k)I_k(x(t_k))\right|_X^2 \\
&\le 3M^2\int_{t-\epsilon}^t p_1(s)ds \\
&\quad + 3M^2\left(c_H H(2H-1)\epsilon^{2H-1}\int_{t-\epsilon}^t (\alpha_1 q+\beta_1 q+c_1)\right)ds \\
&\quad + 3M^2\left(m\sum_{t-\epsilon<t_k<t}\sup_{z\in B(0,q)} E|I_k(z)|_X^2\right).
\end{aligned}$$

Similarly,

$$\begin{aligned}
E|\overline{h}(t)-\overline{h}_\epsilon(t)|_X^2 &\le 3M^2\int_{t-\epsilon}^t p_2(s)ds \\
&\quad + 3M^2\left(c_H H(2H-1)\epsilon^{2H-1}\int_{t-\epsilon}^t (\alpha_2 q+\beta_2 q+c_2)\right)ds \\
&\quad + 3M^2\left(m\sum_{t-\epsilon<t_k<t}\sup_{\overline{z}\in B(0,q)} E|\overline{I}_k(\overline{z})|_X^2\right).
\end{aligned}$$

The right-hand side tends to 0, as $\epsilon \to 0$. Therefore, there are precompact sets arbitrarily close to the set

$$H = \{\widetilde{h}(t) = (h(t),\overline{h}(t)) : \widetilde{h} \in N(x,y)\ (x,y) \in B_q\}.$$

This set is then precompact in $X \times X$.

Step 4. $N = (N_1, N_2)$ has a closed graph. Let

$$u_n = (x_n, y_n) \longrightarrow z_* = (x_*, y_*),$$

$(h_n, \overline{h}_n) \in N(u_n)$, and $(h_n, \overline{h}_n) \longrightarrow (h_*, \overline{h}_*)$ as $n \longrightarrow \infty$. We shall prove that $h_* \in N_1(u_*)$. The fact that $h_n \in N_1(u_n)$ and $\overline{h}_n \in N_2(u_n)$ means that there exists $f_n^i \in S_{F^i,u_n}$ for each $i = 1, 2$ such that

$$\begin{aligned} h_n(t) = S(t)(I - S(2))^{-1}\Bigg(&\sum_{0<t_k<t} S(b - t_k)I_k(x_n(t_k)) + \int_0^b S(b - s)f_n^1(s)ds \\ &+ \sum_{l=1}^{\infty} \int_0^b S(b - s)\sigma_l^1(s, x_n(s), y_n(s))dB_l^H(s)\Bigg) + \int_0^t S(t - s)f_n^1(s)ds \\ &+ \sum_{l=1}^{\infty} \int_0^t S(t - s)\sigma_l^1(s, x_n(s), y_n(s))dB_l^H(s) + \sum_{0<t_k<t} S(t - t_k)I_k(x_n(t_k)). \end{aligned}$$

First, notice that as $n \to \infty$,

$$\begin{aligned} \Bigg\| h_n - S(t)(I - S(2))^{-1}\Bigg(&\sum_{k=1}^{m} S(b - t_k)I_k(x_n(t_k)) \\ &+ \sum_{l=1}^{\infty} \int_0^b S(b - s)\sigma_l^1(s, x_n(s), y_n(s))dB_l^H(s)\Bigg) \\ &- \sum_{l=1}^{\infty} \int_0^t S(t - s)\sigma_l^1(s, x_n(s), y_n(s))dB_l^H(s) - \sum_{0<t_k<k} S(t - t_k)I_k(x_n(t_k)) \\ &- h_* + S(t)(I - S(2))^{-1}\Bigg(\sum_{k=1}^{m} S(b - t_k)I_k(x_*(t_k)) \\ &- \sum_{l=1}^{\infty} \int_0^t S(t - s)\sigma_l^1(s, x(s), y(s))dB_l^H(s) \\ &+ \sum_{l=1}^{\infty} \int_0^b S(b - s)\sigma_l^1(s, x_*(s), y_*(s))dB_l^H(s)\Bigg) \\ &+ \sum_{0<t_k<k} S(t - t_k)I_k(x(t_k))\Bigg\|_{PC} \longrightarrow 0, \quad n \longrightarrow 0. \end{aligned}$$

Now, consider the continuous linear operator

$$\Gamma : L^2(J, X) \longrightarrow PC,$$

defined for each $i = 1, 2$, by

$$\Gamma(v^i)(t) = S(t)(I - S(2))^{-1}\int_0^b S(b-s)v^i(s)ds + \int_0^t S(t-s)v^i(s)ds.$$

From the definition of Γ, we know that

$$\Bigg(h_n(t) - S(t)(I - S(2))^{-1}\Bigg(\sum_{k=1}^m S(b-t_k)I_k(x_n(t_k)) + \sum_{l=1}^\infty \int_0^b S(b-s)\sigma_l^1(s, x_n(s), y_n(s))dB_l^H(s)\Bigg) - \sum_{l=1}^\infty \int_0^t S(t-s)\sigma_l^1(s, x_n(s), y_n(s))dB_l^H(s) - \sum_{0<t_k<t} S(t-t_k)I_k(x_n(t_k))\Bigg) \in \Gamma(S_{F^1,u_n}),$$

and

$$\Bigg(\overline{h}_n(t) - S(t)(I - S(2))^{-1}\Bigg(\sum_{k=1}^m S(b-t_k)\overline{I}_k(y_n(t_k)) + \sum_{l=1}^\infty \int_0^b S(b-s)\sigma_l^2(s, x_n(s), y_n(s))dB_l^H(s)\Bigg) - \sum_{l=1}^\infty \int_0^t S(t-s)\sigma_l^2(s, x_n(s), y_n(s))dB_l^H(s) - \sum_{0<t_k<t} S(t-t_k)\overline{I}_k(y_n(t_k))\Bigg) \in \Gamma(S_{F^2,u_n}).$$

Since

$$u_n = (x_n, y_n) \longrightarrow z_* = (x_*, y_*),$$

and $\Gamma \circ S_{F^i}$ are closed graph operators by Lemma 3.6, there exists $f_*^i \in S_{F^i,u_*}$ for each $i = 1, 2$ such that

$$h_*(t) = S(t)(I - S(2))^{-1}\Bigg(\sum_{k=1}^m S(b-t_k)I_k(x_*(t_k)) + \int_0^b S(b-s)f_*^1(s)ds + \sum_{l=1}^\infty \int_0^b S(b-s)\sigma_l^1(s, x_*(s), y_*(s))dB_l^H(s)\Bigg) + \int_0^t S(t-s)f_*^1(s)ds + \sum_{l=1}^\infty \int_0^t S(t-s)\sigma_l^1(s, x_*(s), y_*(s))dB_l^H(s) + \sum_{0<t_k<t} S(t-t_k)I_k(x_*(t_k)).$$

Similarly,

$$\overline{h}_*(t) = S(t)(I - S(2))^{-1}\left(\sum_{k=1}^{m} S(b - t_k)\overline{I}_k(\overline{y}_*(t_k)) + \int_0^b S(b - s)f_*^2(s)ds\right.$$
$$\left. + \sum_{l=1}^{\infty}\int_0^b S(b - s)\sigma_l^2(s, x_*(s), y_*(s))dB_l^H(s)\right) + \int_0^t S(t - s)f_*^2(s)ds$$
$$+ \sum_{l=1}^{\infty}\int_0^t S(t - s)\sigma_l^2(s, x_*(s), y_*(s))dB_l^H(s) + \sum_{0<t_k<t} S(t - t_k)\overline{I}_k(y_*(t_k)).$$

Hence,

$$(h_*, h_*) \in (N_1(u_*), N_2(u_*)),$$

proves our claim. Lemma 3.8 yields that N is upper semicontinuous. □

Now, we present the first proof of our existence and the compactness of the solution set of problem (9.1).

Theorem 9.1. *Assume that*

$$F^i : [0, b] \times X \times X \longrightarrow \mathcal{P}_{cv,cp}(X),$$

is a Carathèodory map satisfying (H_{41})–(H_{43}) holds. Then (9.1) *has at least one mild solution on J. If further X is a reflexive space, then the solution set is compact in $PC \times PC$.*

Proof. **Part 1.** Existence of solutions.

We transform the problem (9.1) into a fixed-point problem. Consider the multivalued operator

$$N : PC \times PC \to \mathcal{P}(PC \times PC),$$

defined in Lemma 9.2. It is clear that all solutions of problem (9.1) are fixed points of the multivalued operator defined by We shall show that N satisfies the assumptions of Lemma 3.15. Since for each $(x, y) \in PC \times PC$, the nonlinearity F^i takes convex values, the selection set $S_{F^i,u}$ is convex and, therefore, N has convex values. From Lemma 9.2, N is completely continuous and u. s. c.

Step 3. A priori bounds on solutions. Let $(x, y) \in PC \times PC$ be a solution of the abstract nonlinear equation $x \in N_1(x, y)$ and $y \in N_2(x, y)$. Then there exists $f^i \in S_{F^i}$ for $t \in [0, b]$ for each $i = \{1, 2\}$, namely

$$x(t) = S(t)(I - S(2))^{-1}\left(\sum_{k=1}^{m}(S(b - t_k))I_k(x(t_k))\right.$$

$$+\int_0^b S(b-s)f^1(s)ds+\sum_{l=1}^{\infty}\int_0^b S(b-s)\sigma_l^1(s,x(s),y(s))dB_l^H(s)\Bigg)$$
$$+\int_0^t S(t-s)f^1(s)ds+\sum_{l=1}^{\infty}\int_0^t S(t-s)\sigma_l^1(s,x(s),y(s))dB_l^H(s)$$
$$+\sum_{0<t_k<t} S(t-t_k)I_k(x(t_k)),$$

and

$$y(t)=S(t)(I-S(2))^{-1}\Bigg(\sum_{k=1}(S(b-t_k))\bar{I}_k(y(t_k))$$
$$+\int_0^b S(b-s)f^2(s)ds+\sum_{l=1}^{\infty}\int_0^b S(b-s)\sigma_l^2(s,x(s),y(s))dB_l^H(s)\Bigg)$$
$$+\int_0^t S(t-s)f^2(s)ds+\sum_{l=1}^{\infty}\int_0^t S(t-s)\sigma_l^2(s,x(s),y(s))dB_l^H(s)$$
$$+\sum_{0<t_k<t} S(t-t_k)\bar{I}_k(y(t_k)).$$

We first give an estimation for the third part,

$$E|x(t)|_X^2\leq 12M^4\|(I-S(2))^{-1}\|^2\Bigg(m\sum_{k=1}^{m}(d_kE|x(t_k)|_X^2+e_k)+\|p_1\|_{L^1}$$
$$+c_HH(2H-1)b^{2H}(\alpha_1E|x(t)|_X^2+\beta_1E|y(t)|_X^2+c_1)\Bigg)+4M^2\|p_1\|_{L^1}$$
$$+4M^2c_HH(2H-1)b^{2H}(\alpha_1E|x(t)|_X^2+\beta_1E|y(t)|_X^2+c_1)$$
$$+4M^2m\sum_{k=1}^{m}(d_kE|x(t_k)|_X^2+e_k).$$

Similarly,

$$E|y(t)|_X^2\leq 12M^4\|(I-S(2))^{-1}\|^2\Bigg(m\sum_{k=1}^{m}(\bar{d}_kE|y(t_k)|_X^2+\bar{e}_k)+\|p_2\|_{L^1}$$
$$+c_HH(2H-1)b^{2H}(\alpha_2E|x(t)|_X^2+\beta_2E|y(t)|_X^2+c_2)\Bigg)+4M^2\|p_1\|_{L^1}$$
$$+4M^2c_HH(2H-1)b^{2H}(\alpha_2E|x(t)|_X^2+\beta_2E|y(t)|_X^2+c_2)$$
$$+4M^2m\sum_{k=1}^{m}(\bar{d}_kE|y(t_k)|_X^2+\bar{e}_k).$$

Consider the function $\mu, \overline{\mu}$ defined on J by

$$\mu(t) = \sup\{E|x(s)|_X^2 : 0 \le s \le t\},$$

and

$$\overline{\mu}(t) = \sup\{E|y(s)|_X^2 : 0 \le s \le t\}.$$

This implies, for each $t \in J$,

$$\begin{aligned}
\mu(t) &\le 12M^4 \|(I - S(2))^{-1}\|^2 \bigg(m \sum_{k=1}^{m} (d_k \mu(t) + e_k) + \|p_1\|_{L^1} \\
&\quad + c_H H(2H-1) b^{2H} (\alpha_1 \mu(t) + \beta_1 \overline{\mu}(t) + c_1) \bigg) + 4M^2 \|p_1\|_{L^1} \\
&\quad + 4M^2 (c_H H(2H-1) b^{2H} (\alpha_1 \mu(t) + \beta_1 \overline{\mu}(t) + c_1) \\
&\quad + 4M^2 m \sum_{k=1}^{m} (d_k \mu(t) + e_k) \\
&= 12M^4 \|(I - S(2))^{-1}\|^2 \bigg(m \sum_{k=1}^{m} e_k + \|p_1\|_{L^1} + c_H H(2H-1) b^{2H} c_1 \bigg) \\
&\quad + \mu(t) (12M^4 m \|(I - S(2))^{-1}\|^2 \bigg(\sum_{k=1}^{m} d_k + c_H H(2H-1) b^{2H} \alpha_1 \bigg) \\
&\quad + 4M^2 \bigg(c_H H(2H-1) b^{2H} \alpha_1 + 4M^2 m \sum_{k=1}^{m} d_k \bigg) \\
&\quad + \overline{\mu}(t) (12M^4 \|(I - S(2))^{-1}\|^2 c_H H(2H-1) b^{2H} \beta_1 \\
&\quad + 4M^2 c_H H(2H-1) b^{2H} \beta_1) \\
&= K_1 + K_2 \mu(t) + K_3 \overline{\mu}(t).
\end{aligned}$$

There exists a constant $K_j, \overline{K}_j$ for each $j = 1, 2, 3$ defined as follows:

$$K_1 = 12M^4 \|(I - S(2))^{-1}\|^2 \bigg(m \sum_{k=1}^{m} e_k + \|p_1\|_{L^1} + c_H H(2H-1) b^{2H} c_1 \bigg),$$

and

$$\begin{aligned}
K_2 &= 4M^4 \|(I - S(2))^{-1}\|^2 \bigg(m \sum_{k=1}^{m} d_k + c_H H(2H-1) b^{2H} \alpha_1 \bigg) \\
&\quad + 4M^2 (c_H H(2H-1) b^{2H} \alpha_1 + 4M^2 m \sum_{k=1}^{m} d_k,
\end{aligned}$$

and

$$K_3 = 12mM^4\|(I - S(2))^{-1}\|^2 c_H H(2H-1)b^{2H}\beta_1 + 4M^2(c_H H(2H-1)b^{2H}\beta_1.$$

Similarly,

$$\begin{aligned}\overline{\mu}(t) &\le 12M^4\|(I - S(2))^{-1}\|^2\left(m\sum_{k=1}^{m}\overline{e}_k + \|p_2\|_{L^1} + c_H H(2H-1)b^{2H}c_2\right)\\ &\quad + \overline{\mu}(t)(12M^4\|(I - S(2))^{-1}\|^2\left(m\sum_{k=1}^{m}\overline{d}_k + c_H H(2H-1)b^{2H}\beta_2\right)\\ &\quad + 4M^2\left(c_H H(2H-1)b^{2H}\beta_2 + 4M^2\sum_{0<t_k<t}\overline{d}_k\right)\\ &\quad + \mu(t)(4M^4\|(I - S(2))^{-1}\|^2 c_H H(2H-1)b^{2H}\alpha_2 + 4M^4 c_H H(2H-1)b^{2H}\alpha_2)\\ &= \overline{K}_1 + \overline{K}_2\mu(t) + \overline{K}_3\overline{\mu}(t),\end{aligned}$$

where

$$\overline{K}_1 = 12mM^4\|(I - S(2))^{-1}\|^2\left(\sum_{0<t_k<t}\overline{e}_k + \|p_2\|_{L^1} + c_H H(2H-1)b^{2H}c_2\right),$$

and

$$\overline{K}_2 = 12M^4\|(I - S(2))^{-1}\|^2 c_H H(2H-1)b^{2H}\alpha_2 + 4M^4 c_H H(2H-1)b^{2H}\alpha_2,$$

and

$$\begin{aligned}\overline{K}_3 &= \left(12M^4\|(I - S(2))^{-1}\|^2\left(m\sum_{k=1}^{m}\overline{d}_k + c_H H(2H-1)b^{2H}\beta_2\right)\right)\\ &\quad + 4M^2\left(c_H H(2H-1)b^{2H}\beta_2 + 4M^2 m\sum_{k=1}^{m}\overline{d}_k\right).\end{aligned}$$

On the other hand,

$$\mu(t) + \overline{\mu}(t) \le \widetilde{K}_1 + \widetilde{K}_2(\mu(t) + \overline{\mu}(t)).$$

Thus, we have

$$\mu(t) + \overline{\mu}(t) \le \frac{\widetilde{K}_1}{1 - \widetilde{K}_2} = M.$$

The maximum being taken componentwise, and $\widetilde{K}_2$ is a suitable value lower than 1

$$\widetilde{K}_1 = \overline{K}_1 + K_1, \quad \widetilde{K}_2 = \max\{K_2 + \overline{K}_2, K_3 + \overline{K}_3\} < 1,$$
$$E|x(t)|_X^2 + E|y(t)|_X^2 \le M.$$

Consequently,

$$\|x\|_{PC}^2 \le M \quad \text{and} \quad \|y\|_{PC}^2 \le M.$$

Let

$$U = \{(x, y) \in PC \times PC : \|x\|_{PC}^2 < M + 1 \,\text{and}\, \|y\|_{PC}^2 < M + 1\},$$

and consider the operator

$$N : \overline{U} \to \mathcal{P}_{cv,cp}(PC \times PC).$$

From the choice of U, there is no $(x, y) \in \partial U$ such that $x \in \lambda N_1(x, y)$ and $y \in \lambda N_2(x, y)$ for some $\lambda \in (0, 1)$ As a consequence of the Leray and Schauder nonlinear alternative (Lemma 3.15), we deduce that N has a fixed point (x, y) in U, solution of problem (9.1).

Part 2. Compactness of the solution set. Let

$$S_F = \{(x, y) \in PC \times PC : (x, y) \text{ is a solution of problem (9.1)}\}.$$

From Part 1, $S_F \neq \emptyset$ and there exists M, such that for every

$$(x, y) \in S_F, \quad \|x\|_{PC}^2 \le M \quad \text{and} \quad \|y\|_{PC}^2 \le M.$$

Since N is completely continuous, then

$$N(S_F) = (N_1(S_{F^1}), N_2(S_{F^2})),$$

is relatively compact in $PC \times PC$. Let $(x, y) \in S_F$, then $(x, y) \in N(x, y)$ and $S_F \subset \overline{N(S_F)}$. It remains to prove that S_F is a closed set in $PC \times PC$. Let $(x_n, y_n) \in S_F$, such that (x_n, y_n) converge to (x, y). For every $n \in N$, there exists $v_n^i(t) \in F^i(t, x_n, y_n)$ a. e. $t \in J$ for each $i \in \{1, 2\}$, such that

$$x_n(t) = S(t)(I - S(2))^{-1}\Bigg(\sum_{k=1}^{m}(S(b - t_k))I_k(x_n(t_k))$$
$$+ \int_0^b S(b - s)v_n^1(s)ds + \sum_{l=1}^{\infty}\int_0^b S(b - s)\sigma_l^1(t, x_n(s), y_n(s))dB_l^H(s)\Bigg)$$

$$+\int_0^t S(t-s)v_n^1(s)ds + \sum_{l=1}^{\infty}\int_0^t S(t-s)\sigma_l^1(s,x_n(s),y_n(s))dB_l^H(s)$$
$$+\sum_{0<t_k<t} S(t-t_k)I_k(x_n(t_k)),$$

and

$$y_n(t) = S(t)(I-S(2))^{-1}\Bigg(\sum_{k=1}^{m}(S(b-t_k))\bar{I}_k(y_n(t_k))$$
$$+\int_0^b S(-s)v_n^2(s)ds + \sum_{l=1}^{\infty}\int_0^b S(b-s)\sigma_l^2(t,x_n(s),y_n(s))dB_l^H(s)\Bigg)$$
$$+\int_0^t S(t-s)v_n^2(s)ds + \sum_{l=1}^{\infty}\int_0^t S(t-s)\sigma_l^2(s,x_n(s),y_n(s))dB_l^H(s)$$
$$+\sum_{0<t_k<t} S(t-t_k)\bar{I}_k(y_n(t_k)).$$

(H_{43}) implies that for a. e. $t \in J$, $v_n^i \in p_i(t)B(0,1)$, $i = 1,2$; hence, $(v_n^i)_{n\in N}$ is integrably bounded. Note that this still remains true as S_F is a bounded set. Since X is reflexive, by Theorem 3.10, there exists a subsequence, still denoted by $(v_n^i)_{n\in N}$, which converges weakly to some limit $v^i \in L^2(J,X)$. Moreover, the mapping

$$\Gamma : L^2(J,X) \longrightarrow PC,$$

defined by

$$\Gamma(g^i)(t) = \int_0^t S(t-s)g^i(s)ds,$$

is a continuous linear operator. Then it remains continuous if these spaces are endowed with their weak topologies. Therefore for a. e. $t \in J$, the sequence $(x_n(t),y_n(t))$ converges to $(x(t),y(t))$ and by the continuity of $(I_k,\bar{I}_k)$ it follows that

$$x(t) = S(t)(I-S(2))^{-1}\Bigg(\sum_{0<t_k<t}(S(b-t_k))I_k(x(t_k))$$
$$+\int_0^b S(t-s)v^1(s)ds + \sum_{l=1}^{\infty}\int_0^b S(t-s)\sigma_l^1(t,x(s),y(s))dB_l^H(s)\Bigg)$$
$$+\int_0^t S(t-s)v^1(s)ds + \sum_{l=1}^{\infty}\int_0^t S(t-s)\sigma_l^1(s,x(s),y(s))dB_l^H(s)$$

$$+ \sum_{0<t_k<t} S(t-t_k)I_k(x(t_k)),$$

and

$$\begin{aligned} y(t) &= S(t)(I-S(2))^{-1}\bigg(\sum_{0<t_k<t}(S(b-t_k))\bar{I}_k(y(t_k)) \\ &+ \int_0^b S(t-s)v^2(s)ds + \sum_{l=1}^{\infty}\int_0^b S(t-s)\sigma_l^2(t,x(s),y(s))dB_l^H(s)\bigg) \\ &+ \int_0^t S(t-s)v^2(s)ds + \sum_{l=1}^{\infty}\int_0^t S(t-s)\sigma_l^2(s,x(s),y(s))dB_l^H(s) \\ &+ \sum_{0<t_k<t} S(t-t_k)\bar{I}_k(y(t_k)). \end{aligned}$$

Now, we need to prove that

$$v^i(t) \in F^i(t,x(t),y(t)),$$

for a. e. $t \in J$. Lemma 3.11 yields the existence of constants

$$\alpha_i^n \geq 0, \quad j = 1, 2\ldots, k(n),$$

and $i = 1, 2$ such that

$$\sum_{j=1}^{k(n)} \alpha_i^n = 1,$$

and the sequence of convex combinations

$$g_n^i(.) = \sum_{j=1}^{k(n)} \alpha_j^n v_j^i(.),$$

converges strongly to some limit $v^i \in L^2(J,X)$. Since F^i takes convex values, using Lemma 3.10, we obtain that

$$\begin{aligned} v^i(t) &\in \bigcap_{n\geq 1}\overline{\{g_k^i(t) : k \geq n\}}, \quad \text{a. e. } t \in J \\ &\subseteq \bigcap_{n\geq 1}\overline{co}\{v_k^i(t),\ k \geq n\} \\ &\subseteq \bigcap_{n\geq 1}\overline{co}\Big\{\bigcup_{k\geq n} F^i(t,x_k(t),y_k(t))\Big\} \end{aligned}$$

$$\subseteq \overline{co}\Big\{\limsup_{k\to\infty} F^i(t, x_k(t), y_k(t))\Big\}. \tag{9.3}$$

Since F^i is u. s. c. and has compact values, then by Lemma 3.9, we have

$$\limsup_{n\to\infty} F^i(t, x_n(t), y_n(t)) \subseteq F^i(t, x(t), y(t)) \quad \text{for a. e. } t \in J.$$

This and (9.3) imply that

$$v^i(t) \in \overline{co}(F^i(t, x(t), y(t))).$$

Since, for each $i = 1, 2$, $F^i(.,.)$ has closed, convex values, we deduce that

$$v^i(t) \in F^i(t, x(t), y(t)),$$

for a. e. $t \in J$, for each $i = 1, 2$ as claimed. Hence, $(x, y) \in S_{F^i}$, which proves that S_{F^i}, for each $i = 1, 2$, is closed; hence, compact in $PC \times PC$. □

9.1.2 The nonconvex case

Now we present a second result for the problem (9.1) with a nonconvex-valued right-hand side. Let us introduce the following hypotheses:

(H_{44}) $F^i : J \times X \times X \longrightarrow \mathcal{P}_{cp}(X); (t, y) \longrightarrow F^i(t, x, y)$ is measurable for each $(x, y) \in X \times X$.

(H_{45}) There exist functions a_i, $b_i \in L^1([0, T], \mathbb{R}^+)$, such that

$$\begin{cases} H^2_{d_1}(F^1(t, x, y), F^1(t, \overline{x}, \overline{y})) \le a_1(t)|x - \overline{x}|^2_X + b_1(t)|y - \overline{y}|^2_X, \\ H^2_{d_2}(F^2(t, x, y), F^2(t, \overline{x}, \overline{y})) \le a_2(t)|x - \overline{x}|^2_X + b_2(t)|y - \overline{y}|^2_X, \end{cases}$$

with

$$d_i(0, F^i(t, 0, 0) \le a_i(t),$$

for all $x, y, \overline{x}, \overline{y} \in X$ for each $i = 1, 2$.

(H_{46}) There exist functions α_i, $\beta_i \in L^1([0, T], \mathbb{R}^+)$ for each $i = 1, 2$, such that

$$\begin{cases} \|\sigma^1(t, x, y) - \sigma^1(t, \overline{x}, \overline{y})\|^2 \le \alpha_1(t)\|x - \overline{x}\|^2 + \beta_1(t)\|y - \overline{y}\|^2, \\ \|\sigma^2(t, x, y) - \sigma^2(t, \overline{x}, \overline{y})\|^2 \le \alpha_2(t)\|x - \overline{x}\|^2 + \beta_2(t)\|y - \overline{y}\|^2, \end{cases}$$

for all $x, y, \overline{x}, \overline{y} \in X$, and $t \in J$.

(H_{47}) There exist constants $d_k \ge 0$ and $\overline{d}_k \ge 0$, $k = 1, \dots, m$, such that

$$|I_k(x) - I_k(\overline{x})|^2_X \le d_k|x - \overline{x}|^2_X,$$

and

$$|\overline{I}_k(y) - \overline{I}_k(\overline{y})|^2 \leq \overline{d}_k |y - \overline{y}|_X^2,$$

for all $x, y, \overline{x}, \overline{y} \in X$.

Theorem 9.2. *Assume that hypotheses* (H_{44})–(H_{47}) *are fulfilled. If the matrix*

$$M_{\alpha,\beta} = \begin{pmatrix} A_1 & A_2 \\ B_1 & B_1 \end{pmatrix},$$

converges to zero, then problem (9.1) *has at least one mild solution.*

Proof. In order to transform the problem (9.1) into a fixed-point problem, let the multi-valued operator

$$N : PC \times PC \to \mathcal{P}(PC \times PC),$$

be defined as in Lemma 9.2.

Note that (H_{44}) implies that F^i for each $i = 1, 2$ has at most linear growth, that is,

$$E|F^i(t, x, y)|_X^2 \leq a_i(t) E|x|_X^2 + b_i(t) E|y|_X^2,$$

for a. e. $t \in J$ and all $x, y \in X$:

1. $N(x, y) \in \mathcal{P}_{cl}(PC \times PC)$ for each $(x, y) \in PC \times PC$. The proof is similar to that in Theorem 9.1, Part 1, Step 2, and is omitted.
2. There exists $M_{\alpha,\beta} \in \mathcal{M}_{2\times 2}(\mathbb{R}_+)$ a convergent matrix to zero, such that

$$H_d(N(x, y), N(\overline{x}, \overline{y})) \leq M_{\alpha,\beta} \begin{pmatrix} \|x - \overline{x}\|_{PC} \\ \|y - \overline{y}\|_{PC} \end{pmatrix}, \quad \text{for all } x, y, \overline{x}, \overline{y} \in PC.$$

Let $x, y, \overline{x}, \overline{y} \in PC$, and $h_i \in N_i(x, y)$, $i = 1, 2$. Then there exists $f^i(\cdot) \in S_{F^i,x,y}$, such that for each $t \in J$, we have

$$\begin{aligned} h_i(t) = {} & S(t)(I - S(2))^{-1} \Bigg(\sum_{k=1}^{m} (S(b - t_k)) \overline{I}_k(\overline{z^i}(t_k)) \\ & + \int_0^b S(b - s) f^i(s) ds + \sum_{l=1}^{\infty} \int_0^b S(b - s) \sigma_l^i(t, x(s), y(s)) dB_l^H(s) \Bigg) \\ & + \int_0^t S(t - s) f^i(s) ds + \sum_{l=1}^{\infty} \int_0^t S(t - s) \sigma_l^i(t, x(s), y(s)) dB_l^H(s) \\ & + \sum_{0 < t_k < t} S(t - t_k) \overline{I}_k(\overline{z^i}(t_k)), \end{aligned}$$

where

$$\tilde{I}_k(\overline{z^1}(t_k)) = I_k(x(t_k)), \quad \text{and} \quad \tilde{I}_k(\overline{z^2}(t_k)) = \bar{I}_k(y(t_k)), \quad k = 1, \dots, m.$$

From (H_{45}), we learn that

$$\begin{cases} EH_{d_1}^2(F(t,x,y), F(t,\overline{x},\overline{y})) \le a_1(t)E|x-\overline{x}|_X^2 + b_1(t)E|y-\overline{y}|_X^2, & \text{a. e.} t \in J, \\ EH_{d_2}^2(F(t,x,y), F(t,\overline{x},\overline{y})) \le a_2(t)E|x-\overline{x}|_X^2 + b_2(t)E|y-\overline{y}|_X^2, & \text{a. e.} t \in J. \end{cases}$$

Hence, there is

$$(w, \overline{w}) \in F^1(t, \overline{x}(t), \overline{y}(t)) \times F^2(t, \overline{x}(t), \overline{y}(t)),$$

such that

$$E\left|f^1(t) - w\right|_X^2 \le a_1(t)E|x-\overline{x}|_X^2 + b_1(t)E|y-\overline{y}|_X^2, \quad t \in J,$$

and

$$E\left|f^2(t) - \overline{w}\right|^2 \le a_2(t)E|x-\overline{x}|_X^2 + b_2(t)E|y-\overline{y}|_X^2, \quad t \in J.$$

Consider the multivalued maps

$$U_i : J \to \mathcal{P}(X),$$

$i = 1, 2$ defined by

$$U_1(t) = \{w \in F^1(t, \overline{x}(t), \overline{y}(t)) : E\left|f^1(t) - w\right|^2 \le a_1(t)E|x-\overline{x}|_X^2 + b_1(t)E|y-\overline{y}|_X^2 \text{ a. e. } t \in J\},$$

and

$$U_2(t) = \{\overline{w} \in F^2(t, \overline{x}(t), \overline{y}(t)) : E\left|f^2(t) - \overline{w}\right|^2 \le a_2(t)E|x-\overline{x}|_X^2 + b_2(t)E|y-\overline{y}|_X^2, \text{ a. e. } t \in [0,b]\}.$$

That is,

$$U_1 = \overline{B}(f^1(t), a_1(t)E|x-\overline{x}|_X^2 + b_1(t)E|y-\overline{y}|_X^2),$$

and

$$U_2 = \overline{B}(f^2(t), a_2(t)E|x-\overline{x}|_X^2 + b_2(t)E|y-\overline{y}|_X^2).$$

Since

$$f^i, a_i, b_i, x, y, \overline{x}, \overline{y},$$

is measurable for each $i = 1, 2$. Then the closed ball U_i is measurable. In addition, (H_{44}) and (H_{45}) imply that for each

$$(x, y) \in PC \times PC,$$

and

$$F^i(t, x(t), y(t)),$$

the measure is measurable. Finally, the set

$$V_i(.) = U_i(.) \cap F^i(., \overline{x}(.), \overline{y}(.)),$$

is nonempty. Therefore, the intersection multivalued operator V_i is measurable with nonempty, closed values; there exists a function $\overline{f}^i(t)$ that is a measurable selection for $V_i(.)$. Thus,

$$\overline{f}^i(t) \in F^i(t, \overline{x}(t), \overline{y}(t)) \quad \text{for a. e. } t \in J.$$

Hence,

$$E|f^1(t) - \overline{f}^1(t)|_X^2 \le a_1(t)E|x - \overline{x}|_X^2 + b_1(t)E|y - \overline{y}|_X^2, \quad \text{for a. e. } t \in J,$$

and

$$E|f^2(t) - \overline{f}^2(t)|_X^2 \le a_2(t)E|x - \overline{x}|_X^2 + b_2(t)E|y - \overline{y}|_X^2, \quad \text{for a. e. } t \in J.$$

So,

$$\begin{aligned}\overline{h}_i(t) = {} & S(t)(I - S(2))^{-1}\Bigg(\sum_{k=1}^{m}(S(b - t_k))\widetilde{I}_k(\overline{z}^i(t_k)) + \int_0^b S(b - s)\overline{f}^i(s)ds \\ & + \sum_{l=1}^{\infty}\int_0^b S(b - s)\sigma_l^i(t, \overline{x}(s), \overline{y}(s))dB_l^H(s)\Bigg) + \int_0^t S(t - s)\overline{f}^i(s)ds \\ & + \sum_{l=1}^{\infty}\int_0^t S(t - s)\sigma_l^i(t, \overline{x}(s), \overline{y}(s))dB_l^H(s) + \sum_{0<t_k<t} S(t - t_k)\widetilde{I}_k(\overline{z}^i(t_k)).\end{aligned}$$

From (7.5) and (H_{45})–(H_{47}),

$$\begin{aligned}& E|h_1(t) - \overline{h}_1(t)|_X^2 \\ & \le 12M^4m\|(I - S(2))^{-1}\|^2\sum_{k=1}^{m} d_k E|x(t_k) - \overline{x}(t_k)|_X^2\end{aligned}$$

$$+ 12M^4\|(I - S(2))^{-1}\|^2\left(\int_0^b a_1(s)E|x(s) - \overline{x}(s)|_X^2 + b_1(s)E|y(s) - \overline{y}(s)|_X^2 ds\right)$$

$$+ 12M^4\|(I - S(2))^{-1}\|^2 c_H H(2H - 1)b^{2H-1}$$

$$\times\left(\int_0^b a_1(s)E|x(s) - \overline{x}(s)|_X^2 + \beta_1(s)E|y(s) - \overline{y}(s)|_X^2 ds\right)$$

$$+ 4M^2\left(\int_0^t a_1(s)E|x(s) - \overline{x}(s)|_X^2 + b_1(s)E|y(s) - \overline{y}(s)|_X^2 ds\right)$$

$$+ 4M^2 c_H H(2H - 1)b^{2H-1}\int_0^t a_1(s)E|x(s) - \overline{x}(s)|_X^2$$

$$+ \beta_1(s)E|y(s) - \overline{y}(s)|_X^2 ds + 4mM^2\sum_{k=1}^m d_k E|x(t_k) - \overline{x}(t_k)|_X^2.$$

Taking the supremum, we have

$$\sup_{t\in J} E|h_1(t) - \overline{h}_1(t)|_X^2$$

$$\leq 4M^2\Bigg(3m\|(I - S(2))^{-1}\|^2\sum_{k=1}^m d_k + 3M^2\|(I - S(2))^{-1}\|^2\|a_1\|_{L^1}$$

$$+ 3M^2\|(I - S(2))^{-1}\|^2 c_H H(2H - 1)b^{2H-1}\|a_1\|_{L^1}$$

$$+ \|a_1\|_{L^1} + c_H H(2H - 1)b^{2H-1}\|a_1\|_{L^1} + m\sum_{k=1}^m d_k\Bigg)\sup_{t\in J} E|x(t) - \overline{x}(t)|_X^2$$

$$+ 4M^2(\|b_1\|_{L^1} + 3M^2\|(I - S(2))^{-1}\|^2 c_H H(2H - 1)b^{2H-1}\|\beta_1\|_{L^1}$$

$$+ \|b_1\|_{L^1} + c_H H(2H - 1)b^{2H-1}\|\beta_1\|_{L^1})\sup_{t\in J} E|y(t) - \overline{y}(t)|_X^2.$$

Hence,

$$\|h_1 - \overline{h}_1\|_{PC} \leq A_1\|x - \overline{x}\|_{PC} + A_2\|y - \overline{y}\|_{PC},$$

where

$$A_1 = 2M\sqrt{\begin{aligned}&3m\|(I - S(2))^{-1}\|^2\textstyle\sum_{k=1}^m d_k + 3M^2\|(I - S(2))^{-1}\|^2\|a_1\|_{L^1}\\ &\quad + 3M^2\|(I - S(2))^{-1}\|^2 c_H H(2H - 1)b^{2H-1}\|a_1\|_{L^1}\\ &\quad + \|a_1\|_{L^1} + c_H H(2H - 1)b^{2H-1}\|a_1\|_{L^1} + m\textstyle\sum_{k=1}^m d_k,\end{aligned}}$$

and

$$A_2 = 2M\sqrt{\begin{aligned}&\|b_1\|_{L^1} + 3M^2\|(I - S(2))^{-1}\|^2 c_H H(2H - 1)b^{2H-1}\|\beta_1\|_{L^1}\\ &\quad + \|b_1\|_{L^1} + c_H H(2H - 1)b^{2H-1}\|\beta_1\|_{L^1},\end{aligned}}$$

and similarly,

$$\begin{aligned}
&\sup_{t\in J} E\left|h_2(t)-\overline{h}_2(t)\right|_X^2\\
&\leq 4M^2\Bigg(3m\left\|(I-S(2))^{-1}\right\|^2\sum_{k=1}^{m}\bar{d}_k+3M^2\left\|(I-S(2))^{-1}\right\|^2\|a_2\|_{L^1}\\
&\quad+3M^2\left\|(I-S(2))^{-1}\right\|^2 c_H H(2H-1)b^{2H-1}\|\alpha_2\|_{L^1}\\
&\quad+\|a_2\|_{L^1}+c_H H(2H-1)b^{2H-1}\|\alpha_2\|_{L^1}+m\sum_{k=1}^{m}\bar{d}_k\Bigg)\sup_{t\in J} E\left|y(t)-\overline{y}(t)\right|_X^2\\
&\quad+4M^2\big(\|b_2\|_{L^1}+3M^2\left\|(I-S(2))^{-1}\right\|^2 c_H H(2H-1)b^{2H-1}\|\beta_2\|_{L^1}\\
&\quad+\|b_2\|_{L^1}+c_H H(2H-1)b^{2H-1}\|\beta_1\|_{L^1}\big)\sup_{t\in J} E\left|x(t)-\overline{x}(t)\right|_X^2.
\end{aligned}$$

Therefore,

$$\|h_2-\overline{h}_2\|_{PC}\leq B_1\|x-\overline{x}\|_{PC}+B_2\|y-\overline{y}\|_{PC},$$

where

$$B_1=2M\sqrt{\begin{aligned}&3m\|(I-S(2))^{-1}\|^2\textstyle\sum_{k=1}^{m}\bar{d}_k+3M^2\|(I-S(2))^{-1}\|^2\|a_2\|_{L^1}\\&+3M^2\|(I-S(2))^{-1}\|^2 c_H H(2H-1)b^{2H-1}\|\alpha_2\|_{L^1}\\&+\|a_2\|_{L^1}+c_H H(2H-1)b^{2H-1}\|\alpha_2\|_{L^1}+m\textstyle\sum_{k=1}^{m}\bar{d}_k,\end{aligned}}$$

and

$$B_2=2M\sqrt{\begin{aligned}&\|b_2\|_{L^1}+3M^2\|(I-S(2))^{-1}\|^2 c_H H(2H-1)b^{2H-1}\|\beta_2\|_{L^1}\\&+\|b_2\|_{L^1}+c_H H(2H-1)b^{2H-1}\|\beta_1\|_{L^1}.\end{aligned}}$$

By an analogous relation, obtained by interchanging the roles of x, y and $\overline{x}, \overline{y}$, we finally arrive at

$$H_d(N(x,y),N(\overline{x},\overline{y}))\leq M_{\alpha,\beta}\begin{pmatrix}\|x-\overline{x}\|_{PC}\\ \|y-\overline{y}\|_{PC}\end{pmatrix},$$

where

$$M_{\alpha,\beta}=\begin{pmatrix}A_1 & A_2\\ B_1 & B_1\end{pmatrix},$$

since $M_{\alpha,\beta}$ converges to zero. Thanks to Theorem 3.16, we can ensure that N has a fixed point (x, y), which is a mild solution to (9.1). □

9.1.3 Example

We use the abstract result proved in the previous section to study the existence of mild solutions for impulsive Stokes differential inclusions.

Let $D \subset \mathbb{R}^3$ be a bounded open domain with a smooth boundary ∂D, and let $n(x)$ be the outward normal to D at the point $x \in \partial D$. Let

$$X = \{u \in (C_c^\infty(D))^3 : \nabla u = 0 \text{ in } \Omega \text{ and } n.u = 0 \text{ on } \partial D\},$$

and

$$E = \overline{Y}^{(L^2(D))^3},$$

be the closure of Y in $(L^2(D))^3$. It is clear that, endowed with the standard inner product of the space $(L^2(D))^3$, defined by

$$\langle u, v\rangle = \sum_{i=1}^{3} \langle u_i, v_i\rangle_{L^2(D)},$$

E is a Hilbert space. Let

$$P : (L^2(D))^3 \to X,$$

denote the orthogonal projection of $(L^2(D))^3$ in X.

Consider the following system of impulsive stochastic Stokes-type partial differential inclusions:

$$\begin{cases} u_t - P(\Delta u) \in F(t, u(t,x), v(t,x)) + \sigma_1(t)\frac{dB_Q^H}{dt}, & \text{a. e. } \in [0,b],\ x \in D,\\ v_t - P(\Delta v) \in G(t, u(t,x), v(t,x)) + \sigma_2(t)\frac{dB_Q^H}{dt}, & \text{a. e. } t \in [0,b],\ x \in D,\\ u(t_k^+, x) - u(t_k^-, x) = I_k(u(t_k, x)), & \\ v(t_k^+, x) - v(t_k, x) = \overline{I}_k(v(t_k, x)), & k = 1, \dots, m,\\ \nabla u = \nabla v = 0, & (t,x) \in [0,b] \times \partial D,\\ u = v = 0, & (t,x) \in [0,b] \times \partial D,\\ u(0,x) = u(b,x), \quad v(0,x) = v(b,x), & x \in D, \end{cases} \tag{9.4}$$

where $P(\Delta)$ is the Stokes operator. Let

$$A : D(I) \subset X \to X,$$

be defined by

$$\begin{cases} D(I) = (H^2(D) \cap H_0^1(D))^3 \cap X,\\ Au = -P(\Delta u), \quad u \in D(I). \end{cases}$$

Lemma 9.3 (Fujita–Kato, [15, Theorem 7.3.4]). *The operator A, defined as above is the generator of a compact and analytic C_0-semigroup of contractions in X.*

Let us assume that

($\mathcal{K}_1$) Let

$$f_i, g_i : [0,b] \times D \times \mathbb{R} \times \mathbb{R} \to \mathbb{R}, \quad i = 1,2,$$

are functions such that

$$f_1(t,x,u,v) \le f_2(t,x,u,v), \quad g_1(t,x,u,v) \le g_2(t,x,u,v),$$

for all $(t,x,u,v) \in [0,b] \times D \times \mathbb{R} \times \mathbb{R}$.

($\mathcal{K}_2$) There exist

$$\phi_i, \psi_i \in L^1([0,b], \mathbb{R}_+) \cap L^\infty([0,b], \mathbb{R}_+),$$

such that

$$|f_i(t,x,u,v)| \le \phi_i(t) \quad \text{and} \quad |g_i(t,x,u,v)| \le \psi_i(t), \quad i = 1,2,$$

for each

$$(t,x,u,v) \in [0,b] \times D \times \mathbb{R} \times \mathbb{R}.$$

($\mathcal{K}_3$) f_1, g_1 are l.s.c and f_2, g_2 are u. s. c.

($\mathcal{K}_4$) The function

$$\sigma, \sigma_1 : [0,b] \longrightarrow L_Q^2(\mathcal{K}, \mathcal{H}),$$

is bounded, that is, there exists a positive constant L such that

$$\int_0^b \|\sigma_i(s)\|^2_{L_Q^2} ds < L_i, \quad i = 1,2.$$

Lemma 9.4 ([4]). *Let*

$$f_i, g_i : [0,b] \times D \times \mathbb{R} \times \mathbb{R} \to \mathbb{R}, \quad i = 1,2,$$

be functions satisfying ($\mathcal{K}_2$) – ($\mathcal{K}_3$). Let

$$F : G : [0,b] \times L^2(D) \times L^2(D) \to \mathcal{P}(L^2(D)),$$

be a multivalued map defined by

$$F(t,u,v) = \{f \in L^2(D) : f(x) \in [f_1(t,x,u,v), f_2(t,x,u,v)]\},$$

and

$$G(t,u,v) = \{g \in L^2(D) : g(x) \in [g_1(t,x,u,v), g_2(t,x,u,v)]\}.$$

Then F and G are nonempty, u.s.c. with weakly compact and convex values. Moreover,

$$F(.,.,.), G(.,.,.) \in \mathcal{P}_{cl,b,cv}(L^2(D)).$$

Let

$$\begin{aligned}
x(t)(\xi) &= u(t,\xi), \quad t \in J,\ \xi \in D,\\
I_k(x(t_k)) &= K_k \frac{u(t_k^-,\xi)}{1+|u(t_k^-,.)|_X}, \quad \xi \in D,\ k = 1,\dots,m,\\
I_k(y(t_k)) &= \bar{K}_k \frac{v(t_k^-,\xi)}{1+|v(t_k^-,)|_X}, \quad \xi \in \Omega,\ k = 1,\dots,m,\\
x(0)(\xi) &= u(0,\xi) = u(b,\xi) = x(2)(\xi),\\
y(0)(\xi) &= v(0,\xi) = v(b,\xi) = y(2)(\xi),\\
&\xi \in D,
\end{aligned}$$

where $K_k, \bar{K}_k \in \mathbb{R}$, $k = 1,\dots,m$. Assume that $(\mathcal{K}_1) - (\mathcal{K}_4)$ are satisfied. Thus, the problem (9.4) can be written in abstract form:

$$\begin{cases}
x'(t) - A_1 x(t) \in F_1(t, x(t), y(t)) + \sigma_1(t)\frac{dB_Q^H}{dt}, & t \in [0,b]\\
y'(t) - A_2 y(t) \in F_2(t, x(t), y(t)) + \sigma_2(t)\frac{dB_Q^H}{dt}, & t \in [0,b],\\
x(t_k^+) - x(t_k^-) \in I_k(x(t_k)), & \\
y(t_k^+) - y(t_k^-) \in \bar{I}_k(y(t_k)), & k = 1,\dots,m\\
x(0) = x_0, \quad y(0) = y_0, &
\end{cases} \tag{9.5}$$

where $A_1 = A_2 = A$. Since for each $k = 1,\dots,m$, we have

$$\begin{aligned}
\left|I_k(x)\right| &= \left|K_k \frac{x}{1+|x|_X}\right|_X \le |K_k|,\\
\left|\bar{I}_k(x)\right| &= \left|\bar{K}_k \frac{x}{1+|x|_X}\right|_X \le |\bar{K}_k|, \quad \text{for all } x \in X.
\end{aligned}$$

Then, from Theorem 9.1, problem (9.4) has at least one solution.

10 Second-order stochastic differential inclusions

We establish sufficient conditions for the existence of solutions for a class of second-order systems of stochastic impulsive differential inclusions:

$$\begin{cases} d(x'(t) - g^1(t, x(t), y(t))) \in (Ax(t) + F^1(t, x(t), y(t))dt + f^1(t, x(t), y(t))dW(t) \\ \quad + \sum_{l=1}^{\infty} \sigma_l^1(t, x(t), y(t))dB_l^H(t), \quad t \in [0, b], t \neq t_k, \\ d(y'(t) - g^2(t, x(t), y(t))) \in (Ax(t) + F^2(t, x(t), y(t)))dt + f^2(t, x(t), y(t))dW(t) \\ \quad + \sum_{l=1}^{\infty} \sigma_l^2(t, x(t), y(t))dB_l^H(t), \quad t \in [0, b], t \neq t_k, \\ \Delta x(t) = I_k(x(t_k)), \quad \Delta x'(t) = I_k^1(x(t_k)), \quad t = t_k, \ k = 1, 2, \dots, m, \\ \Delta y(t) = \overline{I}_k(y(t_k)), \quad \Delta y'(t) = \overline{I}_k^2(y(t_k)), \quad t = t_k, \ k = 1, 2, \dots, m, \\ x(0) = x_0, \quad y(0) = y_0, \\ x'(0) = x_1, \quad y'(0) = y_1, \end{cases} \tag{10.1}$$

where X is a real separable Hilbert space with inner product $\langle \cdot, \cdot \rangle$ induced by norm $\| \cdot \|$, where A is the infinitesimal generator of a strongly continuous cosine family of linear operators

$$\{C(t), t \geq 0\},$$

B^H is a cylindrical fractional Brownian motion with a Hurst parameter $H \in (1/2, 1)$, and

$$\{W(t) : t \in [0, b]\},$$

is a standard cylindrical Wiener process on Y—an arbitrary separable Hilbert space—and

$$F^i : [0, b] \times X \times X \longrightarrow \mathcal{P}(X),$$
$$g^i : [0, b] \times X \times X \longrightarrow X,$$

are given set-valued functions, where $\mathcal{P}(X)$ denotes the family of nonempty subsets of X:

$$I_k \in C(X, X), \quad (k = 1, 2, \dots, m),$$
$$\sigma_l^1, \sigma_l^2 : J \times X \times X \to L_Q^0(Y, X).$$

Here, $L_Q^0(Y, X)$ denotes the space of all Q-Hilbert–Schmidt operators from Y into X,

$$f^i : J \times X \times X \to L^0(Y, X).$$

Here,

$$L^0(Y, X) = L_2(Q^{1/2}Y, X),$$

https://doi.org/10.1515/9783112251898-010

is a separable Hilbert space with respect to the Hilbert–Schmidt norm $\|.\|_{L^0}$ and Q-Wiener process on $(\Omega, \mathcal{F}, \mathbb{P})$ with a linear bounded covariance operator Q such that $\operatorname{tr} Q < \infty$. Let

$$\{W(t),\ t \in \mathbb{R}\},$$

be a standard cylindrical Wiener process with values in Y and defined on $(\Omega, \mathcal{F}, \mathbb{P})$, which is a complete probability space that will also be defined in the next section. Moreover, the fixed times t_k satisfy

$$0 < t_1 < t_2 < \cdots < t_m < T,$$

$y(t_k^-)$ and $y(t_k^+)$ denote the left and right limits of $y(t)$ at $t = t_k$:

$$\begin{cases} \sigma(.,x) = (\sigma_1(.,x), \sigma_2(.,x), \dots), \\ \|\sigma(.,x)\|^2 = \sum_{l=1}^{\infty} \|\sigma_l(.,x)\|^2_{L_Q^0} < \infty, \end{cases}$$

with $\sigma(.,x) \in \ell^2$ for all $x \in X$, where

$$\ell^2 = \left\{ \phi = (\phi_l)_{l\geq 1} : X \times X \to L_Q^0(Y,X) : \|\phi(x)\|^2 = \sum_{l=1}^{\infty} \|\phi_l(x)\|^2_{L_Q^0} < \infty \right\}.$$

10.1 Stochastic impulsive differential inclusions

The next result ensures the convergence of the series in the previous definition.

Lemma 10.1. *For any*

$$\phi : [0,T] \to L_Q^0(Y,X),$$

such that (8.2) holds, and for any $\alpha, \beta \in [0,T]$ with $\alpha > \beta$,

$$E\left|\int_\alpha^\beta \phi(s)dB^H(s)\right|_X^2 \leq c_H H(2H-1)(\alpha-\beta)^{2H-1} \sum_{n=1}^{\infty} \int_\alpha^\beta |\phi(s)Q^{1/2}e_n|_X^2 ds.$$

If, in addition,

$$\sum_{n=1}^{\infty} |\phi Q^{1/2} e_n|_X \quad \text{is uniformly convergent for } t \in [0,T],$$

then

$$E\left|\int_\alpha^\beta \phi(s)dB^H(s)\right|_X^2 \leq c_H H(2H-1)(\alpha-\beta)^{2H-1} \int_\alpha^\beta \|\phi(s)\|^2_{L_Q^0} ds. \tag{10.2}$$

Notice that d is a generalized metric space on X if and only if d_i, $i = 1, \dots, n$ are metrics on X,

$$H_d(A,B) = \max\Big\{\sup_{a\in A}, d(a,B), \sup_{b\in B} d(A,b)\Big\},$$

where

$$d(A,b) = \inf_{a\in A} d(a,b),$$
$$d(a,B) = \inf_{b\in B} d(a,b).$$

Then $(\mathcal{P}_{b,cl}(X), H_d)$ is a metric space and $(\mathcal{P}_{cl}(X), H_d)$ is a generalized metric space.

A multivalued map

$$F : X \longrightarrow \mathcal{P}(X),$$

is convex (closed) valued if $F(y)$ is convex (closed) for all $y \in X$. F is bounded on bounded sets if

$$F(2) = \bigcup_{y\in B} F(y),$$

is bounded in X for all $B \in \mathcal{P}_b(X)$. F is called upper semicontinuous (u. s. c. for short) on X if for each $y_0 \in X$ the set $F(y_0)$ is a nonempty, closed subset of X, and for each open set $\mathcal{U}$ of X containing $F(y_0)$, there exists an open neighborhood $\mathcal{V}$ of y_0 such that $F(\mathcal{V}) \in \mathcal{U}$. F is said to be completely continuous if $F(2)$ is relatively compact for every $B \in \mathcal{P}_b(X)$.

If the multivalued map F is completely continuous with nonempty compact values, then F is u. s. c. if and only if F has a closed graph, that is, $x_n \longrightarrow x_*, y_n \longrightarrow y_*, y_n \in F(x_n)$ imply $y_* \in F(x_*)$.

A a multivalued map,

$$F : J \longrightarrow \mathcal{P}_{cp,c},$$

is said to be measurable if, for each $y \in X$, the mean-square distance between y and $F(t)$ is measurable.

Lemma 10.2. *Let I be a compact interval and X be a Hilbert space. Let F be an L^2-Carathéodory multivalued map with $S_{F,y} \neq \emptyset$, and let Γ be a linear continuous mapping from $L^2(I,X)$ to $C(I,X)$. Then the operator*

$$\Gamma \circ S_F : C(I,X) \longrightarrow \mathcal{P}_{cp,c}(L^2([0,T],X)),$$
$$y \longmapsto (\Gamma \circ S_F)(y) = \Gamma(S_F, y),$$

is a closed graph operator in $C(I,X)\times C(I,X)$, where $S_{F,y}$ is known as the selectors set from F and is given by

$$S_{F,y} = \{f \in L^2([0,T],X) : f(t) \in F(t,y) \textit{ for a. e. } t \in [0,T]\}.$$

The following two results are easily deduced from the limit properties.

The following so-called nonlinear alternatives of Leray and Schauder will be needed in the proofs of our results.

Lemma 10.3. *Let $(X, \|\cdot\|)$ be a generalized Banach space and*

$$G : X \longrightarrow \mathcal{P}_{cl,cv}(X),$$

be an upper semicontinuous and compact map. Then either:
1. *F has at least one fixed point, or*
2. *The set*

$$\mathcal{M} = \{x \in X \textit{ and } \lambda \in (0,1), \textit{with } x \in \lambda G(u)\},$$

is unbounded.

The single-valued version may be stated as follows.

Lemma 10.4. *Let $(X, \|\cdot\|)$ be a generalized Banach space with $C \subset X$, a closed and convex subset of X. Assume U is an open subset of C, with $0 \in U$, and let $G : \overline{U} \longrightarrow C$ be a compact map. Then either:*
1. *G has a fixed point in $\overline{U}$, or*
2. *There is a point $u \in \partial U$ and $\lambda \in (0,1)$, with $u \in \lambda G(u)$.*

Now, let us recall some facts about cosine families of operators.

Definition 10.1. The one-parameter family

$$\{C(t) : t \in \mathbb{R}\},$$

of operators from X to X satisfying:
1. $C(0) = I$;
2. $C(t)x$ is continuous in t on $\mathbb{R}$, for all $x \in X$;
3. $C(t+s) - C(t-s) = 2C(t)C(s)$, for all $t, s \in \mathbb{R}$;

is called a strongly continuous cosine family.

The corresponding strongly continuous sine family

$$\{S(t) : t \in \mathbb{R}\},$$

associated with the given strongly continuous cosine family

$$\{C(t) : t \in \mathbb{R}\},$$

is defined by

$$S(t)x = \int_0^t C(s)xds \quad \text{for all } x \in X,\ t \in \mathbb{R}.$$

The infinitesimal generator $A : X \to X$ of a cosine family

$$\{C(t) : t \in \mathbb{R}\},$$

is defined by

$$Ax = \frac{d^2}{dt^2}C(t)x\Big|_{t=0} \quad \text{for all } x \in D(I),$$

where

$$D(I) = \{x \in X : C(.)x \in C^2(\mathbb{R}; X)\}.$$

It is well known that the infinitesimal generator A is a closed, densely defined operator on X. Such cosine and its corresponding sine families and their generators satisfy the following properties.

Proposition 10.1. *Suppose that A is the infinitesimal generator of a cosine family of operators*

$$\{C(t) : t \in \mathbb{R}\}.$$

Then:
1. *There exist $M^* \geq 1$ and $\alpha \geq 0$, such that*

   $$\|C(t)\| \leq M^* e^{\alpha|t|},$$

   and hence

   $$\|S(t)\| \leq M^* e^{\alpha|t|}.$$

2. $A \int_s^r S(u)xdu = (C(r) - C(s))x$, *for all* $0 \leq s \leq r < \infty$.
3. *There exists $N^* \geq 1$, such that*

   $$\|S(s) - S(r)\| \leq N^* \int_s^r e^{\alpha|s|}ds,$$

   for all $0 \leq s \leq r < \infty$.

The uniform boundedness principle, together with Proposition 10.1, part (i), implies that both $\{C(t) : t \in [0,b]\}$ and $\{S(t) : t \in [0,b]\}$ are uniformly bounded by a positive constant M.

10.1.1 Mild solutions

Let

$$J_k = (t_k, t_{k+1}], \quad k = 1,2,\dots,m.$$

To define a solution for problem (10.1), consider the following space of piecewise continuous functions.

Let us introduce the spaces

$$\begin{aligned} H_2([0,b];L^2(\Omega,X)) = \{&x : J \to L^2(\Omega,X),\ x|_{(t_k,t_{k+1}]} \in C((t_k,t_{k+1}],L^2(\Omega,X)),\\ &k = 1,2,\dots,m, \text{and there exist } x(t_k^+) \text{ for } k = 1,2,\dots,m\},\end{aligned}$$

and

$$\begin{aligned} H_2'([0,b];L^2(\Omega,X)) = \{&x : J \to L^2(\Omega,X),\ x|_{(t_k,t_{k+1}]} \in C^1((t_k,t_{k+1}],L^2(\Omega,X)),\\ &k = 1,2,\dots,m \text{ and there exist } x(t_k^+) \text{ for } k = 1,2,\dots,m\}.\end{aligned}$$

It is clear that

$$H_2([0,b];L^2(\Omega,X)),$$

is endowed with the norm

$$\|x\|_{H_2} = \sup_{s\in[0,b]} (E\|x(s,.)\|^2)^{\frac{1}{2}}.$$

It is easy to see that H_2' is a Banach space with the norm

$$\|x\|_{H_2'} = \|x\|_{H_2} + \|x'\|_{H_2}.$$

Finally, let the space

$$PC = \Big\{x : [0,b] \to L^2(\Omega,X) \text{ and } x|_J \in H_2' \quad \sup_{t\in[0,b]} E\|x(t,.)\|^2 < \infty \text{ almost surely}\Big\},$$

endowed with the norm

$$\|x\|_{PC} = \sup_{s\in[0,b]} (E\|x(s,.)\|^2)^{\frac{1}{2}}.$$

It is not difficult to check that PC is a Banach space with the norm $\|\cdot\|_{PC}$.

Definition 10.2. An X-valued stochastic process

$$u = (x, y) \in PC \times PC,$$

is said to be a solution of (10.1) with respect to the probability space $(\Omega, \mathcal{F}, \mathbb{P})$, if:
1. $u(t)$ is $\mathcal{F}_t$-adapted for all $t \in J_k = (t_k, t_{k+1}]$, $k = 1, 2, \dots, m$.
2. $u(t)$ is right continuous and has a limit on the left, and there exist selections f^i, $i = 1, 2$, such that $f_*^i(t) \in F^i(t, u(t))$ a. e. $t \in J$.
3. $u(t)$ satisfies

$$\begin{cases} x(t) = C(t)x_0 + S(t)(x_1 - g^1(0, x_0, y_0) + \int_0^t C(t-s)g^1(s, x(s), y(s))ds \\ \quad + \sum_{l=1}^{\infty} \int_0^t S(t-s)\sigma_l^1(t, x(s), y(s))dB^H(s) \\ \quad + \int_0^t S(t-s)f_*^1(s)ds + \int_0^t S(t-s)f^1(t, x(s), y(s))dW(s) \\ \quad + \sum_{0<t_k<t} C(t-t_k)I_k(x(t_k)) + \sum_{0<t_k<t} S(t-t_k)I_k^1(x(t_k)), \quad \mathbb{P}\text{-a. s., } t \in J, \end{cases}$$

and

$$\begin{cases} y(t) = C(t)y_0 + S(t)(y_1 - g^2(0, x_0, x_0) + \int_0^t C(t-s)g^2(s, x(s), y(s))ds \\ \quad + \sum_{l=1}^{\infty} \int_0^t S(t-s)\sigma_l^2(t, x(s), y(s))dB^H(s) \\ \quad + \int_0^t S(t-s)f_*^2(s)ds + \int_0^t S(t-s)f^2(t, x(s), y(s))dW(s) \\ \quad + \sum_{0<t_k<t} C(t-t_k)\bar{I}_k(y(t_k)) + \sum_{0<t_k<t} S(t-t_k)I_k^2(y(t_k)), \quad \mathbb{P}\text{-a. s., } t \in J. \end{cases} \tag{10.3}$$

First, we will list the following hypotheses that will be imposed in our main theorem. In this section, we assume that there exists $M > 0$ such that the cosine family of operators

$$\{C(t) : t \in [0, b]\},$$

on X and the corresponding sine family

$$\{S(t) : t \in [0, b]\},$$

satisfy

$$\begin{aligned} \|S(t)\| &\leq M, \\ \|C(t)\| &\leq M, \quad \text{for every } t \in [0, b], \end{aligned}$$

and $S(t)$, $c(t)$ are compact in X for $t \geq 0$.

(H_{48}) The function

$$\sigma_l^i : J \times X \times X \to L_Q^0(Y, X),$$

and there exists positive constants α_i and β_i, c_i for each $i = 1, 2$,

$$\left\|\sigma^i(t,x,y)\right\|^2 \le \alpha_i|x|_X^2 + \beta_i|y|_X^2 + c_i,$$

and

$$\sum_{l=1}^{\infty}\int_0^b \|\sigma_l^i(t,x,y)\|_{L_Q^0}^2\,dt < \infty,$$

for all $x, y \in X$, and $t \in J$.

(H_{49}) $F^i : [0,b]\times X\times X \longrightarrow \mathcal{P}_{cv,cp}(X)$ is an integrable bounded multivalued map, that is, there exists $p_i \in L^2(J,X)$, such that

$$|F^i(t,x,y)|_X^2 = \sup_{f_*^i\in F^i(t,x,y)} |f_*^i(t)|_X^2 \le p_i(t), \quad t\in J,\ (x,y)\in X\times X,$$

for each $i = 1,2$

(H_{50}) The function

$$f^i : J\times X\times X \to L^0(Y,X),$$

and there exists constants $a_i, b_i, c_i, \overline{a}_i, \overline{b}_i, \overline{c}_i \in \mathbb{R}^+$, such that

$$|f^i(t,x,y)|_X^2 \le a_i|x|_X^2 + b_i|y|_X^2 + c_i,$$

and

$$|g^i(t,x,y)|_X^2 \le \overline{a}_i|x|_X^2 + \overline{b}_i|y|_X^2 + \overline{c}_i,$$

and

$$\int_0^b \|f^i(t,x,y)\|_{L^0}^2\,dt < \infty,$$

for all $x, y \in X$, and $t \in J$, for each $i = 1,2$.

(H_{51}) There exist constants $d_k, \overline{d}_k \ge 0$, and $e_k, \overline{e}_k \ge 0$ for each $k = 1,\dots,m$ such that

$$\begin{cases} |I_k(x)|_X^2 \vee |I_k^1(x)|_X^2 \le d_k|x|_X^2 + e_k,\\ |\overline{I}_k(y)|_X^2 \vee |I_k^2(x)|_X^2 \le \overline{d}_k|y|_X^2 + \overline{e}_k,\end{cases}$$

for all $x, y \in X$, and $t \in J$, where $\vee$ denotes the maximum.

Theorem 10.1. *Assume that*

$$F^i : [0,b]\times X\times X \longrightarrow \mathcal{P}_{cv,cp}(X), \quad i = 1,2,$$

is a Caratheodory map and assume that (H_{48})–(H_{51}) *holds. Then* (10.1) *has at least one mild solution on* J.

Proof. Existence of solutions.

We transform the problem (10.1) into a fixed-point one. Consider the multivalued operator

$$N : PC \times PC \to \mathcal{P}(PC \times PC).$$

It is clear that all solutions of problem (10.1) are fixed points of the multivalued operator

$$N(x,y) = (N_1(x,y), N_2(x,y)), \quad (x,y) \in PC \times PC,$$

defined by

$$N(x,y) = \{(h,\overline{h}) \in PC \times PC\},$$

where

$$\begin{aligned}
h(t) &= C(t)x_0 + S(t)(x_1 - g^1(0,x_0,y_0)) + \int_0^t C(t-s)g^1(s,x(s),y(s))ds \\
&\quad + \sum_{l=1}^{\infty}\int_0^t S(t-s)\sigma_l^1(t,x(s),y(s))dB^H(s) + \int_0^t S(t-s)f_*^1(s)ds \\
&\quad + \int_0^t S(t-s)f^1(t,x(s),y(s))dW(s) + \sum_{0<t_k<t} C(t-t_k)I_k(x(t_k)) \\
&\quad + \sum_{0<t_k<t} S(t-t_k)I_k^1(x(t_k)), \quad \text{if } t \in [0,b], \\
\overline{h}(t) &= C(t)y_0 + S(t)(y_1 - g^2(0,x_0,y_0)) + \int_0^t C(t-s)g^2(s,x(s),y(s))ds \\
&\quad + \int_0^t S(t-s)\sigma_l^2(t,x(s),y(s))dB^H(s) + \int_0^t S(t-s)f_*^2(s)ds \\
&\quad + \sum_{l=1}^{\infty}\int_0^t S(t-s)f^2(t,x(s),y(s))dW(s) + \sum_{0<t_k<t} C(t-t_k)\overline{I}_k(x(t_k)) \\
&\quad + \sum_{0<t_k<t} S(t-t_k)I_k^2(x(t_k)), \quad \text{if } t \in [0,b],
\end{aligned}$$

and

$$f_*^i \in S_{F^i,u} = \{f_*^i \in L^2(J,X) : f_*^i(t) \in F^i(t,x,y) \text{ for a. e., } t \in J, \text{ for each } i = 1,2\}.$$

We will prove that

$$N = (N_1(x,y), N_2(x,y)),$$

has a fixed point.

We shall show that N satisfies the assumptions of Lemma 10.3. Since for each $(x,y) \in PC \times PC$, the nonlinearity F^i takes convex values for each $i = 1,2$, the selection set $S_{F^i,u}$ is convex; therefore,

$$N = (N_1(x,y), N_2(x,y)),$$

has convex values.We now divide the proof into several steps.

Step 1. $N = (N_1, N_2)$ is completely continuous.

1. N maps bounded sets into bounded sets in $PC \times PC$.
   Indeed, it is enough to show that for any $q > 0$ there exists a positive constant $l = (l_1, l_2)$, such that for each

   $$(x,y) \in B_q = \{(x,y) \in PC \times PC : E|x|_X^2 \le q,\ E|y|_X^2 \le q\},$$

   one has

   $$E|h|_X^2 \le l_1,$$
   $$E|\overline{h}|_X^2 \le l_2.$$

   Let $(h,\overline{h}) \in (N_1,N_2)$. There exists $f^i(t) \in F^i(t,x,y)$ for each $t \in [0,b]$, and we have

   $$\begin{aligned} E|h(t)|_X^2 = E\Big|&C(t)x_0 + S(t)(x_1 - g^1(0,x_0,y_0)) + \int_0^t C(t-s)g^1(s,x(s),y(s))ds \\ &+ \sum_{l=1}^{\infty}\int_0^t S(t-s)\sigma_l^1(t,x(s),y(s))dB^H(s) + \int_0^t S(t-s)f^1(t,x(s),y(s))dW(s) \\ &+ \int_0^t S(t-s)f_*^1(s)ds + \sum_{0<t_k<t} C(t-t_k)I_k(x(t_k)) + \sum_{0<t_k<t} S(t-t_k)I_k^1(x(t_k))\Big|_X^2. \end{aligned}$$

   Consequently,

   $$\begin{aligned} E|h(t)|_X^2 \le{}& 7E|C(t)x_0 + S(t)(x_1 - g^1(0,x_0,y_0)|_X^2 + 7E\Big|\int_0^t C(t-s)g^1(s,x(s),y(s))ds\Big|_X^2 \\ &+ 7E\Big|\sum_{l=1}^{\infty}\int_0^t S(t-s)\sigma_l^1(t,x(s),y(s))dB^H(s)\Big|_X^2 + 7E\Big|\int_0^t S(t-s)f_*^1(s)ds\Big|_X^2 \end{aligned}$$

$$
\begin{aligned}
&+7E\left|\int_0^t S(t-s)f^1(t,x(s),y(s))dW(s)\right|_X^2\\
&+7E\left|\sum_{0<t_k<t} C(t-t_k)I_k(x(t_k))\right|_X^2\\
&+7E\left|\sum_{0<t_k<t} S(t-t_k)I_k^1(x(t_k))\right|_X^2.
\end{aligned}
$$

Using (H_{48})–(H_{51}) and (10.2),

$$
\begin{aligned}
E|h(t)|_X^2 \le\ & 7M^2E|x_0|_X^2+7M^2(E|x_1|_X^2+\overline{a}_1E|x_0|_X^2+\overline{b}_1E|y_0|_X^2+\overline{c}_1)\\
&+7M^2\overline{a}_1bq+7M^2\overline{b}_1bq+7M^2\overline{c}_1b\\
&+7M^2c_HH(2H-1)b^{2H}(\alpha_1q+\beta_1q+c_1)\\
&+7M^2\|p_1\|_{L^1}+7M^2C_2a_1qb+7M^2C_2b_1qb+7M^2c_1b\\
&+14M^2m\sum_{0<t_k<t}d_kq+14M^2m\sum_{0<t_k<t}e_k\\
=\ & l_1.
\end{aligned}
$$

Similarly,

$$
\begin{aligned}
E|\overline{h}(t)|_X^2 \le\ & 7M^2E|y_0|_X^2+7M^2(E|y_1|_X^2+\overline{a}_1E|x_0|_X^2+\overline{b}_1E|y_0|_X^2+\overline{c}_1)\\
&+7M^2\overline{a}_2q+(7M^2\overline{b}_2q+7M^2\overline{c}_2)b\\
&+7M^2c_HH(2H-1)b^{2H}(\alpha_2q+\beta_2q+c_2)\\
&+7M^2\|p_2\|_{L^1}+7M^2C_2a_2qb+7M^2C_2b_2qb+7M^2c_2b\\
&+14M^2m\sum_{0<t_k<t}\overline{d}_kq+14M^2m\sum_{0<t_k<t}\overline{e}_k\\
=\ & l_2.
\end{aligned}
$$

Therefore,

$$
\begin{pmatrix} E|h(t)|_X^2 \\ E|\overline{h}(t)|_X^2 \end{pmatrix} \le \begin{pmatrix} l_1 \\ l_2 \end{pmatrix}.
$$

2. $N=(N_1,N_2)$ maps bounded sets into equicontinuous sets of $PC\times PC$. Let B_q be a bounded set in $PC\times PC$ as in (1). Let $t_1,t_2\in[0,b]$, $t_1<t_2$ and $(x,y)\in B_q$, there exists $f_*^i(t)\in F^i(t,x,y)$ for each $i=1,2$, and we have

$$
\begin{aligned}
E|h(t_2)-h(t_1)|_X^2 \le\ & 12E|(C(t_2)-C(t_1))x_0|_X^2\\
&+12E|(S(t_2)-S(t_1))(x_1-g^1(0,x_0,y_0))|_X^2
\end{aligned}
$$

$$+ 12E\left|\int_0^{t_2}(C(t_2-s)-C(t_1-s))g^1(s,x(s),y(s))ds\right|_X^2$$
$$+ 12E\left|\int_{t_1}^{t_2}C(t_1-s)g^1(s,x(s),y(s))ds\right|_X^2$$
$$+ 12E\left|\sum_{l=1}^{\infty}\int_0^{t_2}(S(t_2-s)-S(t_1-s))\sigma_l^1(t,x(s),y(s))dB^H(s)\right|_X^2$$
$$+ 12E\left|\sum_{l=1}^{\infty}\int_{t_1}^{t_2}S(t_1-s)\sigma_l^1(t,x(s),y(s))dB^H(s)\right|_X^2$$
$$+ 12E\left|\int_0^{t_2}(S(t_2-s)-S(t_1-s))f_*^1(s)ds\right|_X^2$$
$$+ 12E\left|\int_{t_1}^{t_2}S(t_1-s)f_*^1(s)ds\right|_X^2$$
$$+ 12E\left|\int_0^{t_2}(S(t_2-s)-S(t_1-s))f^1(t,x(s),y(s))dW(s)\right|_X^2$$
$$+ 12E\left|\int_{t_1}^{t_2}S(t_1-s)f^1(t,x(s),y(s))dW(s)\right|_X^2$$
$$+ 12E\left|\sum_{0<t_k<t}(C(t_2-t_k)-C(t_1-t_k))I_k(x(t_k))\right|_X^2$$
$$+ 12E\left|\sum_{0<t_k<t}(S(t_2-t_k)-S(t_1-t_k))I_k^1(x(t_k))\right|_X^2.$$

Using (10.2),

$$\begin{aligned}
&E|h(t_2)-h(t_1)|_X^2\\
&\quad\le 12(\|C(t_2)-C(t_1)\|^2E|x_0|_X^2+\|S(t_2)-S(t_1)\|^2(E|x_1|_X^2+\overline{a}_1E|x_0|_X^2+\overline{b}_1E|y_0|_X^2+\overline{c}_1))\\
&\qquad+12\int_0^{t_2}\|C(t_2-s)-C(t_1-s)\|^2(\overline{a}_1E|x(s)|_X^2+\overline{b}_1E|y(s)|_X^2+\overline{c}_1)ds\\
&\qquad+12\int_{t_1}^{t_2}\|C(t_1-s)\|^2(\overline{a}_1E|x(s)|_X^2+\overline{b}_1E|y(s)|_X^2+\overline{c}_1)ds
\end{aligned}$$

$$+ 12M^2 c_H H(2H-1)t_2^{2H-1}\int_0^{t_2}\|(S(t_2-s)-S(t_1-s))\|^2(\alpha_1 q+\beta_1 q+c_1)ds$$
$$+ 12M^2 c_H H(2H-1)(t_2-t_1)^{2H-1}\int_{t_1}^{t_2}\|S(t_1-s)\|^2(\alpha_1 q+\beta_1 q+c_1)ds$$
$$+ 12\int_0^{t_2}\|(S(t_2-s)-S(t_1-s))\|^2 p_1(s)ds + 12\int_{t_1}^{t_2}\|S(t_1-s)\|p_1(s)ds$$
$$+ 12C_2\int_0^{t_2}\|S(t_2-s)-S(t_1-s)\|^2(a_1 q+b_1 q+c_1)ds$$
$$+ 12C_2\int_{t_1}^{t_2}\|S(t_1-s)\|^2(a_1 q+b_1 q+c_1)ds$$
$$+ 12m\sum_{t_1<t_k<t_2}\|C(t_2-t_k)-C(t_2-t_k)\|^2\sup_{z\in B(0,q)}E|I_k(z)|_X^2$$
$$+ 12m\sum_{t_1<t_k<t_2}\|S(t_2-t_k)-S(t_2-t_k)\|^2\sup_{z\in B(0,q)}E|I_k^1(z)|_X^2.$$

Similarly,

$$\begin{aligned}
E|\overline{h}(t_2)-\overline{h}(t_1)|_X^2 &\\
\le 12(&\|C(t_2)-C(t_1)\|^2 E|y_0|_X^2\\
&+\|S(t_2)-S(t_1)\|^2(E|y_1|_X^2+\overline{a}_2E|x_0|_X^2+\overline{b}_2E|y_0|_X^2+\overline{c}_2))\\
&+\int_0^{t_2}\|C(t_2-s)-C(t_1-s)\|^2(\overline{a}_2E|x(s)|_X^2+\overline{b}_2E|y(s)|_X^2+\overline{c}_2)ds\\
&+12\int_{t_1}^{t_2}\|C(t_1-s)\|^2(\overline{a}_2E|x(s)|_X^2+\overline{b}_2E|y(s)|_X^2+\overline{c}_2)ds\\
&+12M^2c_HH(2H-1)t_2^{2H-1}\int_0^{t_2}\|(S(t_2-s)-S(t_1-s))\|^2(\alpha_2q+\beta_2q+c_2)ds\\
&+12M^2c_HH(2H-1)(t_2-t_1)^{2H-1}\int_{t_1}^{t_2}\|S(t_1-s)\|^2(\alpha_2q+\beta_2q+c_2)ds\\
&+12\int_0^{t_2}\|(S(t_2-s)-S(t_1-s))\|^2p_2(s)ds+\int_{t_1}^{t_2}\|S(t_1-s)\|p_2(s)ds
\end{aligned}$$

$$
\begin{aligned}
&+ 12C_2 \int_0^{t_2} \|S(t_2 - s) - S(t_1 - s)\|^2 (a_2 q + b_2 q + c_2) ds \\
&+ 12C_2 \int_{t_1}^{t_2} \|S(t_1 - s)\|^2 (a_2 q + b_2 q + c_2) ds \\
&+ 12m \sum_{t_1 < t_k < t_2} \|C(t_2 - t_k) - C(t_2 - t_k)\|^2 \sup_{z \in B(0,q)} E|\bar{I}_k(z)|_X^2 \\
&+ 12m \sum_{t_1 < t_k < t_2} \|S(t_2 - t_k) - S(t_2 - t_k)\|^2 \sup_{z \in B(0,q)} E|I_k^2(z)|_X^2 .
\end{aligned}
$$

The right-hand term tends to zero as $|t_2 - t_1| \to 0$ since $(S(t), C(t))$ are strongly continuous operators, and the compactness of $(S(t), C(t))$ for $t > 0$ implies continuity in the uniform operator topology. This proves the equicontinuity.

3. $N(B_q)(t)$ is pre-compact in $X \times X$. As a consequence of steps (1) to (2), together with the Arzelá–Ascoli theorem, it suffices to show that N maps B_q into a pre-compact set in $X \times X$. Let $0 < t < b$ be fixed and let ϵ be a real number satisfying $0 < \epsilon < t$. For $(x, y) \in B_q$, we define

$$
\begin{aligned}
h_\epsilon(t) = {} & C(\epsilon) C(t - \epsilon) x_0 + S(\epsilon) S(t - \varepsilon)(x_1 - g^1(0, x_0, y_0)) \\
&+ C(\epsilon) \int_0^{t-\epsilon} C(t - s) g^1(s, x(s), y(s)) ds \\
&+ \sum_{l=1}^{\infty} S(\epsilon) \int_0^{t-\epsilon} S(t - s) \sigma_l^1(t, x(s), y(s)) dB^H(s) \\
&+ S(\epsilon) \int_0^{t-\epsilon} S(t - s) f_*^1(s) ds + S(\epsilon) \int_0^{t-\epsilon} S(t - s) f^1(t, x(s), y(s)) dW(s) \\
&+ C(\epsilon) \sum_{0 < t_k < t - \epsilon} C(t - \epsilon - t_k) I_k(x(t_k)) + S(\epsilon) \sum_{0 < t_k < t - \epsilon} S(t - \epsilon - t_k) I_k^1(x(t_k)).
\end{aligned}
$$

Since $S(t)$, $c(t)$ are compact, the set

$$
H_\epsilon = \{\tilde{h}_\epsilon(t) = (h_\epsilon(t), \bar{h}_\epsilon(t)) : \tilde{h}_\epsilon \in N_\epsilon(x, y) \ \ (x, y) \in B_q\},
$$

is precompact, and

$$
\begin{aligned}
E|h(t) - h_\epsilon(t)|_X^2 \le {} & 6E \left| \int_{t-\epsilon}^{t} C(t - s) g^1(s, x(s), y(s)) ds \right|_X^2 \\
&+ 6E \left| \sum_{l=1}^{\infty} \int_{t-\epsilon}^{t} S(t - s) \sigma_l^1(t, x(s), y(s)) dB^H(s) \right|_X^2
\end{aligned}
$$

$$+ 6E\left|\int_{t-\epsilon}^{t} S(t-s)f_*^1(s)ds\right|_X^2 + 6E\left|\int_{t-\epsilon}^{t} S(t-s)f^1(t,x(s),y(s))dW(s)\right|_X^2$$
$$+ 6E\left|\sum_{t-\epsilon<t_k<t} C(t-t_k)I_k(x(t_k))\right|_X^2 + 6E\left|\sum_{t<t_k<t-\epsilon} S(t-t_k)I_k^1(x(t_k))\right|_X^2.$$

From (H_{48})–(H_{50}) and (10.2), we have

$$\begin{aligned}E|h(t)-h_\epsilon(t)|_X^2 &\le 6M^2\int_{t-\epsilon}^{t}(\overline{a}_1q+\overline{b}_1q+\overline{c}_1)ds\\ &+ 6M^2c_HH(2H-1)(\epsilon)^{2H-1}\int_{t-\epsilon}^{t}(\alpha_1q+\beta_1q+c_1)ds\\ &+ 6M^2\int_{t-\epsilon}^{t}p_1(s)ds + 6M^2C_2\int_{t-\epsilon}^{t}(a_1q+b_1q+c_1)ds\\ &+ 6M^2m\Big(\sup_{z\in B(0,\epsilon)}E|I_k(z)|_X^2\Big) + 6M^2m\Big(\sup_{z\in B(0,\epsilon)}E|I_k^1(z)|_X^2\Big).\end{aligned}$$

Similarly,

$$\begin{aligned}E|\overline{h}(t)-\overline{h}_\epsilon(t)|_X^2 &\le 6M^2\int_{t-\epsilon}^{t}(\overline{a}_2q+\overline{b}_2q+\overline{c}_2)ds\\ &+ 6M^2c_HH(2H-1)(\epsilon)^{2H-1}\int_{t-\epsilon}^{t}(\alpha_2q+\beta_2q+c_2)ds\\ &+ 6M^2\int_{t-\epsilon}^{t}p_2(s)ds + 8M^2C_2\int_{t-\epsilon}^{t}(a_2q+b_2q+c_2)ds\\ &+ 6M^2m\Big(\sup_{z\in B(0,\epsilon)}E|\overline{I}_k(z)|_X^2\Big) + 6M^2m\Big(\sup_{z\in B(0,\epsilon)}E|I_k^2(z)|_X^2\Big).\end{aligned}$$

The right-hand side tends to 0, as $\epsilon \to 0$. Therefore, there are precompact sets arbitrarily close to the set

$$H = \{\widetilde{h}(t) = (h(t),\overline{h}(t)) : \widetilde{h} \in N(x,y)\ \ (x,y) \in B_q\}.$$

This set is then precompact in $X \times X$.

Step 2. $N = (N_1, N_2)$ has a closed graph. Let

$$u_n = (x_n, y_n) \longrightarrow z_* = (x_*, y_*),$$

$$(h_n, \overline{h}_n) \in N(u_n),$$

and

$$(h_n, \overline{h}_n) \longrightarrow (h_*, \overline{h}_*),$$

as $n \longrightarrow \infty$. We shall prove that $h_* \in N_1(u_*)$. The fact that $h_n \in N_1(u_n)$ and $\overline{h}_n \in N_2(u_n)$ mean that there exists $f^i_{*,n} \in S_{F^i,u_n}$ for each $i = 1, 2$, such that

$$\begin{aligned} h_n &= C(t)x_0 + S(t)(x_1 - g^1(0, x_0, y_0)) + \int_0^t C(t-s)g^1(s, x_n(s), y_n(s))ds \\ &\quad + \sum_{l=1}^{\infty} \int_0^t S(t-s)\sigma_l^1(t, x_n(s), y_n(s))dB^H(s) + \int_0^t S(t-s)f^1_{*,n}(s)ds \\ &\quad + \int_0^t S(t-s)f^1(t, x_n(s), y_n(s))dW(s) + \sum_{0<t_k<t} C(t-t_k)I_k(x_n(t_k)) \\ &\quad + \sum_{0<t_k<t} S(t-t_k)I_k^1(x_n(t_k)). \end{aligned}$$

First, notice that, as $n \to \infty$,

$$\begin{aligned} &\Bigg\| h_n - C(t)x_0 - S(t)(x_1 - g^1(0, x_0, y_0)) - \int_0^t C(t-s)g^1(s, x_n(s), y_n(s))ds \\ &- \sum_{l=1}^{\infty} \int_0^t S(t-s)\sigma_l^1(t, x_n(s), y_n(s))dB^H(s) - \int_0^t S(t-s)f^1(t, x_n(s), y_n(s))dW(s) \\ &- \sum_{0<t_k<t} C(t-t_k)I_k(x_n(t_k)) - \sum_{0<t_k<t} S(t-t_k)I_k^1(x_n(t_k)) \\ &- (h_* - C(t)x_0 - S(t)(x_1 - g^1(0, x_0, y_0)) \\ &- \int_0^t C(t-s)g^1(s, x_*(s), y_*(s))ds - \sum_{l=1}^{\infty} \int_0^t S(t-s)\sigma_l^1(t, x_*(s), y_*(s))dB^H(s) \\ &- \int_0^t S(t-s)f^1(t, x_*(s), y_*(s))dW(s) \\ &- \sum_{0<t_k<t} C(t-t_k)I_k(x_*(t_k)) - \sum_{0<t_k<t} S(t-t_k)I_k^1(x_*(t_k))\Bigg\|_{PC} \longrightarrow 0, \quad n \longrightarrow 0. \end{aligned}$$

Now, consider the continuous linear operator

$$\Gamma : L^2(J,X) \longrightarrow PC,$$

defined, for each $i = 1,2$, by

$$\Gamma(v^i)(t) = \int_0^t S(t-s)v^i(s)ds.$$

From the definition of Γ, we know that

$$h_n - \Bigg(C(t)x_0 + S(t)(x_1 - g^1(0,x_0,y_0)) + \int_0^t C(t-s)g^1(s,x_n(s),y_n(s))ds$$
$$+ \sum_{l=1}^{\infty} \int_0^t S(t-s)\sigma_l^1(t,x_n(s),y_n(s))dB^H(s) + \int_0^t S(t-s)f^1(t,x_n(s),y_n(s))dW(s)$$
$$+ \sum_{0<t_k<t} C(t-t_k)I_k(x_n(t_k)) - \sum_{0<t_k<t} S(t-t_k)I_k^1(x_n(t_k)) \Bigg) \in \Gamma(S_{F^1,u_n}),$$

and

$$\overline{h}_n - \Bigg(C(t)x_0 + S(t)(y_1 - g^2(0,x_0,y_0)) + \int_0^t C(t-s)g^2(s,x_n(s),y_n(s))ds$$
$$+ \sum_{l=1}^{\infty} \int_0^t S(t-s)\sigma_l^2(t,x_n(s),y_n(s))dB^H(s) + \int_0^t S(t-s)f^2(t,x_n(s),y_n(s))dW(s)$$
$$+ \sum_{0<t_k<t} C(t-t_k)\overline{I}_k(y_n(t_k)) + \sum_{0<t_k<t} S(t-t_k)I_k^2(y_n(t_k)) \Bigg) \in \Gamma(S_{F^2,u_n}).$$

Since

$$u_n = (x_n,y_n) \longrightarrow z_* = (x_*,y_*),$$

and

$$\Gamma \circ S_{F^i},$$

are closed graph operators by Lemma 10.2, there exists $f_{**}^i \in S_{F^i,u_*}$ for each $i = 1,2$, such that

$$h_* = C(t)x_0 + S(t)(x_1 - g^1(0,x_0,y_0)) + \int_0^t C(t-s)g^1(s,x_*(s),y_*(s))ds$$

$$+\sum_{l=1}^{\infty}\int_0^t S(t-s)\sigma_l^1(t,x_*(s),y_*(s))dB^H(s)+\int_0^t S(t-s)f_{**}^1(s)ds$$
$$+\int_0^t S(t-s)f^1(t,x_*(s),y_*(s))dW(s)+\sum_{0<t_k<t}C(t-t_k)I_k(x_*(t_k))$$
$$+\sum_{0<t_k<t}S(t-t_k)I_k^1(x_*(t_k)).$$

Similarly,

$$\overline{h}_*=C(t)y_0+S(t)(y_1-g^2(0,x_0,y_0))+\int_0^t C(t-s)g^2(s,x_*(s),y_*(s))ds$$
$$+\sum_{l=1}^{\infty}\int_0^t S(t-s)\sigma_l^2(t,x_*(s),y_*(s))dB^H(s)+\int_0^t S(t-s)f_{**}^2(s)ds$$
$$+\int_0^t S(t-s)f^2(t,x_*(s),y_*(s))dW(s)+\sum_{0<t_k<t}C(t-t_k)\overline{I}_k(x_*(t_k))$$
$$+\sum_{0<t_k<t}S(t-t_k)I_k^2(x_*(t_k)).$$

Hence,

$$(h_*,h_*)\in(N_1(u_*),N_2(u_*)),$$

proving our claim yields that $N=(N_1,N_2)$ is u. s. c.

Step 3. A priori bounds on solutions. Let $(x,y)\in PC\times PC$ be a solution of the abstract nonlinear equation $x\in\lambda N_1(x,y)$ and $y\in\lambda N_2(x,y)$. Then there exists $f_*^i\in S_{F^i}$ for $t\in[0,b]$ for each $i=\{1,2\}$, such that

$$x(t)=C(t)x_0+S(t)(x_1-g^1(0,x_0,y_0))+\int_0^t C(t-s)g^1(s,x(s),y(s))ds$$
$$+\sum_{l=1}^{\infty}\int_0^t S(t-s)\sigma_l^1(t,x(s),y(s))dB^H(s)+\int_0^t S(t-s)f_*^1(s)ds$$
$$+\int_0^t S(t-s)f^1(t,x(s),y(s))dW(s)+\sum_{0<t_k<t}C(t-t_k)I_k(x(t_k))$$
$$+\sum_{0<t_k<t}S(t-t_k)I_k^1(x(t_k)),$$

and

$$
\begin{aligned}
y(t) = C(t)y_0 + S(t)(y_1 - g^2(0,x_0,y_0)) + \int_0^t C(t-s)g^2(s,x(s),y(s))ds \\
+ \sum_{l=1}^{\infty}\int_0^t S(t-s)\sigma_l^2(t,x(s),y(s))dB^H(s) + \int_0^t S(t-s)f_*^2(s)ds \\
+ \int_0^t S(t-s)f^2(t,x(s),y(s))dW(s) + \sum_{0<t_k<t} C(t-t_k)\bar{I}_k(x(t_k)) \\
+ \sum_{0<t_k<t} S(t-t_k)I_k^2(x(t_k)).
\end{aligned}
$$

Now, we have

$$
\begin{aligned}
E|x(t)|_X^2 \le\ & 7M^2E|x_0|_X^2 + 7M^2(E|x_1|_X^2 + \overline{a}_1E|x_0|_X^2 + \overline{b}_1E|y_0|_X^2 + \overline{c}_1) \\
& + 7M^2\overline{a}_1\int_0^t E|x(s)|_X^2ds + 7M^2\overline{b}_1\int_0^t E|y(s)|_X^2ds + 7M^2\overline{c}_1b \\
& + 14M^2m\sum_{0<t_k<t} d_kE|x(t_k)|_X^2 + 14M^2m\sum_{0<t_k<t} e_k \\
& + 7M^2c_HH(2H-1)b^{2H-1}\left(\alpha_1\int_0^t E|x(s)|_X^2ds + \beta_1\int_0^t E|y(s)|_X^2ds + c_1b\right) \\
& + 7M^2\int_0^t p_1(s)ds + 7M^2C_2a_1\int_0^t E|x(s)|_X^2ds + 7M^2C_2b_1\int_0^t E|y(s)|_X^2ds + 7C_2M^2c_1b.
\end{aligned}
$$

Similarly,

$$
\begin{aligned}
E|y(t)|_X^2 \le\ & 7M^2E|y_0|_X^2 + 7M^2(E|x_1|_X^2 + \overline{a}_2E|x_0|_X^2 + \overline{b}_2E|y_0|_X^2 + \overline{c}_2) \\
& + 7M^2\overline{a}_2\int_0^t E|x(s)|_X^2ds + 7M^2\overline{b}_2\int_0^t E|y(s)|_X^2ds + 7M^2\overline{c}_2b \\
& + 7M^2c_HH(2H-1)b^{2H-1}\left(\alpha_2\int_0^t E|x(s)|_X^2ds + \beta_2\int_0^t E|y(s)|_X^2ds + c_2b\right) \\
& + 7M^2\int_0^t p_2(s)ds + 7M^2C_2a_2\int_0^t E|x(s)|_X^2ds + 7M^2C_2b_2\int_0^t E|y(s)|_X^2ds \\
& + 7M^2C_2c_2b + 14M^2m\sum_{0<t_k<t}\overline{d}_kE|y(t_k)|_X^2 + 14M^2m\sum_{0<t_k<t}\overline{e}_k.
\end{aligned}
$$

Consider functions $\mu, \overline{\mu}$ defined on $[0,+\infty)$ by

$$
\begin{aligned}
\mu(t) &= \sup\{E|x(s)|_X^2 : 0 \le s \le t\},\\
\overline{\mu}(t) &= \sup\{E|y(s)|_X^2 : 0 \le s \le t\},\\
\mu(t) &\le 7M^2E|x_0|_X^2 + 7M^2(E|x_1|_X^2 + \overline{a}_1E|x_0|_X^2 + \overline{b}_1E|y_0|_X^2 + \overline{c}_1)\\
&\quad + 7M^2\overline{a}_1\int_0^t \mu(s)ds + 7M^2\overline{b}_1\int_0^t \overline{\mu}(s)ds + 7M^2\overline{c}_1 b\\
&\quad + 7M^2c_H H(2H-1)b^{2H-1}\left(\alpha_1\int_0^t \mu(s)ds + \beta_1\int_0^t \overline{\mu}(s)ds + c_1 b\right)\\
&\quad + 7M^2\int_0^t p_1(s)ds + 7M^2C_2a_1\int_0^t \mu(s)ds + 7M^2C_2b_1\int_0^t \overline{\mu}(s)ds + 7M^2C_2c_1b\\
&\quad + 14M^2m\sum_{0<t_k<t} e_k + 14M^2m\sum_{0<t_k<t} d_k\mu(t)\\
&\le K_1 + K_2\int_0^t \mu(s)ds + K_3\int_0^t \overline{\mu}(s)ds.
\end{aligned}
$$

With a similar manner,

$$
\begin{aligned}
\overline{\mu}(t) &\le 7M^2E|y_0|_X^2 + 7M^2(E|y_1|_X^2 + \overline{a}_2E|x_0|_X^2 + \overline{b}_2E|y_0|_X^2 + \overline{c}_2) + 7M^2C_2c_2b\\
&\quad + 7M^2\overline{a}_2\int_0^t \mu(s)ds + 7M^2\overline{b}_2\int_0^t \overline{\mu}(s)ds + 7M^2\overline{c}_2 b\\
&\quad + 7M^2c_H H(2H-1)b^{2H-1}\left(\alpha_2\int_0^t \mu(s)ds + \beta_2\int_0^t \overline{\mu}(s)ds + c_2 b\right)\\
&\quad + 7M^2\int_0^t p_1(s)ds + 7M^2C_2a_2\int_0^t \mu(s)ds + 7M^2C_2b_2\int_0^t \overline{\mu}(s)ds\\
&\quad + 14M^2m\sum_{0<t_k<t} \overline{d}_k\mu(t) + 14M^2m\sum_{0<t_k<t} \overline{e}_k\\
&\le \overline{K}_1 + \overline{K}_2\int_0^t \mu(s)ds + \overline{K}_3\int_0^t \overline{\mu}(s)ds.
\end{aligned}
$$

We can write

$$
\overline{\mu}(t) \le \overline{K}_1 + \overline{K}_2\int_0^t \mu(s)ds + \overline{K}_3\int_0^t \overline{\mu}(s)ds.
$$

Therefore,

$$\mu(t) + \overline{\mu}(t) \leq \overline{K}_{**} + \overline{K}_{*} \int_0^t (\mu(s) + \overline{\mu}(s))ds,$$

where

$$K_1 = \frac{\overline{\xi}_1 + \overline{\xi}_2}{1 - 14M^2 m \sum_{0<t_k<t} d_k},$$

and

$$\overline{\xi}_1 = 7M^2 E|x_0|_X^2 + 7M^2(E|x_1|_X^2 + \overline{a}_1 E|x_0|_X^2 + \overline{b}_1 E|y_0|_X^2 + \overline{c}_1),$$

$$\overline{\xi}_2 = 7M^2\overline{c}_1 b + 7M^2 c_H H(2H-1)b^{2H-1}c_1 b + 7M^2 C_2 c_1 + 14M^2 m \sum_{0<t_k<t} e_k + 7M^2\|p_2\|_{L^1},$$

$$\overline{K}_1 = \frac{\xi_1 + \xi_2}{1 - 14M^2 m \sum_{0<t_k<t} \overline{d}_k},$$

and

$$\xi_1 = 7M^2 E|y_0|_X^2 + 7M^2(E|y_1|_X^2 + \overline{a}_2 E|x_0|_X^2 + \overline{b}_2 E|y_0|_X^2 + \overline{c}_2),$$

and

$$\xi_2 = 7M^2\|p_2\|_{L^1} + 7M^2\overline{c}_2 b + 7M^2 c_H H(2H-1)b^{2H-1}c_1 b + 7M^2 C_2 c_2 + 14M^2 m \sum_{0<t_k<t} \overline{e}_k,$$

$$K_2 = \frac{7M^2\overline{a}_1 + 7M^2 c_H H(2H-1)b^{2H-1}a_1 + 7M^2 C_2 a_1}{1 - 14M^2 m \sum_{0<t_k<t} d_k},$$

and

$$K_3 = \frac{7M^2\overline{b}_1 + 7M^2 c_H H(2H-1)b^{2H-1}\beta_1 + 7M^2 C_2 b_1}{1 - 14M^2 m \sum_{0<t_k<t} d_k},$$

$$\overline{K}_2 = \frac{7M^2\overline{a}_2 + 7M^2 c_H H(2H-1)b^{2H-1}a_2 + 7M^2 C_2 a_2}{1 - 14M^2 m \sum_{0<t_k<t} \overline{d}_k},$$

and

$$\overline{K}_3 = \frac{7M^2\overline{b}_2 + 7M^2 c_H H(2H-1)b^{2H-1}\beta_2 + 7M^2 C_2 b_2}{1 - 14M^2 m \sum_{0<t_k<t} \overline{d}_k},$$

where

$$1 - 14M^2 m \sum_{0<t_k<t} \overline{d}_k > 0, \quad 1 - 14M^2 m \sum_{0<t_k<t} d_k > 0,$$

$$\overline{K}_{**} = K_1 + \overline{K}_1,$$

and

$$K_* = \max\{K_2 + \overline{K}_2, K_3 + \overline{K}_3\},$$

where $\overline{K}_{**}$ is a suitable positive constant. Thanks to the Gronwall inequality, we derive

$$\mu(t) + \overline{\mu}(t) \le \overline{K}_{**} e^{\overline{K}_* t} = M.$$

So, we have

$$E|x(t)|_X^2 + E|y(t)|_X^2 \le M$$

and, consequently,

$$\|x\|_{PC}^2 \le M \quad \text{and} \quad \|y\|_{PC}^2 \le M.$$

Let

$$U = \{(x,y) \in PC \times PC : \|x\|_{PC}^2 < M + 1 \text{ and } \|y\|_{PC}^2 < M + 1\},$$

and consider the operator

$$N : \overline{U} \to \mathcal{P}_{cv,cp}(PC \times PC).$$

From the choice of U, there is no $(x,y) \in \partial U$, such that $x \in \lambda N_1(x,y)$ and $y \in \lambda N_2(x,y)$ for some $\lambda \in (0,1)$. As a consequence of the Leray and Schauder nonlinear alternative (Lemma 10.3), we deduce that N has a fixed point (x,y) in U, which is a solution to problem (10.1). The proof is therefore complete. □

10.1.2 Nonconvex case

In this section, we present a result for the problem (10.1) in the spirit of the nonlinear alternative of the Leray–Schauder type (Lemma 10.4) for single-valued maps, combined with a selection theorem due to Bressan and Colombo,[1] for lower semicontinuous multivalued maps with decomposable values.

Let $\mathcal{A}$ be a subset of $J \times \mathcal{D}$. $\mathcal{A}$ is $\mathcal{L} \otimes \mathcal{B}$ measurable if $\mathcal{A}$ belongs to the σ-algebra generated by all sets of the form $\mathcal{J} \times \mathcal{D}$ where $\mathcal{J}$ is Lebesgue measurable in J and $\mathcal{D}$ is Borel measurable in X. A subset $\mathcal{A}$ of $L^2(J,X)$ is decomposable if, for all $w, v \in \mathcal{A}$, and $\mathcal{J} \in J$ measurable,

1 A. Bressan and G. Colombo, Extensions and selections of maps with decomposable values, *Studia Math.* **90** (1988), 69–86.

$$w\chi_{\mathcal{J}} + v\chi_{J-\mathcal{J}} \in A,$$

where χ stands for the characteristic function.

Let

$$F^i : J \times X \times X \longrightarrow \mathcal{P}(X),$$

be a multivalued operator with nonempty closed values. F is lower semicontinuous (l. s. c.) if the set

$$\{x, y \in X : F^i(x, y) \cap B \neq \emptyset\},$$

is open for any open set B in $X \times X$.

Definition 10.3. Let Y be a separable metric space, and let

$$L : Y \longrightarrow \mathcal{P}(L^2(J, X)),$$

be a multivalued operator. We say that L has property (BC) if:
1. L is lower semicontinuous (l. s. c.).
2. L has nonempty closed and decomposable values.

Let

$$F^i : J \times X \times X \longrightarrow \mathcal{P}(X),$$

be a multivalued map with nonempty compact values. Assign to F^i the multivalued operator

$$\Psi^i : PC \times PC \longrightarrow \mathcal{P}(L^2(J, X)),$$

by letting

$$\Psi^i(x, y) = \{f_*^i \in L^2(J, X) : f_*^i(t) \in F^i(t, x, y) \text{ for a. e. } t \in J\}.$$

The operator Ψ is called the Nemytskii operator associated with F^i.

Definition 10.4. Let

$$F^i : J \times X \times X \longrightarrow \mathcal{P}(X),$$

be a multivalued function with nonempty compact values. We say F^i is of lower semicontinuous type (l. s. c. type) if its associated Nemytskii operator Ψ is lower semicontinuous and has nonempty closed and decomposable values.

Next, we state a selection lemma due to Bressan and Colombo.

Lemma 10.5. *Let Y be a separable metric space, and let*

$$L : Y \longrightarrow \mathcal{P}(L^2(J,X)),$$

be a multivalued operator that has property (BC). Then L has a continuous selection, that is, there exists a continuous (single-valued) function

$$\tilde{g} : Y \longrightarrow L^2(J,X),$$

such that

$$\tilde{g}(y) \in L(y),$$

for every $y \in Y$.

Let us introduce the following hypotheses, which are assumed hereafter:

(H_{52}) $F^i : J \times X \times X \longrightarrow \mathcal{P}(X)$ is a nonempty compact valued multivalued map such that:
  (a) $(t,x,y) \longmapsto F^i(t,x,y)$ is $\mathcal{L} \otimes \mathcal{B}$ measurable, and for every $t \in J$, the multifunction $t \to F^i(t,x,y)$ is measurable.
  (b) $(t,x,y) \longmapsto F^i(t,x,y)$ is lower semicontinuous for almost every $t \in J$.

(H_{53}) $E|F^i(t,x,y)|^2 = \sup\{E(|f_*^i|)^2 : f_* \in F^i(t,x,y)\} \le p(t)$, $t \in J$, $p : \mathbb{R}^+ \to (0,\infty)$ is continuous and increasing for all $x,y \in X$.

The following lemma is crucial to the proof of our main theorem.

Lemma 10.6.[2] *Let*

$$F^i : J \times X \times X \longrightarrow \mathcal{P}_{cp}(X),$$

be a multivalued map. Assume that (H_{52}) and (H_{53}) hold. Then F^i is of l. s. c. type.

Theorem 10.2. *Assume that hypotheses (H_{48})–(H_{51}) and (H_{52})–(H_{53}) hold. Then the impulsive initial value problem (10.1) has at least one solution.*

Proof. (H_{52})–(H_{53}) imply, by Lemma 10.6, that F^i is of lower semicontinuous type. From Lemma 10.5, there is a continuous selection

$$f_*^i : PC \times PC \to L^2([0,b],X),$$

such that $f_*^i \in \Psi(x,y)$ for all $(x,y) \in PC \times PC$. Consider the problem

2 M. Frigon and A. Granas, Théorèmes d'existence pour des inclusions différentielles sans convexité, *C. R. Acad. Sci. Paris, Ser. 1* **310** (1990), 819–822.

$$\begin{cases} d(x'(t) - g^1(t,x(t),y(t))) = (Ax(t) + f^1_*(x,y)(t))dt + f^1(t,x(t),y(t)))dW(t) \\ \quad + \sum_{l=1}^{\infty} \sigma_l^1(t,x(t),y(t))dB^H(t), \quad t \in [0,b],\ t \neq t_k, \\ d(y'(t) - g^2(t,x(t),y(t))) = (Ax(t) + f^2_*(x,y)(t))dt + f^1(t,x(t),y(t)))dW(t) \\ \quad + \sum_{l=1}^{\infty} \sigma_l^2(t,x(t),y(t))dB^H(t), \quad t \in [0,b],\ t \neq t_k, \\ \Delta x(t) = I_k(x(t_k)), \quad \Delta x'(t) = I_k^1(x(t_k)), \quad t = t_k,\ k = 1,2,\dots,m, \\ \Delta y(t) = \overline{I}_k(y(t_k)), \quad \Delta y'(t) = \overline{I}_k^2(y(t_k)), \quad t = t_k,\ k = 1,2,\dots,m, \\ x(0) = x_0, \quad y(0) = y_0, \\ x'(0) = x_1, \quad y'(0) = y_1. \end{cases}$$

Consider the operator

$$G^i : PC \times PC \to \mathcal{P}(PC),$$

defined by

$$\begin{aligned} G^1(x,y)(t) = {} & C(t)x_0 + S(t)(x_1 - g^1(0,x_0,y_0)) + \int_0^t C(t-s)g^1(s,x(s),y(s))ds \\ & + \sum_{l=1}^{\infty} \int_0^t S(t-s)\sigma_l^1(s,x(s),y(s))dB^H(s) + \int_0^t S(t-s)f^1_*(x,y)(s)ds \\ & + \int_0^t S(t-s)f^1(s,x(s),y(s))dW(s) + \sum_{0<t_k<t} C(t-t_k)I_k(x(t_k)) \\ & + \sum_{0<t_k<t} S(t-t_k)I_k^1(x(t_k)), \end{aligned}$$

and

$$\begin{aligned} G^2(x,y)(t) = {} & C(t)y_0 + S(t)(y_1 - g^2(0,x_0,y_0)) + \int_0^t C(t-s)g^2(s,x(s),y(s))ds \\ & + \sum_{l=1}^{\infty} \int_0^t S(t-s)\sigma_l^2(s,x(s),y(s))dB^H(s) + \int_0^t S(t-s)f^2_*(x,y)(s)ds \\ & + \int_0^t S(t-s)f^1(s,x(s),y(s))dW(s) + \sum_{0<t_k<t} C(t-t_k)\overline{I}_k(y(t_k)) \\ & + \sum_{0<t_k<t} S(t-t_k)I_k^2(y(t_k)). \end{aligned}$$

We define the Nemytskii operator associated with F^i,

$$\Psi^i : PC \times PC \longrightarrow \mathcal{P}(L^2(J,X)),$$

by letting

$$\Psi^i(x,y) = \{f_*^i \in L^2(J,X) : f_*^i(t) \in F^i(t,x,y) \text{ for a. e. } t \in J, \text{ for each } i = 1,2\}.$$

As in Theorem 10.1, we can prove that the single-valued operator G^i is compact and there exists $M_* > 0$ such that for all possible solutions (x,y), we have $\|x\|_{PC} < M_*$ and $\|y\|_{PC} < M_*$. Now, we only check that G^i is continuous. Let (x_n, y_n) be a sequence such that

$$(x_n, y_n) \to (x_*, y_*),$$

in $PC \times PC$, as $n \to \infty$. Then

$$\begin{aligned}
&E|G^1(x_n,y_n)(t) - G^1(x_*,y_*)(t)|_X^2 \\
&\quad = 6M^2 b \int_0^t E|g^1(s,x_n(s),y_n(s)) - g^1(s,x_*(s),y_*(s))|_X^2 ds \\
&\qquad + 6M^2 c(H)H(2H-1)b^{2H-1} \int_0^t E|\sigma^1(t,x_n(s),y_n(s)) - \sigma^1(t,x_*(s),y_*(s))|_X^2 ds \\
&\qquad + 6M^2 \int_0^t E|f_{*,n}^1(x_n(s),y_n)(s) - f_{*,n}^1(x_*(s),y_*)(s)|_X^2 ds \\
&\qquad + 6M^2 C_2 \int_0^t E|f^1(t,x_n(s),y_n(s)) - f^1(t,x_*(s),y_*(s))|_X^2 ds \\
&\qquad + 6mM^2 \sum_{0<t_k<t} E|I_k(x_n(t_k)) - I_k(x_*(t_k))|_X^2 + 6mM^2 \sum_{0<t_k<t} E|I_k^1(x_n(t_k)) - I_k^1(x_*(t_k))|_X^2.
\end{aligned}$$

Similarly,

$$\begin{aligned}
&E|G^2(x_n,y_n)(t) - G^2(x_*,y_*)(t)|_X^2 \\
&\quad = 6M^2 b \int_0^t E|g^2(s,x_n(s),y_n(s)) - g^2(s,x_*(s),y_*(s))|_X^2 ds \\
&\qquad + 6M^2 c(H)H(2H-1)b^{2H-1} \int_0^t E|\sigma^2(t,x_n(s),y_n(s)) - \sigma^2(t,x_*(s),y_*(s))|_X^2 ds \\
&\qquad + 6M^2 \int_0^t E|f_{*,n}^2(x_n(s),y_n)(s) - f_{*,n}^2(x_*(s),y_*)(s)|_X^2 ds
\end{aligned}$$

$$
\begin{aligned}
&+ 6M^2C_2 \int_0^t E|f^2(t,x_n(s),y_n(s)) - f^2(t,x_*(s),y_*(s))|_X^2 ds \\
&+ 6mM^2 \sum_{0<t_k<t} E|\bar{I}_k(y_n(t_k)) - \bar{I}_k(y_*(t_k))|_X^2 \\
&+ 6mM^2 \sum_{0<t_k<t} E|I_k^2(y_n(t_k)) - I_k^2(y_*(t_k))|_X^2.
\end{aligned}
$$

Since f^i, g^i, σ^i, $f_{*,n}^i$ are Carathéodory functions and I_k^i, $\bar{I}_k$, I_k are continuous functions. By the Lebesgue dominated convergence theorem,

$$
\begin{aligned}
&\|G^1(x_n,y_n)(t) - G^1(x_*,y_*)(t)\|_{PC} \\
&\quad = 6M^2b\|g^1(s,x_n(s),y_n(s)) - g^1(s,x_*(s),y_*(s))\|_{L^2}^2 \\
&\qquad + 6M^2c(H)H(2H-1)b^{2H-1}\|\sigma^1(t,x_n(s),y_n(s)) - \sigma^1(t,x_*(s),y_*(s))\|_{L^2}^2 \\
&\qquad + 6M^2 \int_0^t \|f_{*,n}^1(x_n(s),y_n)(s) - f_{*,n}^1(x_*(s),y_*)(s)\|_{L^2}^2 \\
&\qquad + 6M^2C_2\|f^1(t,x_n(s),y_n(s)) - f^1(t,x_*(s),y_*(s))\|_{L^2}^2 \\
&\qquad + 6mM^2 \sum_{0<t_k<t} E|I_k(x_n(t_k)) - I_k(x_*(t_k))|_X^2 \\
&\qquad + 6mM^2 \sum_{0<t_k<t} E|I_k^1(x_n(t_k)) - I_k^1(x_*(t_k))|_X^2 \to 0 \quad \text{as } n \to \infty,
\end{aligned}
$$

and

$$
\begin{aligned}
&\|G^2(x_n,y_n)(t) - G^2(x_*,y_*)(t)\|_{PC} \\
&\quad = 6M^2b\|g^2(s,x_n(s),y_n(s)) - g^2(s,x_*(s),y_*(s))\|_{L^2}^2 \\
&\qquad + 6M^2c(H)H(2H-1)b^{2H-1}\|\sigma^2(t,x_n(s),y_n(s)) - \sigma^2(t,x_*(s),y_*(s))\|_{L^2}^2 \\
&\qquad + 6M^2 \int_0^t \|f_{*,n}^2(x_n(s),y_n)(s) - f_{*,n}^2(x_*(s),y_*)(s)\|_{L^2}^2 \\
&\qquad + 6M^2C_2\|f^2(t,x_n(s),y_n(s)) - f^2(t,x_*(s),y_*(s))\|_{L^2}^2 \\
&\qquad + 6mM^2 \sum_{0<t_k<t} E|I_k(y_n(t_k)) - I_k(y_*(t_k))|_X^2 \\
&\qquad + 6mM^2 \sum_{0<t_k<t} E|I_k^2(y_n(t_k)) - I_k^2(y_*(t_k))|_X^2 \to 0 \quad \text{as } n \to \infty.
\end{aligned}
$$

Set

$$
U = \{(x,y) \in PC \times PC : \|y\|_{PC} < M_*,\ \|y\|_{PC} < M_*\}.
$$

From the choice of U, there is no $y \in \partial U$ such that $(x,y) = \lambda N(x,y)$ for in $\lambda \in (0,1)$. As a consequence of the nonlinear alternative of the Leray–Schauder Lemma 10.3, we deduce that G has a fixed point $(x,y) \in U$, which is a solution to the problem (10.1). This completes the proof. □

10.1.3 A further result

Now, we present a second result for the problem (10.1) with a nonconvex valued right-hand side. Our considerations are based on a fixed-point theorem for contraction multivalued operators given by Covitz and Nadler in 1970, Theorem 11.1.[3] In this section, we will provide a multivalued version of Perov's fixed-point theorem.

Definition 10.5. Let (X,d) be a generalized metric space. An operator

$$N : X \to X,$$

is said to be contractive if there exists a convergent matrix M to zero such that

$$d(N(x), N(y)) \le Md(x,y) \quad \text{for all } x, y \in X.$$

Theorem 10.3. *(X,d) is a complete generalized metric space, and*

$$F : X \to \mathcal{P}_{cl,b}(X),$$

is a contractive multivalued operator with a Lipschitz matrix M. Then N has at least one fixed point.

Let us introduce the following hypotheses:

(H_{54}) $F^i : J \times X \times X \longrightarrow \mathcal{P}_{cp}(X); (t,y) \longrightarrow F^i(t,x,y)$ is measurable for each $(x,y) \in X \times X$.

(H_{55}) There exist constants $a_i, b_i \in \mathbb{R}^+$ such that

$$H^2_{d_i}(F^1(t,x,y), F^1(t,\overline{x},\overline{y})) \le a_i|x-\overline{x}|^2_X + b_i|y-\overline{y}|^2_X,$$

for all $x, y, \overline{x}, \overline{y} \in X$ for each $i = 1, 2$.

(H_{56}) There exist functions $\alpha_i, \beta_i \in \mathbb{R}^+$ for each $i = 1, 2$, such that

$$\|\sigma^1(t,x,y) - \sigma^1(t,\overline{x},\overline{y})\|^2 \le \alpha_i|x-\overline{x}|^2_X + \beta_i|y-\overline{y}|^2_X,$$

for all $x, y, \overline{x}, \overline{y} \in X$ and $t \in J$.

3 K. Deimling, *Multi-Valued Differential Equations*, Walter De Gruyter, Berlin–New York, 1992.

(H_{57}) There exist constants $a_i, b_i, \overline{a}_i, \overline{b}_i \in \mathbb{R}^+$, such that

$$|f^i(t,x,y) - f^i(t,\overline{x},\overline{y})|_X^2 \le a_{i+2}|x-\overline{x}|_X^2 + b_{i+2}|y-\overline{y}|_X^2,$$

and

$$|g^i(t,x,y) - g^i(t,\overline{x},\overline{y})|_X^2 \le \overline{a}_i|x-\overline{x}|_X^2 + \overline{b}_i|y-\overline{y}|_X^2,$$

for all $x, y, \overline{x}, \overline{y} \in X$.

(H_{58}) There exist constants $d_k \ge 0$ and $\overline{d}_k \ge 0$, $k = 1, \dots, m$, such that

$$|I_k^1(x) - I_k^1(\overline{x})|_X^2 \vee |I_k(x) - I_k(\overline{x})|_X^2 \le d_k|x-\overline{x}|_X^2,$$
$$|I_k^2(x) - I_k^2(\overline{x})|_X^2 \vee |\overline{I}_k(y) - \overline{I}_k(\overline{y})|^2 \le \overline{d}_k|y-\overline{y}|_X^2,$$

for all $x, y, \overline{x}, \overline{y} \in X$.

Theorem 10.4. *Assume that hypotheses (H_{54})–(H_{58}) are satisfied. If the matrix,*

$$M = \begin{pmatrix} L_1 & L_2 \\ L_3 & L_4 \end{pmatrix},$$

where

$$L_3 = \sqrt{\left\{(6M^2\overline{a}_2 + 6M^2c_H H(2H-1)b^{2H-1}a_2 + 6M^2a_2 + 6M^2a_4C_2)b + 12M^2m\sum_{k=1}^m \overline{d}_k\right\}},$$
$$L_4 = \sqrt{\{b(6\overline{b}_2 6M^2 + 6M^2c_H H(2H-1)b^{2H-1}\beta_2 + 6M^2b_2 + 6M^2b_4C_2)\}},$$

and

$$L_1 = \sqrt{\left\{(6M^2\overline{a}_1 + 6M^2c_H H(2H-1)b^{2H-1}a_1 + 6M^2a_1 + 6M^2a_3C_2)b + 12M^2m\sum_{k=1}^m d_k\right\}},$$
$$L_2 = \sqrt{\{(\overline{b}_1 6M^2 + 6M^2c_H H(2H-1)b^{2H-1}\beta_1 + 6M^2b_3C_2 + 6M^2b_1)b\}},$$

converges to zero, then problem (10.1) has at least one mild solution.

Proof. In order to transform the problem (10.1) into a fixed-point problem, let the multivalued operator

$$N : PC \times PC \to \mathcal{P}(PC \times PC),$$

be defined as in Theorem 10.1. We shall show that N satisfies the assumptions of Theorem 10.3. Note that (H_{50}) implies that F^i, for each $i = 1, 2$ has at most linear growth, that is,

$$E|F^i(t,x,y)|_X^2 \le a_i E|x|_X^2 + b_i E|y|_X^2,$$

for a. e. $t \in J$ and all $x,y \in X$

(1) $N(x,y) \in \mathcal{P}_{cl}(PC \times PC)$ for each $(x,y) \in PC \times PC$. The proof is similar to that in Theorem 10.1 in (Step 2) and is omitted.

(2) There exists a matrix M with $\|M\| < 1$ such that

$$EH_d(N(x,y), N(\overline{x},\overline{y})) \le M \begin{pmatrix} \|x-\overline{x}\|_{PC} \\ \|y-\overline{y}\|_{PC} \end{pmatrix},$$

for any $x,y,\overline{x},\overline{y} \in PC$ and $h_i \in N_i(x,y)$ there exists $f_*^i(\cdot) \in S_{F^i,x,y}$ (f_*^i is a measurable selection) such that for each $t \in [0,b]$, $i = 1,2$ let

$$\begin{aligned} h_1(t) = {} & C(t)x_0 + S(t)(x_1 - g^1(0,x_0,y_0)) + \int_0^t C(t-s)g^1(s,x(s),y(s))ds \\ & + \sum_{l=1}^{\infty} \int_0^t S(t-s)\sigma_l^1(t,x(s),y(s))dB^H(s) + \int_0^t S(t-s)f_*^1(s)ds \\ & + \int_0^t S(t-s)f^1(t,x(s),y(s))dW(s) + \sum_{0<t_k<t} C(t-t_k)I_k(x(t_k)) \\ & + \sum_{0<t_k<t} S(t-t_k)I_k^1(x(t_k)), \end{aligned}$$

and

$$\begin{aligned} h_2(t) = {} & C(t)y_0 + S(t)(y_1 - g^2(0,x_0,y_0)) + \int_0^t C(t-s)g^2(s,x(s),y(s))ds \\ & + \sum_{l=1}^{\infty} \int_0^t S(t-s)\sigma_l^2(t,x(s),y(s))dB^H(s) + \int_0^t S(t-s)f_*^2(s)ds \\ & + \int_0^t S(t-s)f^2(t,x(s),y(s))dW(s) + \sum_{0<t_k<t} C(t-t_k)\overline{I}_k(y(t_k)) \\ & + \sum_{0<t_k<t} S(t-t_k)I_k^2(y(t_k)). \end{aligned}$$

From (H_{52}),

$$EH_{d_i}^2(F(t,x,y), F(t,\overline{x},\overline{y})) \le a_i E|x-\overline{x}|_X^2 + b_i E|y-\overline{y}|_X^2, \quad \text{a. e. } t \in J.$$

Hence, there is

$$(w, \overline{w}) \in F^1(t, \overline{x}(t), \overline{y}(t)) \times F^2(t, \overline{x}(t), \overline{y}(t)),$$

such that

$$E|f_*^1(t) - w|_X^2 \leq a_1 E|x - \overline{x}|_X^2 + b_1 E|y - \overline{y}|_X^2, \quad t \in [0, b],$$

and

$$E|f_*^2(t) - \overline{w}|^2 \leq a_2 E|x - \overline{x}|_X^2 + b_2 E|y - \overline{y}|_X^2, \quad t \in [0, b]$$

Consider the multivalued maps

$$U_i : [0, b] \to \mathcal{P}(X),$$

$i = 1, 2$ defined by

$$U_1(t) = \{w \in F^1(t, \overline{x}(t), \overline{y}(t)) : E|f^1(t) - w|^2 \leq a_1 E|x - \overline{x}|_X^2 + b_1 E|y - \overline{y}|_X^2 \text{ a.e. } t \in [0, b]\},$$

and similarly,

$$U_2(t) = \{\overline{w} \in F^2(t, \overline{x}(t), \overline{y}(t)) : E|f^2(t) - \overline{w}|^2 \leq a_2 E|x - \overline{x}|_X^2 + b_2 E|y - \overline{y}|_X^2, \text{ a.e. } t \in [0, b]\},$$

that is,

$$U_1 = \overline{B}(f_*^1(t), a_1 E|x - \overline{x}|_X^2 + b_1 E|y - \overline{y}|_X^2),$$

and

$$U_2 = \overline{B}(f_*^2(t), a_2 E|x - \overline{x}|_X^2 + b_2 E|y - \overline{y}|_X^2).$$

Since $f^i, a_i, b_i, x, y, \overline{x}, \overline{y}$, are measurable for each $i = 1, 2$, then the closed ball U_i is measurable. In addition (H_{51}) and (H_{52}) imply that for each $(x, y) \in PC \times PC$ and $F^i(t, x(t), y(t))$ is measurable. Finally, the set

$$V_i(.) = U_i(.) \cap F^i(., \overline{x}(.), \overline{y}(.)),$$

is nonempty since it contains w, $\overline{w}$. Therefore, the intersection multivalued operator V_i is measurable with nonempty, closed values, there exists a function $\overline{f}^i(t)$, which is a measurable selection for $V_i(.)$. Thus,

$$\overline{f}_*^i(t) \in F^i(t, \overline{x}(t), \overline{y}(t)),$$

and

$$E|f_*^1(t) - \overline{f}_*^1(t)|_X^2 \leq a_1 E|x - \overline{x}|_X^2 + b_1 E|y - \overline{y}|_X^2, \quad \text{for a.e. } t \in J,$$

and

$$E|f_*^2(t) - \overline{f}_*^2(t)|_X^2 \le a_2 E|x - \overline{x}|_X^2 + b_2 E|y - \overline{y}|_X^2, \quad \text{for a. e. } t \in J,$$

$$\begin{aligned}\overline{h}_1(t) &= C(t)x_0 + S(t)(x_1 - g^1(0, x_0, y_0) + \int_0^t C(t-s)g^1(s, \overline{x}(s), \overline{y}(s))ds \\ &+ \sum_{l=1}^{\infty} \int_0^t S(t-s)\sigma_l^1(t, \overline{x}(s), \overline{y}(s))dB^H(s) + \int_0^t S(t-s)\overline{f}_*^1(s)ds \\ &+ \int_0^t S(t-s)f^1(t, \overline{x}(s), \overline{y}(s))dW(s) + \sum_{0<t_k<t} C(t-t_k)I_k(\overline{x}(t_k)) \\ &+ \sum_{0<t_k<t} S(t-t_k)I_k^1(\overline{x}(t_k)),\end{aligned}$$

and

$$\begin{aligned}\overline{h}_2(t) &= C(t)y_0 + S(t)(y_1 - g^2(0, x_0, y_0) + \int_0^t C(t-s)g^2(s, \overline{x}(s), \overline{y}(s))ds \\ &+ \sum_{l=1}^{\infty} \int_0^t S(t-s)\sigma_l^2(t, \overline{x}(s), \overline{y}(s))dB^H(s) + \int_0^t S(t-s)\overline{f}_*^2(s)ds \\ &+ \int_0^t S(t-s)f^1(t, \overline{x}(s), \overline{y}(s))dW(s) + \sum_{0<t_k<t} C(t-t_k)\overline{I}_k(\overline{y}(t_k)) \\ &+ \sum_{0<t_k<t} S(t-t_k)I_k^2(\overline{y}(t_k)).\end{aligned}$$

Let us define for a. e. $t \in J$, which implies that

$$\begin{aligned}E|h_1(t) - \overline{h}_1(t)|_X^2 &\le 6E\left|\int_0^t C(t-s)(g^1(s, x(s), y(s)) - g^1(s, \overline{x}(s), \overline{y}(s)))ds\right|_X^2 \\ &+ 6E\left|\sum_{l=1}^{\infty}\int_0^t S(t-s)(\sigma_l^1(t, x(s), y(s)) - \sigma_l^2(t, \overline{x}(s), \overline{y}(s)))dB^H(s)\right|_X^2 \\ &+ 6E\left|\int_0^t S(t-s)(f_*^1(s) - \overline{f}_*^1(s))ds\right|_X^2 \\ &+ 6E\left|\int_0^t S(t-s)(f^1(t, x(s), y(s)) - f^1(t, \overline{x}(s), \overline{y}(s)))dW(s)\right|_X^2\end{aligned}$$

$$+ 6E\Big|\sum_{0<t_k<t} C(t-t_k)(I_k(x(t_k)) - I_k(\overline{x}(t_k)))\Big|_X^2$$
$$+ 6E\Big|\sum_{0<t_k<t} S(t-t_k)(I_k^1(x(t_k)) - I_k^1(\overline{x}(t_k)))\Big|_X^2.$$

First, applying inequality (10.2) and $(H6)$–$(H9)$ become

$$\begin{aligned}
&\sup_{t\in J} E|h_1(t) - \overline{h}_1(t)|_X^2 \\
&\quad \le 6M^2\overline{a}_1 \int_0^t \sup_{0\le\theta\le s} E|x(\theta) - \overline{x}(\theta)|_X^2 + \overline{b}_1 6M^2 \int_0^t \sup_{0\le\theta\le s} E|y(\theta) - \overline{y}(\theta)|_X^2 \\
&\quad + 6M^2 c_H H(2H-1) b^{2H-1} \alpha_1 \int_0^t \sup_{0\le\theta\le s} E|x(\theta) - \overline{x}(\theta)|_X^2 \\
&\quad + 6M^2 c_H H(2H-1) b^{2H-1} \beta_1 \int_0^t \sup_{0\le\theta\le s} E|y(\theta) - \overline{y}(\theta)|_X^2 ds \\
&\quad + 6M^2 a_1 \int_0^t \sup_{0\le\theta\le s} E|x(\theta) - \overline{x}(\theta)|_X^2 ds \\
&\quad + 6M^2 b_1 \int_0^t \sup_{0\le\theta\le s} E|y(\theta) - \overline{y}(\theta)|_X^2 ds \\
&\quad + 6M^2 a_3 C_2 \int_0^t \sup_{0\le\theta\le s} E|x(\theta) - \overline{x}(\theta)|_X^2 ds \\
&\quad + 6M^2 b_3 C_2 \int_0^t \sup_{0\le\theta\le s} E|y(\theta) - \overline{y}(\theta)|_X^2 ds \\
&\quad + 12M^2 m \sum_{0<t_k<t} d_k \sup_{0\le\theta\le t} E|x(\theta) - \overline{x}(\theta)|_X^2.
\end{aligned}$$

Similarly,

$$\begin{aligned}
\sup_{t\in J} E|h_2(t) - \overline{h}_2(t)|_X^2 &\le 6M^2\overline{a}_2 \int_0^t \sup_{0\le\theta\le s} E|x(\theta) - \overline{x}(\theta)|_X^2 \\
&\quad + 6\overline{b}_2 M^2 \int_0^t \sup_{0\le\theta\le s} E|y(\theta) - \overline{y}(\theta)|_X^2
\end{aligned}$$

$$
\begin{aligned}
&+ 6M^2 c_H H(2H-1) b^{2H-1} \alpha_2 \int_0^t \sup_{0\le\theta\le s} E|x(\theta) - \overline{x}(\theta)|_X^2 \\
&+ 6M^2 c_H H(2H-1) b^{2H-1} \beta_2 \int_0^t \sup_{0\le\theta\le s} E|y(\theta) - \overline{y}(\theta)|_X^2 ds \\
&+ 6M^2 a_2 \int_0^t \sup_{0\le\theta\le s} E|x(\theta) - \overline{x}(\theta)|_X^2 ds \\
&+ 6M^2 b_2 \int_0^t \sup_{0\le\theta\le s} E|y(\theta) - \overline{y}(\theta)|_X^2 ds \\
&+ 6M^2 a_4 C_2 \int_0^t \sup_{0\le\theta\le s} E|x(\theta) - \overline{x}(\theta)|_X^2 ds \\
&+ 6M^2 b_4 C_2 \int_0^t \sup_{0\le\theta\le s} E|y(\theta) - \overline{y}(\theta)|_X^2 ds \\
&+ 12M^2 \sum_{0<t_k<t} \overline{d}_k \sup_{0\le\theta\le t} E|x(\theta) - \overline{x}(\theta)|_X^2,
\end{aligned}
$$

such that we can prove that

$$
\sup_{t\in J} E|h_1(t) - \overline{h}_1(t)|_X \le L_1 \|x - \overline{x}\|_{PC} + L_2 \|y - \overline{y}\|_{PC}.
$$

On the other hand,

$$
\sup_{t\in J} E|h_1(t) - \overline{h}_1(t)|_X \le L_3 \|x - \overline{x}\|_{PC} + L_4 \|y - \overline{y}\|_{PC}.
$$

By an analogous relation obtained by exchanging the roles of x, y and $\overline{x}, \overline{y}$,

$$
\begin{aligned}
EH_d(N(t,x,y), N(t,\overline{x},\overline{y})) &= \begin{pmatrix} EH_{d_1}(N_1((x,y) - N_1(\overline{x},\overline{y})) \\ EH_{d_2}(N_2(x,y) - N_2(\overline{x},\overline{y})) \end{pmatrix} \\
&\le M \begin{pmatrix} \|x - \overline{x}\|_{PC} \\ \|y - \overline{y}\|_{PC} \end{pmatrix},
\end{aligned}
$$

where

$$
M = \begin{pmatrix} L_1 & L_2 \\ L_3 & L_4 \end{pmatrix}.
$$

Therefore, combining the above relations, we obtain

$$EH_d(N(t,x,y),N(t,\overline{x},\overline{y})) \leq M\begin{pmatrix} \|x-\overline{x}\|_{PC} \\ \|y-\overline{y}\|_{PC} \end{pmatrix}.$$

As M converges to zero, and thus by Theorem 10.3, N has a fixed point (x,y), which is a mild solution to (10.1). This completes the proof. □

10.1.4 Example

We consider the following stochastic nonlinear wave equation with impulsive effects:

$$\begin{cases} d(\frac{\partial u(t,\xi)}{\partial t} - g^1(t,u(t,\xi),v(t,\xi))) \in (\frac{\partial^2 u(t,\xi)}{\partial \xi^2} + F^1(t,u(t,\xi),v(t,\xi)))dt + \sigma^1(t)dB_Q^H(t) \\ \quad +f^1(t)dW(t), \quad t \geq 0,\ t \neq t_k,\ 0 \leq \xi \leq \pi, \\ d(\frac{\partial v(t,\xi)}{\partial t} - g^2(t,u(t,\xi),v(t,\xi))) \in (\frac{\partial^2 u(t,\xi)}{\partial \xi^2} + F^2(t,u(t,\xi),v(t,\xi)))dt + \sigma^2(t)dB_Q^H(t) \\ \quad +f^2(t)dW(t), \quad t \geq 0,\ t \neq t_k,\ 0 \leq \xi \leq \pi, \\ u(t_k^+,\xi) - u(t_k^-,\xi) = I_k(u(t_k,\xi)), \\ v(t_k^+,\xi) - v(t_k,\xi) = \overline{I}_k(v(t_k,\xi)), \quad k = 1,\dots,m, \\ u'(t_k^+,\xi) - u'(t_k^-,x) = I_k^1(u(t_k,\xi)), \\ v'(t_k^+,\xi) - v'(t_k,x) = \overline{I}_k^2(v(t_k,\xi)), \quad k = 1,\dots,m, \\ u(t,0) = u(t,\pi) = 0, \quad t \geq 0, \\ u'(0,\xi) = \eta_1(\xi), \\ v'(0,\xi) = \eta_2(\xi), \\ u(0,\xi) = u_0(\xi), \quad 0 \leq \xi \leq \pi, \\ v(0,\xi) = v_0(\xi), \quad 0 \leq \xi \leq \pi. \end{cases} \tag{10.4}$$

Take

$$\mathcal{K} = \mathcal{H} = L^2([0,\pi]).$$

We define the operator A by

$$(A_1u)(\xi) = \frac{d^2u(\xi)}{d\xi^2},$$

and

$$(A_2v)(\xi) = \frac{d^2v(\xi)}{d\xi^2},$$

with domain

$$D(I) = \{u \in \mathcal{H}, u'' \in \mathcal{H} \text{ and } u(0) = u(\pi) = 0\}.$$

The spectrum of A consists of e^{-n^2t} for $n \in \mathbb{N}$, with associated eigenvectors

$$u_n(\xi) = (2/\pi)^{1/2}\sin(n\xi).$$

Furthermore, the set $\{u_n, n \in \mathbb{N}\}$ is an orthonormal basis of $\mathcal{H}$. In particular,

$$A_i x = \sum_{n=1}^{\infty} e^{-n^2t}\langle x, z_n\rangle z_n, \quad x \in \mathcal{H}, \text{ for each } i = 1, 2.$$

Moreover, the operators $C(t)$ are defined by

$$C(t)x = \sum_{n=1}^{\infty} \cos(nt)\langle x, z_n\rangle z_n, \quad t \in \mathbb{R}$$

a cosine function on $\mathcal{H}$, with an associated sine function.

$$S(t)x = \sum_{n=1}^{\infty} \frac{\sin(nt)}{n}\langle x, z_n\rangle z_n, \quad t \in \mathbb{R}.$$

Notice that

$$\{C(t)\},\ \{S(t)\},\ t \in J,$$

is compact, and there exists a constant $M \geq 1$ such that

$$\|S(t)\| \leq M,$$

and

$$\|C(t)\| \leq M.$$

In order to define the operator

$$Q : \mathcal{K} \longrightarrow \mathcal{K},$$

we choose a sequence

$$\{\sigma_n\}_{n\geq 1} \subset \mathbb{R}^+$$

in the set $Qe_n = \sigma_n e_n$ and assume that

$$\operatorname{tr}(Q) = \sum_{n=1}^{\infty} \sqrt{\sigma_n} < \infty.$$

Define the process $B_Q^H(s)$ by

$$B_Q^H = \sum_{n=1}^{\infty} \sqrt{\sigma_n} \gamma_n^H(t) e_n,$$

where $H \in (1/2, 1)$, and $\{\gamma_n^H\}_{n \in \mathbb{N}}$ is a sequence of two-sided, one-dimensional fractional Brownian motions that are mutually independent.

Here, $W(t)$ is a standard cylindrical Wiener process in $\mathbb{R}$ defined on the probability space $(\Omega, \mathcal{F}, P)$, where

$$F^i(t, u, v) = [h_i(t, u, v), \overline{h}_i(t, u, v)],$$

and

$$g^i : [0, b] \times \mathbb{R} \times \mathbb{R} \to \mathbb{R},$$

for each $i = 1, 2$

($\mathcal{K}_1$) Let $h_i, \overline{h}_i : [0, b] \times \mathbb{R} \times \mathbb{R} \to \mathbb{R}$, $i = 1, 2$ are functions such that

$$h_1(t, u, v) \leq \overline{h}_1(t, u, v),$$
$$h_2(t, u, v) \leq \overline{h}_2(t, u, v), \quad \text{for all } (t, u, v) \in [0, b] \times \mathbb{R} \times \mathbb{R}.$$

($\mathcal{K}_2$) There exist

$$\phi_i, \psi_i \in L^1([0, b], \mathbb{R}_+) \cap L^\infty([0, T], \mathbb{R}_+),$$

such that

$$|h_i(t, u, v)| \leq \phi_i(t) \quad \text{and} \quad |\overline{h}_i(t, u, v)| \leq \psi_i(t), \quad i = 1, 2$$

for each $(t, u, v) \in [0, b] \times \mathbb{R} \times \mathbb{R}$.

($\mathcal{K}_3$) h_1, h_2 are l. s. c. and $\overline{h}_1, \overline{h}_2$ are u. s. c.

($\mathcal{K}_4$) The function

$$\sigma_i : [0, b] \longrightarrow L_Q^0(\mathcal{K}, \mathcal{H}),$$

is bounded, that is, there exists a positive constant L_i such that

$$\int_0^b \|\sigma_i(s)\|_{L_Q^0}^2 ds < L_i, \quad i = 1, 2.$$

The function

$$f_i : [0, T] \longrightarrow L^0(\mathcal{K}, \mathcal{H}),$$

is bounded, that is, there exists a positive constant M_i such that

$$\int_0^b \|f^i(s)\|^2_{L^0} ds < M_i, \quad i = 1, 2.$$

($\mathcal{K}_5$) There exist some positive number $d_k, \overline{d}_k$ $k \in \{1, \ldots, m\}$, such that

$$\max\{|I_k u(t_k, \xi)|, |I^1_k(u)(t_k, \xi)|\} = d_k,$$

and

$$\max\{\overline{I}_k v(t_k, \xi), \overline{I}^2_k(v)(t_k, \xi)\} = \overline{d}_k,$$

where

$$\begin{aligned} x(t)(\xi) &= u(t, \xi), \quad t \in J,\ 0 \le \xi \le \pi, \\ I^1_k(u)(t_k, \xi = I_k(x(t_k))) &= d_k \frac{u(t_k^-, \xi)}{1 + |u(t_k^-, .)|_X}, \quad 0 \le \xi \le \pi,\ k = 1, \ldots, m, \\ \overline{I}_k v(t_k, \xi) = \overline{I}^2_k(v)(t_k, \xi) &= \bar{d}_k \frac{v(t_k^-, \xi)}{1 + |v(t_k^-,)|_X}, \quad 0 \le \xi \le \pi,\ k = 1, \ldots, m, \\ x(0)(\xi) = u(0, \xi) = x_0(\xi), &\quad y(0)(\xi) = v(0, \xi) = y_0(\xi), \quad 0 \le \xi \le \pi, \\ x'(0)(\xi) = u'(0, \xi) = x_1(\xi), &\quad y'(0)(\xi) = v'(0, \xi) = y_1(\xi), \quad 0 \le \xi \le \pi, \end{aligned}$$

where $d_k, \overline{d}_k \in \mathbb{R}^+, k = 1, \ldots, m$. Assume that $(\mathcal{K}_1) - (\mathcal{K}_5)$ are satisfied. Thus, problem (10.4) can be written in the abstract form

$$\begin{cases} d(x'(t) - g^1(t, x(t), y(t))) \in (Ax(t) + F^1(t, x(t), y(t))dt + f^1(t, x(t), y(t))dW(t) \\ \quad + \sum_{l=1}^{\infty} \sigma^1_l(t, x(t), y(t))dB^H_l(t), \quad t \in [0, b],\ t \neq t_k, \\ d(y'(t) - g^2(t, x(t), y(t))) \in (Ax(t) + F^2(t, x(t), y(t)))dt + f^1(t, x(t), y(t))dW(t) \\ \quad + \sum_{l=1}^{\infty} \sigma^2_l(t, x(t), y(t))dB^H_l(t), \quad t \in [0, b],\ t \neq t_k, \\ \Delta x(t) = I_k(x(t_k)) \quad \Delta x'(t) = I^1_k(x(t_k)), \quad t = t_k,\ k = 1, 2, \ldots, m, \\ \Delta y(t) = \overline{I}_k(y(t_k)), \quad \Delta y'(t) = \overline{I}^2_k(y(t_k)), \quad t = t_k,\ k = 1, 2, \ldots, m, \\ x(0) = x_0, \quad y(0) = y_0, \\ x'(0) = x_1, \quad y'(0) = y_1, \end{cases} \tag{10.5}$$

where $A_1 = A_2 = A$. Since for each $k = 1, \ldots, m$, then Theorem 10.1 ensures that problem (10.4) possesses at least one solution.

11 Sweeping process with fractional Brownian by time delay

We study a convex stochastic sweeping process with fractional Brownian motion and time delay. The approach is based on discretizing stochastic functional differential inclusions. The so-called sweeping process is a particular differential inclusion of the general form

$$\begin{aligned}-x'(t) &\in N_{C(t)}(x(t)) \quad \text{a. e. } t \in [0,T],\\ x(0) &\in C(0),\end{aligned}$$

where $C(t)$ is a convex time-dependent set, and $N_C(t)(x(t))$ is the normal cone to $C(t)$ at $x(t)$. The sweeping process was introduced by Moreau in the early 1970s. We consider the problem formally expressed by

$$\begin{cases}-dx(t) \in N_{C_1(t)}(x(t))dt + G^1(t,x_t,y_t)dB^{H_1}, & \text{a. e. } t \in J = [0,T],\\ -dy(t) \in N_{C_2(t)}(y(t))dt + G^2(t,x_t,y_t)dB^{H_2}, & \text{a. e. } t \in J = [0,T],\\ x(t) = \phi(t), & t \in [-r,0],\ x(0) \in C_1(0),\\ y(t) = \overline{\phi}(t), & t \in [-r,0],\ y(0) \in C_2(0),\end{cases} \tag{11.1}$$

where $C_1(t)$, $c_2(t)$ is convex for all t, X is a real separable Hilbert space with an inner product $\langle\cdot,\cdot\rangle$ induced by the norm $\|\cdot\|$, and

$$G^j : M_2([-r,0],X) \times M_2([-r,0],X) \to L^0_{Q_{H_j}}(Y,X),$$

are given functions. Here, $L^0_{Q_{H_j}}(Y,X)$ denotes the space of all Q_{H_j}-Hilbert–Schmidt operators from Y into X; B^{H_j} is a sequence of mutually independent fractional Brownian motions with $H_1 \neq H_2$, i. e. ($B^{H_1} \neq B^{H_2}$) for each $j = 1,2$, with Hurst parameter $H_j > \frac{1}{2}$. Here,

$$y(\cdot,\cdot) : [-r,T] \times \Omega \to X,$$

then for any $t \geq 0$,

$$y_t(\cdot,\cdot) : [-r,0] \times \Omega \to X,$$

is given by

$$y_t(\theta,\omega) = y(t+\theta,\omega), \quad \text{for } \theta \in [-r,0],\ \omega \in \Omega.$$

Here, $y_t(\cdot)$ represents the history of the state from time $t-r$ up to the present time t. Let $M^2([-r,0],X)$ be the following space defined by

$$M^2([-r,0],X) = \{\phi,\overline{\phi}\colon [-r,0] \times \Omega \to X,\ \phi,\overline{\phi} \in C([-r,0],L^2(\Omega,X))\},$$

https://doi.org/10.1515/9783112251898-011

endowed with the norm

$$\|\phi(t)\|_{M^2_{\mathcal{F}_0}} = \int_{-r}^{0} |\phi(t)|^2 dt.$$

Now, for a given $T > 0$, we define

$$M^2([-r,T],X) = \Big\{ y\colon [-r,T] \times \Omega \to X,\ \phi, \overline{\phi} \in C([-r,T], L^2(\Omega,X)) \text{ and } \sup_{t\in[0,T]} E(|y(t)|^2) < \infty, \int_{-r}^{0} |\phi(t)|^2 dt < \infty \Big\},$$

endowed with the norm

$$\|y\|_{M^2_{\mathcal{F}_b}} = \sup_{-r\le s\le T} (E\|y(s)\|^2)^{\frac{1}{2}}.$$

11.1 Existence of unique solution

Definition 11.1. A function $x, y \in M^2([-r,b],X)$ is said to be a solution of (11.1) if x, y satisfies the equation

$$\begin{cases} dx(t) \in N^p(x(t), c_1(t))dt + G^1(t, x_t, y_t)dB^{H_1} & \text{a.e. } t \in [0,T], \\ dy(t) \in N^p(y(t), c_2(t))dt + G^2(t, x_t, y_t)dB^{H_2} & \text{a.e. } t \in [0,T], \end{cases}$$

and the conditions

$$(x(t), y(t)) \in (C_1(t), C_2(t)),$$

for all $t \in [0,T]$.

First, we will list the following hypotheses that will be imposed in our main theorem.

(H_{59}) $C_j(t)$ is convex for every $t \in [0,T]$ and there exists $\lambda > 0$ such that

$$H_{d_j}(C_j(t), c_j(s)) \le \lambda |t - s|,$$

for all $t, s \in [0,T]$.

(H_{60}) There exists a positive constant α_j, β_j for each $j = 1, 2$ such that

$$E|G^j(t,x,y) - G^j(t,\overline{x},\overline{y})| \le \alpha_j \|x - \overline{x}\|_{M^2_{\mathcal{F}_0}} + \beta_j \|y - \overline{y}\|_{M^2_{\mathcal{F}_0}},$$

for all $t \in [0,T]$ and $x, y, \overline{x}, \overline{y} \in M^2([-r,0],X)$.

Theorem 11.1. *Assume that* (H_{59}) *and* (H_{60}) *hold. Then problem* (11.1) *possesses a unique solution on* $[0,T]$.

Proof. We now have the existence part. Therefore, we pass immediately to uniqueness. We shall obtain the solution through a well-established discretization procedure. The following discretization scheme lies at the heart of many proofs for sweeping processes. Consider, for every $n \in \mathbb{N}$, the following partition of $[0,T]$:

$$t_{n,i} = \frac{iT}{2^n}, \quad 0 \le i \le 2^n \quad \text{and} \quad I_{n,i} = (t_{n,i}, t_{n,i+1}], \quad \text{if } 0 \le i \le 2^n - 1,\ n \ge 0,$$

$$x_{n,0} = \begin{cases} \phi(t), & t \in [-r,0], \\ \phi(0), & t \in [0,t_{n,0}], \end{cases}$$

for any $I_{n,0} = (t_{n,0}, t_{n,1}]$, we have

$$x_{n,1} = \begin{cases} x_{n,0}(t), & t \in [-r,t_{n,0}], \\ \operatorname{proj}(\phi(0)+G^1(t_{n,0},(x_{n,0})_{t_{n,0}},(y_{n,0})_{t_{n,0}})(B^{H_1}(t_{n,1}) - B^{H_1}(t_{n,0}), c_1(t_{n,1})), & t \in [t_{n,0}, t_{n,1}], \end{cases}$$

for any $I_{n,1} = (t_{n,1}, t_{n,2}]$, we have

$$x_{n,2} = \begin{cases} x_{n,1}(t), \quad t \in [-r,t_{n,1}], \\ \operatorname{proj}(x_{n,1}(t_{n,1}) + G^1(t_{n,1},(x_{n,1})_{t_{n,1}},(y_{n,1})_{t_{n,1}})(B^{H_1}(t_{n,2}) - B^{H_1}(t_{n,1}), c_1(t_{n,2})), \\ \quad t \in [t_{n,1}, t_{n,2}]. \end{cases}$$

With the same argument, we can define recursively

$$x_{n,i+1} = \begin{cases} x_{n,i}(t), \quad t \in [-r,t_{n,i}], \\ \operatorname{proj}(x_{n,i}(t_{n,i}) + G^1(t_{n,i},(x_{n,i})_{t_{n,1}},(y_{n,i})_{t_{n,1}})(B^{H_1}(t_{n,i+1}) - B^{H_1}(t_{n,i}), c_1(t_{n,i+1})), \\ \quad t \in [t_{n,i}, t_{n,i+1}]. \end{cases}$$

Estimate (x_n, y_n) by norm

$$M^2([-r,T],X) \times M^2([-r,T],X),$$

since (x_n, y_n) is piecewise affine, by direct calculations,

$$\sup\left\{ \sqrt{E\left|x_{n,i+1}(t) - x_{n,i}(t)\right|^2} : t \in [-r,T] \right\} \le \lambda \frac{T}{2^n}. \tag{11.2}$$

Observe that $(x_{n,i}(t), y_{n,i}(t)) \in (C_1(t_{n,i}), c_2(t_{n,i}))$, and

$$E\left|x_{n,i+1}(t) - x_{n,i}(t)\right| \le EH_{d_1}(C_1(t_{n,i}), c_1(t_{n,i+1})) \le \lambda \frac{T}{2^n}, \tag{11.3}$$

and

$$E\left|y_{n,i+1}(t)-y_{n,i}(t)\right| \le EH_{d_2}(C_2(t_{n,i}),c_2(t_{n,i+1})) \le \lambda\frac{T}{2^n}, \tag{11.4}$$

for all $t \in (t_{n,i-1},t_{n,i}]$, for every $0 \le i \le 2^n$.

By affine interpolation, we define a corresponding sequence of approximate solutions $x_n, y_n \in M^2([-r,T],X)$; for $t \in I_{n,i}$, the explicit formula is

$$x_n(t) = \begin{cases} x_{n,i}(t), & t \in [-r,t_{n,i}] \\ x_{n,i}(t_{n,i}) + \frac{t-t_{n,i}}{\epsilon_n}(x_{n,i+1}(t)-x_{n,i}(t)) & \\ \quad + G^1(t_{n,i},x(n,i)_{t_{n,i}})(B^{H_1}(t)-B^{H_1}(t_{n,1})), & t \in [t_{n,i},t_{n,i+1}], \end{cases}$$

and

$$y_n(t) = \begin{cases} y_{n,i}(t), & t \in [-r,t_{n,i}] \\ y_{n,i}(t_{n,i}) + \frac{t-t_{n,i}}{\epsilon_n}(y_{n,i+1}(t)-y_{n,i}(t)) & \\ \quad + G^2(t_{n,i},x(n,i)_{t_{n,i}},y(n,i)_{t_{n,i}})(B^{H_2}(t)-B^{H_2}(t_{n,1})), & t \in [t_{n,i},t_{n,i+1}], \end{cases}$$

where $\epsilon_n = \frac{T}{2^n}$ and for every $0 \le i \le 2^n - 1$.

From the definition of the normal proximal cone, we have

$$dx_n(t) \in -N(x_{n,i+1},c_1(t_{n,i+1}))dt + G^1(t_{n,i},x(n,i)_{t_{n,i}},y(n,i)_{t_{n,i}})(B^{H_1}(t)-B^{H_1}(t_{n,1})), \tag{11.5}$$

and

$$dy_n(t) \in -N(y_{n,i+1},c_2(t_{n,i+1}))dt + G^2(t_{n,i},x(n,i)_{t_{n,i}},y(n,i)_{t_{n,i}})(B^{H_2}(t)-B^{H_2}(t_{n,1})). \tag{11.6}$$

Now, we prove that

$$\{x_n, y_n,\ n \in \mathbb{N}\}$$

is compact in

$$M^2([-r,T],X),$$

for each $z_n = (x_n, y_n)$ in

$$M^2([-r,T],X) \times M^2([-r,T],X).$$

Step 1. $\{(x_n,y_n),\ n \in \mathbb{N}\}$ are bounded sets in

$$M^2([-r,T],X) \times M^2([-r,T],X).$$

We obtain

$$\left|x_n(t)\right| \le \left|x_{n,i}(t)\right| + \left|x_{n,i+1}(t)-x_{n,i}(t)\right| + b\left|G^1(t_{n,i},x(n,i)_{t_{n,i}},y(n,i)_{t_{n,i}})\right|\left|(B^{H_1}(t)-B^{H_1}(t_{n,1}))\right|$$

$$
\begin{aligned}
&\leq |x_{n,0}(t)| + \sum_{k=1}^{i+1} |x_{n,k-1}(t) - x_{n,k}(t)| \\
&\quad + T|G^1(t_{n,i}, x(_{n,i}, y(_{n,i})_{t_{n,i}}, y(_{n,i})_{t_{n,i}})||(B^{H_1}(t) - B^{H_1}(t_{n,1}))| \\
&\leq \|\phi\| + 2T + T(|G^1(t_{n,i}, x(_{n,i})_{t_{n,i}}, y(_{n,i})_{t_{n,i}}) - G^1(t_{n,i}, 0, 0)| \\
&\quad + |G^1(t_{n,i}, 0, 0)|)|(B^{H_1}(t) - B^{H_1}(t_{n,1}))| \\
&\leq \|\phi\| + 2T + T(\alpha_1 \|(x_{n,i})_{t_{n,i}}\|_{M^2_{\mathcal{F}_0}} + \beta_1 \|(y_{n,i})_{t_{n,i}}\|_{M^2_{\mathcal{F}_0}} \\
&\quad + |G^1(t_{n,i}, 0, 0)|)|(B^{H_1}(t) - B^{H_1}(t_{n,1}))|.
\end{aligned}
$$

By definition $(x_{n,i}, y_{n,i})$, we can prove that there exist $M, \overline{M} > 0$, such that

$$\sup\{E|x_{n,i}(t)| : t \in [-r, T]\} \leq M,$$

and

$$\sup\{E|y_{n,i}(t)| : t \in [-r, T]\} \leq \overline{M}.$$

Hence, by using (11.3) and (11.4), we have

$$
\begin{aligned}
&E|x_n(t)|^2 \\
&\quad \leq 2E\|\phi\|^2 + 4T^2 + 2T^2\Big(\alpha_1 E\|(x_{n,i})_{t_{n,i}}\|^2 + \beta_1 E\|(y_{n,i})_{t_{n,i}}\|^2 \\
&\qquad + \sup_{t\in[0,b]} |G^1(t, 0, 0)|^2\Big)E|(B^{H_1}(t) - B^{H_1}(t_{n,1}))|^2 \\
&\quad \leq 2E\|\phi\|^2 + 4T^2 + 2T^2\Big(\alpha_1 E\|(x_{n,i})_{t_{n,i}}\|^2 + \beta_1 E\|(y_{n,i})_{t_{n,i}}\|^2 + \sup_{t\in[0,T]} |G^1(t, 0, 0)|^2\Big)|t - t_{n,1}|^{2H_1} \\
&\quad \leq 2E\|\phi\|^2 + 4T^2 + 2T^2\Big(\alpha_1 M + \beta_1 \overline{M} + \sup_{t\in[0,T]} |G^1(t, 0, 0)|^2\Big)|t - t_{n,1}|^{2H_1} \\
&\quad \leq 2E\|\phi\|^2 + 4T^2 + 2T^2\Big(\alpha_1 M + \beta_1 \overline{M} + \sup_{t\in[0,T]} |G^1(t, 0, 0)|^2\Big)T^{2H_1} = l_1.
\end{aligned}
$$

Similarly, we have

$$E|y_n(t)|^2 \leq 2E\|\overline{\phi}\|^2 + 4T^2 + 2T^2\Big(\alpha_2 \overline{M} + \beta_2 \overline{M} + \sup_{t\in[0,T]} |G^2(t, 0, 0)|^2\Big)T^{2H_2} = l_2,$$

which implies that

$$\begin{pmatrix} E|x_n(t)|^2 \\ E|y_n(t)|^2 \end{pmatrix} \leq \begin{pmatrix} l_1 \\ l_2 \end{pmatrix}.$$

Step 2. $\{(x_n, y_n),\ n \in \mathbb{N}\}$ are equicontinuous sets in

$$M^2([-r, T], X) \times M^2([-r, T], X).$$

Let $\tau_1, \tau_2 \in [t_{n,i}, t_{n,i+1}]$, $\tau_1 < \tau_2$. Thus,

$$\begin{aligned}\mathbb{E}|x_n(\tau_2) - x_n(\tau_1)|^2 &= \mathbb{E}\left|\frac{\tau_2 - \tau_1}{\epsilon_n}(x_{n,i+1} - x_{n,i}) + G^1(t_{n,i}, x(_{n,i})_{t_{n,i}}, y(_{n,i})_{t_{n,i}})(B^{H_1}(\tau_2) - B^{H_1}(\tau_1))\right|^2 \\ &\leq 2|\tau_2 - \tau_1|^2 + 2\Big(\alpha_1 M + \beta_1 \overline{M} + \sup_{t\in[0,T]} |G^1(t,0,0)|^2\Big)|\tau_2 - \tau_1|^{2H_1}.\end{aligned}$$

Similarly,

$$E\big|y_n(\tau_2) - y_n(\tau_1)\big|^2 \leq 2|\tau_2 - \tau_1|^2 + 2\Big(\alpha_2 M + \beta_2 \overline{M} + \sup_{t\in[0,T]} |G^2(t,0,0)|^2\Big)|\tau_2 - \tau_1|^{2H_2}.$$

The right-hand side tends to zero as $\tau_2 - \tau_1 \to 0$ and ϵ are sufficiently small, from Steps 1 and 2. By the Arzelà–Ascoli theorem, we conclude that there is a subsequence of (x_n, y_n), again denoted (x_n, y_n), which converges to $(x, y) \in M^2([-r, T], X)$.

Now, we prove that

$$(x(t), y(t)) \in (C_1(t), c_2(t)).$$

Let $\rho_n(t)$, $\mu_n(t)$ be two functions from $[0, T]$ into $[0, T]$ defined by

$$\begin{aligned}\rho_n(t) &= t_{n,i}, \quad \text{if } t \in [t_{n,i}, t_{n,i+1}), \quad \rho_n(0) = 0,\\ \mu_n(t) &= t_{n,i+1}, \quad \text{if } t \in [t_{n,i}, t_{n,i+1}), \quad \mu_n(0) = 0,\end{aligned}$$

for all $t \in [0, T]$. From (11.5) and (11.6), we have

$$dx_n(t) \in -N(x_n(\mu_n(t)), c_1(\mu_n(t)))dt + G^1(t_{\rho_n(t)}, x_{\rho_n(t)}, y_{\rho_n(t)})dB^{H_1}(\rho_n(t)), \quad \text{a. e. } t \in [0, T],$$

and

$$dy_n(t) \in -N(x_n(\mu_n(t)), c_2(\mu_n(t)))dt + G^2(t_{\rho_n(t)}, x_{\rho_n(t)}, y_{\rho_n(t)})dB^{H_2}(\rho_n(t)), \quad \text{a. e. } t \in [0, T].$$

Moreover, for all n large enough, we have

$$\rho_n(t) \to t, \quad \mu_n(t) \to t \text{ uniformly on } [0, b],$$

since

$$|\rho_n(t) - t| \leq \frac{T}{2^n},$$

and

$$|\mu_n(t) - t| \leq \frac{T}{2^n}.$$

Thus,

$$|y_n(\rho_n(t)) - y_n(t)| \leq H_{d_1}(C_1(\rho_n(t)), c_1(t)) \leq \lambda|\rho_n(t) - t|,$$

which immediately yields

$$\sup\left\{\sqrt{E|y_n(\rho_n(t)) - y_n(t)|^2} : t \in [0,T]\right\} \leq \lambda\sqrt{E|\rho_n(t) - t|^2} \to 0 \quad \text{as } n \to \infty.$$

Let $t \in [0,T]$. From (11.2) for each $n \in \mathbb{N}$, $t_{n,i} \in I_{n,i}$ for some i,

$$\begin{aligned}|x_n(t) - C_1(t)| &\leq |x_n(t) - x_n(t_{n,i})| + d(x_n(t_{n,i}), c_1(t)) \\ &\leq \lambda\frac{T}{2^n} + H_{d_1}(C_1(t_{n,i}), c_1(t)).\end{aligned}$$

Thus,

$$|x_n(t) - C_1(t)| \leq \lambda\frac{T}{2^{n-1}}. \tag{11.7}$$

Since (x_n, y_n) is defined by linear interpolation, we obtain

$$|x_n'(t)| \leq \frac{1}{\epsilon_n}\sup_i|x_{n,i+1}(t) - x_{n,i}(t)|,$$

and

$$|y_n'(t)| \leq \frac{1}{\epsilon_n}\sup_i|y_{n,i+1}(t) - y_{n,i}(t)|.$$

By letting $n \to \infty$ in (11.7) for all $t \in [0,T]$, we obtain that

$$(x(t), y(t)) \in (C_1, c_2).$$

Now, we prove that the sequences of composition mappings

$$(x_n \circ \mu_n, y \circ \mu_n),$$

and

$$(x_n \circ \rho_n, y \circ \rho_n),$$

converge uniformly to (x_t, y_t) in $M^2([-r,0],X)$,

$$\begin{aligned}E|x_n(\rho_n(t) + \tau) - x(t+\tau)|^2 \leq\ & 3E|x_n(\rho_n(t) + \tau) - x_n(t+\tau)|^2 \\ &+ 3E|x_n(\rho_n(t) + \tau) - x_n(\mu_n(t) + \tau)|^2 \\ &+ 3E|x_n(\mu_n(t) + \tau) - x_n(t+\tau)|^2.\end{aligned}$$

Thus,

$$\sup_{\tau\in[-r,0]} E\left|(x_n)_{\rho_n(t)} - x_t\right|^2 \leq 3\lambda^2 E\left|\rho_n(t) - t\right|^2 + 3E\left|\rho_n(t) - \mu_n(t)\right|^2$$
$$+ 3 \sup_{\tau\in[-r,T]} E\left|x_n(\mu_n(t)) - x(t)\right|^2 \to 0 \quad \text{as} \quad n \to \infty.$$

Since

$$\left|(\rho_n(t) - \tau) - (t - \tau)\right| \leq \frac{T}{2^n},$$

and

$$\left|\mu_n(t) - \rho_n(t)\right| \leq \frac{T}{2^{n-1}}.$$

We can pass to the limit when $n \to \infty$, we deduce from

$$(x_{\rho_n(t)}, y_{\rho_n(t)}) \to (x_t, y_t) \in M^2([-r,0], X),$$

and the fact that $G^i(.,.,.)$ is a continuous function, then we have

$$G^i(\rho_n(t), x_{\rho_n(t)}, y_{\rho_n(t)}) \to G^i(t, x_t, y_t).$$

Now, we show that

$$dx(t) \in -N(x(t), c_1(t))dt + G^1(t, x_t, y_t)dB^{H_1}(t), \quad \text{a. e. } t \in [0,T],$$

and

$$dy(t) \in -N(y(t), c_2(t))dt + G^2(t, x_t, y_t)dB^{H_2}(t), \quad \text{a. e. } t \in [0,T].$$

Since (x_n, y_n) is bounded in $X \times X$, there exists a subsequence of (x_n, y_n) converging to (x, y). Then

$$\int_0^T \sigma(-x_n'(t) + G^1(t,(x_n)_t,(y_n)_t)dB^{H_1}(t), c_1(\mu_n(t)))dt$$
$$\leq \int_0^T (-x_n'(t) + G^1(t,(x_n)_t,(y_n)_t)dB^{H_1}(t), x(\mu_n(t)))dt. \tag{11.8}$$

Using the fact that $\sigma(., c_j(t))$ is lower semicontinuous, then

$$\liminf_{n\to\infty} \int_0^T \sigma(-x_n'(t) + G^1(t,(x_n)_t,(y_n)_t)dB^{H_1}(t), c_1(\mu_n(t)))dt$$

$$\geq \int_0^T (-x'(t) + G^1(t, x_t, y_t)dB^{H_1}(t), c_1(t))dt. \tag{11.9}$$

By (11.8) and (11.9), we obtain

$$\int_0^T (-x'(t) + G^1(t, x_t, y_t)dB^{H_1}(t), c_1(t))dt$$
$$\geq \int_0^T \sigma(-x'(t) + G^1(t, x_t, y_t)dB^{H_1}(t), c_1(t))dt.$$

Thus,

$$dx(t) \in -N(x(t), c_1(t))dt + G^1(t, x_t, y_t)dB^{H_1}(t), \quad \text{a. e. } t \in [0, T],$$

and

$$dy(t) \in -N(y(t), c_2(t))dt + G^2(t, x_t, y_t)dB^{H_2}(t), \quad \text{a. e. } t \in [0, T].$$

Finally, we prove the uniqueness of solutions to the problem (11.1). Let us assume that (x, y) and $(\overline{x}, \overline{y})$ are two solutions of (11.1):

$$d\overline{x}(t) \in -N(\overline{x}(t), c_1(t))dt + G^1(t, \overline{x}_t, \overline{y}_t)dB^{H_1}(t), \quad \text{a. e. } t \in [0, T],$$

and

$$d\overline{y}(t) \in -N(\overline{y}(t), c_2(t))dt + G^2(t, \overline{x}_t, \overline{y}_t)dB^{H_2}(t), \quad \text{a. e. } t \in [0, T].$$

Since $C(t) = (C_1(t), c_2(t))$ is a convex set, then

$$T_{C_j}(z) = \bigcup_{h>0} \overline{\frac{C_j(t) - z}{h}},$$

for all $t \in [0, T]$,

$$T_{C_j}(z) \subset \{v \in X : \langle v, \xi \rangle \leq 0 \text{ for all } \xi \in N^p(z, \xi)\},$$

which immediately yields

$$\langle x'(t) - \overline{x}'(t) + (G^1(t, x_t, y_t) - G^1(t, \overline{x}_t, \overline{y}_t))dB^{H_1}(t), x(t) - \overline{x}(t) \rangle \leq 0.$$

Thus, we deduce

$$\langle x'(t) - \overline{x}'(t), x(t) - \overline{x}(t) \rangle + \langle (G^1(t, x_t, y_t) - G^1(t, \overline{x}_t, \overline{y}_t))dB^{H_1}(t), x(t) - \overline{x}(t) \rangle \leq 0.$$

By assumptions (H_1), (H_2) imply

$$\frac{1}{2}\cdot\frac{d}{dt}\left|x(t)-\overline{x}(t)\right|^2 \le \alpha_1\|x_t-\overline{x}_t\|_{M^2_{\mathcal{F}_0}}\left|x(t)-\overline{x}(t)\right|dB^{H_1}(t)+\beta_1\|y_t-\overline{y}_t\|_{M^2_{\mathcal{F}_0}}\left|x(t)-\overline{x}(t)\right|dB^{H_1}(t), \tag{11.10}$$

and

$$\frac{1}{2}\cdot\frac{d}{dt}\left|y(t)-\overline{y}(t)\right|^2 \le \alpha_2\|x_t-\overline{x}_t\|_{M^2_{\mathcal{F}_0}}\left|y(t)-\overline{y}(t)\right|dB^{H_1}(t)+\beta_2\|y_t-\overline{y}_t\|_{M^2_{\mathcal{F}_0}}\left|y(t)-\overline{y}(t)\right|dB^{H_1}(t). \tag{11.11}$$

Integrating (11.10) and (11.11) over $(0,t)$, we arrive at

$$\begin{aligned}\left|x(t)-\overline{x}(t)\right|^2 &\le \alpha_1\int_0^t\|x_s-\overline{x}_s\|_{M^2_{\mathcal{F}_0}}\left|x(s)-\overline{x}(s)\right|dB^{H_1}(s)\\&\quad+\beta_1\int_0^t\|y_s-\overline{y}_s\|_{M^2_{\mathcal{F}_0}}\left|x(s)-\overline{x}(s)\right|dB^{H_1}(s)\\&\le \alpha_1\int_0^t\sup_{s\in[0,t]}\sqrt{E\left|x(s)-\overline{x}(s)\right|^2}\left|x(s)-\overline{x}(s)\right|dB^{H_1}(s)\\&\quad+\beta_1\int_0^t\sup_{s\in[0,t]}\sqrt{E\left|y(s)-\overline{y}(s)\right|^2}\left|x(s)-\overline{x}(s)\right|dB^{H_1}(s).\end{aligned}$$

Then, for each $t\in[0,T]$ and thanks to Lemma 4.1,

$$\begin{aligned}E\left|x(t)-\overline{x}(t)\right|^4 &\le 2\alpha_1E\left|\int_0^t\sup_{s\in[0,t]}\sqrt{E\left|x(s)-\overline{x}(s)\right|^2}\left|x(s)-\overline{x}(s)\right|dB^{H_1}(s)\right|^2\\&\quad+2\beta_1E\left|\int_0^t\sup_{s\in[0,t]}\sqrt{E\left|y(s)-\overline{y}(s)\right|^2}\left|x(s)-\overline{x}(s)\right|dB^{H_1}(s)\right|^2\\&\le 2c_2(H_1)H_1(2H_1-1)T^{2H_1-1}\alpha_1\int_0^t\sup_{s\in[0,t]}E\left|x(s)-\overline{x}(s)\right|^4ds\\&\quad+2c_2(H_1)H_1(2H_1-1)T^{2H_1-1}\beta_1\int_0^t\sup_{s\in[0,t]}E\left|x(s)-\overline{x}(s)\right|^2E\left|y(s)-\overline{y}(s)\right|^2ds.\end{aligned}$$

Thus,

$$E|x(t)-\overline{x}(t)|^4 \le A_1\int_0^t \sup_{s\in[0,t]} E|x(s)-\overline{x}(s)|^4 ds + B_1\int_0^t \sup_{s\in[0,t]} E|y(s)-\overline{y}(s)|^4 ds,$$

where

$$A_1 = 2c_2(H_1)H_1(2H_1-1)T^{2H_1-1}(2\alpha_1+\beta_1),$$

and

$$B_1 = c_2(H_1)H_1(2H_1-1)T^{2H_1-1}\beta_1.$$

In the same way, we also have

$$\begin{aligned} E|y(t)-\overline{y}(t)|^4 \le\ & 2c_2(H_2)H_2(2H_2-1)T^{2H_2-1}\alpha_2\int_0^t \sup_{s\in[0,t]} E|y(s)-\overline{y}(s)|^4 ds \\ & + 2c_2(H_2)H_2(2H_2-1)T^{2H_2-1}\beta_2\int_0^t \sup_{s\in[0,t]} E|x(s)-\overline{x}(s)|^2 E|y(s)-\overline{y}(s)|^2 ds \end{aligned}$$

and consequently,

$$E|y(t)-\overline{y}(t)|^4 \le A_2\int_0^t \sup_{s\in[0,t]} E|y(s)-\overline{y}(s)|^4 ds + B_2\int_0^t \sup_{s\in[0,t]} E|x(s)-\overline{x}(s)|^4 ds,$$

where

$$A_3 = c_2(H_2)H_2(2H_2-1)T^{2H_2-1}(2\alpha_2+\beta_2),$$

and

$$A_4 = c_2(H_2)H_2(2H_2-1)T^{2H_2-1}\beta_2.$$

Adding these, we obtain

$$\begin{aligned} & E|x(t)-\overline{x}(t)|^4 + E|y(t)-\overline{y}(t)|^4 \\ & \le A_*\int_0^t \sup_{s\in[0,t]} E|x(s)-\overline{x}(s)|^4 ds + B_*\int_0^t \sup_{s\in[0,t]} E|y(s)-\overline{y}(s)|^4 ds, \end{aligned}$$

where

$$A_* = A_1 + B_2, \quad B_* = A_2 + B_1.$$

Then

$$\sup_{s\in[0,t]} E|x(t)-\overline{x}(t)|^4 + E|y(t)-\overline{y}(t)|^4$$
$$\leq A_{**}\int_0^t \sup_{s\in[0,t]}(E|x(s)-\overline{x}(s)|^4 + E|y(s)-\overline{y}(s)|^4)ds,$$

where

$$A_{**} = \max\{A_*, b_*\}.$$

By a generalization of Grönwall's inequality, we have

$$\sup_{s\in[0,t]} E|x(t)-\overline{x}(t)|^4 + E|y(t)-\overline{y}(t)|^4 = 0 \quad\Longrightarrow\quad (x(t),y(t)) = (\overline{x}(t),\overline{y}(t)), \quad \text{a. e. } t\in[0,T].$$

The proof is therefore complete. □

11.2 Perturbation problem

To prove the main result, we will need the following auxiliary inclusion:

$$\begin{cases} -dx(t)\in N_{C_1(t)}(x(t))dt + F^1(t,x_t,y_t)dt + G^1(t,x_t,y_t)dB^{H_1}, & \text{a. e. } t\in[0,T],\\ -dy(t)\in N_{C_2(t)}(y(t))dt + F^2(t,x_t,y_t)dt + G^2(t,x_t,y_t)dB^{H_2}, & \text{a. e. } t\in[0,T],\\ x(t) = \phi(t), & t\in[-r,0],\ x(0)\in C_1(0),\\ y(t) = \overline{\phi}(t), & t\in[-r,0],\ y(0)\in C_2(0). \end{cases} \tag{11.12}$$

The goal of this section is to study the existence results of the problem (11.12).

Theorem 11.2. *Assume that (H_1) and (H_2) and the condition*

(H_{61}) $F^j : [0,T]\times M^2([-r,0],X)\times M^2([-r,0],X)\to \mathcal{P}_{cp,cv}(X)$ *be a u. s. c. Carathéodory multi-map, and for each $t\in[0,T]$, scalar*

$$\mathcal{L}([0,T])\otimes\mathcal{B}(M^2([-r,0],X),X),$$

measurable, where $\mathcal{L}([0,T])$ is the σ-algebra of Lebesgue measurable sets of $[0,T]$ and $\mathcal{B}(M^2)$ is the Borel tribe of M^2 and $|F^j(t,x,y)|\leq k_j$ for all

$$(t,x,y)\in[0,T]\times M^2([-r,0],X)\times M^2([-r,0],X),$$

or some constant $k_j>0$.

Then problem (11.12) has at least one solution on $[0,T]$.

Proof. Consider, for every $n \in \mathbb{N}$, the following partition of $[0, T]$:

$$t_{n,i} = \frac{iT}{2^n}, \quad 0 \le i \le 2^n,$$

and

$$I_{n,i} = (t_{n,i}, t_{n,i+1}], \quad \text{if } 0 \le i \le 2^n - 1,\ n \ge 0.$$

$$x_{n,0} = \begin{cases} \phi(t), & t \in [-r, 0], \\ \phi(0), & t \in [0, t_{n,0}]. \end{cases}$$

For any $I_{n,0} = (t_{n,0}, t_{n,1}]$, we have

$$x_{n,1} = \begin{cases} x_{n,0}(t), & t \in [-r, t_{n,0}], \\ \operatorname{proj}(\phi(0) + g_0^1(t_{n,0}) \\ \quad + G^1(t_{n,0}, (x_{n,0})_{t_{n,0}}, (y_{n,0})_{t_{n,0}})(B^{H_1}(t_{n,1}) - B^{H_1}(t_{n,0}), c(t_{n,1})), & t \in [t_{n,0}, t_{n,1}]. \end{cases}$$

Similarly, for any $I_{n,1} = (t_{n,1}, t_{n,2}]$, we have

$$x_{n,2} = \begin{cases} x_{n,1}(t), & t \in [-r, t_{n,1}], \\ \operatorname{proj}(x_{n,1}(t_{n,1}) + g_0^1(t_{n,1}) \\ \quad + G^1(t_{n,1}, (x_{n,1})_{t_{n,1}}, (y_{n,1})_{t_{n,1}})(B^{H_1}(t_{n,2}) - B^{H_1}(t_{n,1}), c(t_{n,2})), & t \in [t_{n,1}, t_{n,2}]. \end{cases}$$

With the same argument, we can define recursively, for any $I_{n,i} = (t_{n,i}, t_{n,i+1}]$,

$$x_{n,i+1} = \begin{cases} x_{n,i}(t), & t \in [-r, t_{n,i}], \\ \operatorname{proj}(x_{n,i}(t_{n,i}) + g_0^1(t_{n,i}) \\ +G^1(t_{n,i}, (x_{n,i})_{t_{n,1}}, (y_{n,i})_{t_{n,1}})(B^{H_1}(t_{n,i+1}) - B^{H_1}(t_{n,i}), c(t_{n,i+1})), & t \in [t_{n,i}, t_{n,i+1}], \end{cases}$$

where

$$g_0^j(t, u) = \min\{|x| : x \in F^j(t, u)\}.$$

By construction, we have $(x_{n,i}, y_{n,i}) \in (C_1, C_2)$, for all $t \in [t_{n,i-1}, t_{n,i}]$. Then, for every $0 \le i \le 2^n$,

$$|x_{n,i+1}(t) - x_{n,i}(t)| \le H_{d_1}(C_1(t_{n,i}), c_1(t_{n,i+1})) \le \lambda \frac{T}{2^n},$$

and

$$|y_{n,i+1}(t) - y_{n,i}(t)| \le H_{d_2}(C_1(t_{n,i}), c_1(t_{n,i+1})) \le \lambda \frac{T}{2^n}$$

and, consequently,

$$\sup\left\{\sqrt{E\left|x_{n,i+1}(t)-x_{n,i}(t)\right|^2} : t\in[-r,T]\right\} \le \lambda\frac{T}{2^n},$$

and

$$\sup\left\{\sqrt{E\left|y_{n,i+1}(t)-y_{n,i}(t)\right|^2} : t\in[-r,T]\right\} \le \lambda\frac{T}{2^n}.$$

Put

$$x_n(t) = \begin{cases} x_{n,i}(t), & t\in[-r,t_{n,i}], \\ x_{n,i}(t_{n,i}) + \frac{t-t_{n,i}}{\epsilon_n}(x_{n,i+1}(t)-x_{n,i}(t)) + (t-t_{n,i})g_0^1(t_{n,i}) & \\ \quad + G^1(t_{n,i}, x_{t_{n,i}}, y_{t_{n,i}})(B^{H_1}(t)-B^{H_1}(t_{n,1})), & t\in[t_{n,i},t_{n,i+1}], \end{cases}$$

and

$$y_n(t) = \begin{cases} y_{n,i}(t), & t\in[-r,t_{n,i}], \\ y_{n,i}(t_{n,i}) + \frac{t-t_{n,i}}{\epsilon_n}(y_{n,i+1}(t)-y_{n,i}(t)) + (t-t_{n,i})g_0^2(t_{n,i}) & \\ \quad + G^2(t_{n,i}, x_{t_{n,i}}, y_{t_{n,i}})(B^{H_2}(t)-B^{H_2}(t_{n,1})), & t\in[t_{n,i},t_{n,i+1}]. \end{cases}$$

Since (x_n, y_n) is defined by linear interpolation, we have

$$\left|x_n'(t)\right| \le \frac{1}{\epsilon_n}\sup_i\left|x_{n,i+1}(t)-x_{n,i}(t)\right|,$$

and

$$\left|y_n'(t)\right| \le \frac{1}{\epsilon_n}\sup_i\left|y_{n,i+1}(t)-y_{n,i}(t)\right|.$$

Using the fact that the projections are nonexpansive, we

$$\left|x_{n,i+1}(t)-\mathrm{proj}(x_{n,i}(t), c_1(t_{n,i+1}))\right| \le \epsilon_n\left|g_0^1(t_{n,i})\right| \le \epsilon_n k_1,$$

and

$$\left|y_{n,i+1}(t)-\mathrm{proj}(y_{n,i}(t), c_2(t_{n,i+1}))\right| \le \epsilon_n\left|g_0^2(t_{n,i})\right| \le \epsilon_n k_2.$$

Hence,

$$\left|x_{n,i+1}(t)-x_{n,i}(t)\right| \le \epsilon_n(k_1+\lambda). \tag{11.13}$$

Thus,

$$\left|x_n'(t)\right| \le k_1+\lambda \quad \text{and} \quad \sup_{t\in J}\left|x_n'(t)\right|^2 \le (k_1+\lambda)^2. \tag{11.14}$$

From the definition of the normal proximal cone, we have

$$\begin{aligned} dx_n(t) &\in -N(x_{n,i+1}, c_1(t_{n,i+1}))dt + g_0^1(t_{n,i})dt \\ &\quad + G^1(t_{n,i}, x(_{n,i})_{t_{n,i}}, y(_{n,i})_{t_{n,i}})(B^{H_1}(t) - B^{H_1}(t_{n,1})), \quad \text{a. e. } t \in [0,T], \end{aligned}$$

and

$$\begin{aligned} dy_n(t) &\in -N(y_{n,i+1}, c_2(t_{n,i+1}))dt + g_0^2(t_{n,i})dt \\ &\quad + G^2(t_{n,i}, x(_{n,i})_{t_{n,i}}, y(_{n,i})_{t_{n,i}})(B^{H_2}(t) - B^{H_2}(t_{n,1})), \quad \text{a. e. } t \in [0,T]. \end{aligned}$$

Now, we prove that

$$\{(x_n, y_n),\ n \in \mathbb{N}\},$$

is compact in

$$M^2([-r,T],X) \times M^2([-r,T],X).$$

Step 1. $\{(x_n, y_n),\ n \in \mathbb{N}\}$ are bounded sets in

$$M^2([-r,T],X) \times M^2([-r,T],X).$$

We have

$$\begin{aligned} |x_n(t)| &\leq |x_{n,i}(t)| + |x_{n,i+1}(t) - x_{n,i}(t)| + T|g_0^1(t_{n,i}, x(_{n,i})_{t_{n,i}}, y(_{n,i})_{t_{n,i}})| \\ &\quad + |G^1(t_{n,i}, x(_{n,i})_{t_{n,i}}, y(_{n,i})_{t_{n,i}})||(B^{H_1}(t) - B^{H_1}(t_{n,1}))| \\ &\leq |x_{n,0}(t)| + 2\sum_{k=1}^{i+1} |x_{n,k-1}(t) - x_{n,k}(t)| + Tk_1 \\ &\quad + |G^1(t_{n,i}, x(_{n,i}, y(_{n,i})_{t_{n,i}}, y(_{n,i})_{t_{n,i}})||(B^{H_1}(t) - B^{H_1}(t_{n,1}))| \\ &\leq \|\phi\| + 2T + (|G^1(t_{n,i}, x(_{n,i})_{t_{n,i}}, y(_{n,i})_{t_{n,i}}) - G^1(t_{n,i}, 0, 0)| \\ &\quad + |G^1(t_{n,i}, 0, 0)|)|(B^{H_1}(t) - B^{H_1}(t_{n,1}))| \\ &\leq \|\phi\| + 2T + Tk_1 \\ &\quad + T(\alpha_1\|(x_{n,i})_{t_{n,i}}\|_{M^2_{\mathcal{F}_0}} + \beta_1\|(y_{n,i})_{t_{n,i}}\|_{M^2_{\mathcal{F}_0}} + |G^1(t_{n,i}, 0, 0)|)|(B^{H_1}(t) - B^{H_1}(t_{n,1}))|. \end{aligned}$$

Then

$$\begin{aligned} E|x_n(t)|^2 &\leq 2(\|\phi\|^2 + 2T + Tk_1)^2 + 2T^2\Big(\alpha_1 M + \beta_1\overline{M} \\ &\quad + \sup_{t\in[0,T]} |G^1(t,0,0)|^2\Big)E|(B^{H_1}(t) - B^{H_1}(t_{n,1}))|^2 \\ &\leq 2(\|\phi\|^2 + 2T + Tk_1)^2 + 2T^2\Big(\alpha_1 M + \beta_1\overline{M} + \sup_{t\in[0,T]} |G^1(t,0,0)|^2\Big)T^{2H_1} = \bar{l}_1. \end{aligned}$$

Hence,

$$\sup\left\{\sqrt{E|x_n(t)|^2} : t \in [-r, T]\right\} \le \bar{l}_1,$$

and

$$\sup\left\{\sqrt{E|y_n(t)|^2} : t \in [-r, T]\right\} \le \bar{l}_2,$$

which implies that

$$\begin{pmatrix} E|x_n(t)|^2 \\ E|y_n(t)|^2 \end{pmatrix} \le \begin{pmatrix} \bar{l}_1 \\ \bar{l}_2 \end{pmatrix}.$$

Step 2. $\{(x_n, y_n),\ n \in \mathbb{N}\}$ are equicontinuous sets in

$$M^2([-r, T], X).$$

Let $\tau_1, \tau_2 \in [t_{n,i}, t_{n,i+1}]$, $\tau_1 < \tau_2$. Thus,

$$\begin{aligned}
&E|x_n(\tau_2) - x_n(\tau_1)|^2 \\
&\quad = E\left|\frac{\tau_2 - \tau_1}{\epsilon_n}(x_{n,i+1} - x_{n,i}) + (\tau_2 - \tau_1)g_0^1(t_{n,i}, x(n,i)_{t_{n,i}}, y(n,i)_{t_{n,i}})\right. \\
&\qquad \left. + G^1(t_{n,i}, x(n,i)_{t_{n,i}}, y(n,i)_{t_{n,i}})(B^{H_1}(\tau_2) - B^{H_1}(\tau_1))\right|^2 \\
&\quad \le 3|\tau_2 - \tau_1|^2 + 3\Big(\alpha_1 M + \beta_1 \overline{M} + \sup_{t\in[0,T]} |G^1(t, 0, 0)|^2\Big)|\tau_2 - \tau_1|^{2H_1} + 3k_1^2|\tau_2 - \tau_1|^2.
\end{aligned}$$

Similarly,

$$\begin{aligned}
&E|y_n(\tau_2) - y_n(\tau_1)|^2 \\
&\quad \le 3|\tau_2 - \tau_1|^2 + 3\Big(\alpha_2 M + \beta_2 \overline{M} + \sup_{t\in[0,T]} |G^2(t, 0, 0)|^2\Big)|\tau_2 - \tau_1|^{2H_2} \\
&\qquad + 3k_2^2|\tau_2 - \tau_1|^2.
\end{aligned}$$

The right-hand side tends to zero as $\tau_2 - \tau_1 \to 0$ and ϵ are sufficiently small. From Steps 1 and 2. By the Arzelà–Ascoli theorem, we conclude that there is a subsequence of (x_n, y_n), again denoted (x_n, y_n), which converges to (x, y) in

$$M^2([-r, T], X) \times M^2([-r, T], X).$$

It remains to prove that

$$(x(t), y(t)) \in (C_1(t), C_2(t)).$$

Let $t \in [0, T]$. From (11.14), we obtain

$$\begin{aligned} 0 \le |x_n(t) - C_1(t)| &= d(x_n(t), c_1(t)) \\ &\le |x_n(t) - x_n(t_{n,i})| + d(x_n(t_{n,i}), c_1(t)) \\ &\le (k_1 + \lambda)|t - t_{n,i}| + H_{d_1}(C_1(t_{n,i}), c_1(t)) \\ &\le \frac{(k_1 + \lambda)b}{2^{n-1}}. \end{aligned}$$

Then

$$|x_n(t) - C_1(t)| \le \frac{(k_1 + \lambda)T}{2^{n-1}}, \tag{11.15}$$

and

$$|y_n(t) - C_2(t)| \le \frac{(k_2 + \lambda)T}{2^{n-1}}. \tag{11.16}$$

By letting $n \to \infty$ in (11.15) and (11.16), we obtain that

$$(x(t), y(t)) \in (C_1, c_2). \tag{11.17}$$

Now, we define, for $t \in [0, T]$,

$$\rho_n(t) = t_{n,i}, \quad \mu_n(t) = t_{n,i+1} \quad \text{if } t \in [t_{n,i}, t_{n,i+1}).$$

Hence, by using (11.5) and (11.6), we have

$$\begin{aligned} dx_n(t) \in -N(x_n(\mu_n(t)), c_1(\mu_n(t)))dt + g_0^1(t_{\rho_n(t)}, x_{\rho_n(t)}, y_{\rho_n(t)}) \\ + G^1(t_{\rho_n(t)}, x_{\rho_n(t)}, y_{\rho_n(t)})dB^{H_1}(\rho_n(t)) \quad \text{a. e. } t \in [0, T], \end{aligned}$$

and

$$\begin{aligned} dy_n(t) \in -N(x_n(\mu_n(t)), c_2(\mu_n(t)))dt + g_0^2(t_{\rho_n(t)}, x_{\rho_n(t)}, y_{\rho_n(t)}) \\ + G^2(t_{\rho_n(t)}, x_{\rho_n(t)}, y_{\rho_n(t)})dB^{H_2}(\rho_n(t)), \quad t \in \text{a. e. } t \in [0, T]. \end{aligned}$$

Hence,

$$\begin{aligned} \rho_n(t) &\to t, \\ \mu_n(t) &\to t, \end{aligned}$$

uniformly on $[0, b]$.

Since

$$|\rho_n(t) - t| \le \frac{T}{2^n},$$

and

$$|\mu_n(t) - t| \le \frac{T}{2^n}.$$

Moreover,

$$|x_n(\rho_n(t)) - x_n(t)| \le H_{d_1}(C_1(\rho_n(t)), c_1(t)) \le \lambda|\rho_n(t) - t|.$$

Similarly,

$$|y_n(\rho_n(t)) - y_n(t)| \le H_{d_2}(C_2(\rho_n(t)), c_2(t)) \le \lambda|\rho_n(t) - t|.$$

Therefore,

$$\sup\Big\{\sqrt{E|x_n(\rho_n(t)) - x_n(t)|^2} : t \in [0,T]\Big\} \le \lambda\sqrt{E|\rho_n(t) - t|^2} \to 0 \quad \text{as } n \to \infty,$$

and

$$\sup\Big\{\sqrt{E|y_n(\rho_n(t)) - y_n(t)|^2} : t \in [0,T]\Big\} \le \lambda\sqrt{E|\rho_n(t) - t|^2} \to 0 \quad \text{as } n \to \infty.$$

In Theorem 11.1, it was proved that $(x_{\rho_n(t)}, y_{\rho_n(t)})$ converges to (x_t, y_t) in $M^2([-r,T],X)$.

Let

$$v_n^j(t) = g_0^j(\rho_n(t), (x_n)_{\rho_n(t)}, (y_n)_{\rho_n(t)}).$$

From (H_3), we have $|v_n^j(t)| \le k_j$ for $n \in \mathbb{N}$ implies that $v_n^j(t) \in lB(0,1)$; hence, $(v_n^j)_{n\in\mathbb{N}}$, which converges weakly to some limit $v^j \in L^2(J,X)$. Since $F(.,x,y)$ is u. s. c. with closed and convex values and $F^j(.,.,.)$ is bounded for each $j = 1,2$, then there exists a sequence $\{F_m\}_{m\in\mathbb{N}}$ of globally u. s. c. set-valued mappings on

$$J \times M^2([-r,0],X) \times M^2([-r,0],X),$$

with convex compact values in $X \times X$ satisfying the following conditions:

$$\|F_m^j(t,x,y)\| \le k_j, \quad j = 1,2,$$

for all

$$\begin{aligned} (t,x,y) &\in J \times M^2([-r,0],X) \times M^2([-r,0],X), \\ F_{m+1}^j(t,x,y) &\subset F_m^j(t,x,y), \\ F(t,x,y) &= \bigcap_{m\ge1} F_m^j(t,x,y). \end{aligned}$$

Now we need to prove that

$$v^j(t) \in F^j(t, x_t, y_t),$$

for a. e. $t \in J$. Lemma 3.11 yields the existence of constants $\alpha_l^n \geq 0$, $l = 1, 2, \ldots, k(n)$, and $j = 1, 2$ such that

$$\sum_{l=1}^{k(n)} \alpha_l^n = 1,$$

and the sequence of convex combinations

$$\psi_n^j(.) = \sum_{l=1}^{k(n)} \alpha_l^n v_l^j(.),$$

converges strongly to some limit $v^j \in L^2(J, X)$. Since F^j takes convex values, using Lemma 3.10, we obtain that

$$\begin{aligned} v^j(t) &\in \bigcap_{n\geq 1} \overline{\{\psi_n^j(t)\}}, \quad \text{a. e. } t \in J \\ &\subset \bigcap_{n\geq 1} \overline{co}\{v_k^j(t),\ k \geq n\} \\ &\subset \bigcap_{n\geq 1} \overline{co}\Big\{\bigcup_{k\geq n} F_m^j(\rho_k(t), (x_k)_{\rho_k(t)}, (y_k)_{\mu_k(t)})\Big\} \\ &= \overline{co}\Big\{\limsup_{k\to\infty} F_m^j(\mu_k(t), (x_k)_{\mu_k(t)}, (y_k)_{\mu_k(t)})\Big\}. \end{aligned}$$

Since F_m^j is u. s. c. and has compact values, then by Lemma 3.9, we have

$$\limsup_{n\to\infty} F_m^j(\rho_n(t), (x_n)_{\rho_n(t)}, (y_n)_{\rho_n(t)}) = F_m^j(t, x_t, y_t) \quad \text{for a. e. } t \in J.$$

This and (9.3) imply that

$$v^j(t) \in \overline{co}(F^j(t, x_t, y_t)).$$

Since, for each $j = 1, 2$, $F_m^j(., ., .)$ has closed, convex values, we deduce that

$$v^j(t) \in F_m^j(t, x_t, y_t),$$

for a. e. $t \in J$; then

$$v^j(t) \in F^j(t, x_t, y_t).$$

We can pass to the limit when $n \to \infty$; we deduce from

$$(x_{\rho_n(t)}, y_{\rho_n(t)}) \to (x_t, y_t) \in M^2([-r, 0], X) \quad \text{as } n \to \infty.$$

Using the fact that $G^j(., ., .)$ is a continuous function, we have

$$G^j(\rho_n(t), x_{\rho_n(t)}, y_{\rho_n(t)}) \to G^j(t, x_t, y_t) \quad \text{as } n \to \infty.$$

Now, we show that

$$dx(t) \in -N(x(t), c_1(t))dt + v^1(t)dt + G^1(t, x_t, y_t)dB^{H_1}(t) \quad \text{a. e. } t \in [0, T],$$

and

$$dy(t) \in -N(y(t), c_2(t))dt + v^2(t)dt + G^2(t, x_t, y_t)dB^{H_2}(t) \quad \text{a. e. } t \in [0, T].$$

Since (x_n, y_n) is bounded in $X \times X$, there exists a subsequence of (x_n, y_n) converging to (x, y). Then

$$\begin{aligned} &\int_0^T \sigma(-x_n'(t) + v_n^1(t) + G^1(t, (x_n)_t, (y_n)_t)dB^{H_1}(t), c_1(\mu_n(t)))dt \\ &\quad \le \int_0^T (-x_n'(t) + v_n^1(t) + G^1(t, (x_n)_t, (y_n)_t)dB^{H_1}(t), x(\mu_n(t)))dt. \end{aligned} \tag{11.18}$$

Using the fact that $\sigma(., c_1(t))$ is lower semicontinuous, we can conclude that

$$\begin{aligned} &\liminf_{n\to\infty} \int_0^T \sigma(-x_n'(t) + v_n^1(t) + G^1(t, (x_n)_t, (y_n)_t)dB^{H_1}(t), c_1(\mu_n(t)))dt \\ &\quad \ge \int_0^T (-x'(t) + v^1(t) + G^1(t, x_t, y_t)dB^{H_1}(t), c_1(t))dt. \end{aligned} \tag{11.19}$$

By (11.18) and (11.19), we obtain

$$\begin{aligned} &\int_0^T (-x'(t) + v^1(t) + G^1(t, x_t, y_t)dB^{H_1}(t), c_1(t))dt \\ &\quad \ge \int_0^T \sigma(-x'(t) + v^1(t) + G^1(t, x_t, y_t)dB^{H_1}(t), c_1(t))dt. \end{aligned}$$

Thus,

$$dx(t) \in -N(x(t), c_1(t))dt + F^1(t, x_t, y_t)dt + G^1(t, x_t, y_t)dB^{H_1}(t), \quad \text{a. e. } t \in [0, T],$$

and

$$dy(t) \in -N(y(t), c_2(t))dt + F^1(t, x_t, y_t)dt + G^2(t, x_t, y_t)dB^{H_2}(t), \quad \text{a. e. } t \in [0, T],$$

and the proof is complete. □

Bibliography

[1] J. P. Aubin and H. Frankowska, *Set-Valued Analysis*, Birkhauser, Boston, 1990.

[2] M. Benchohra, J. Henderson and S. K. Ntouyas, *Impulsive Differential Equations and Inclusions*, vol. 2, Hindawi Publishing Corporation, New York, 2006.

[3] F. Biagini et al., *Stochastic Calculus for Fractional Brownian Motion and Applications*, Springer, 2008. Print.

[4] O. Cârjă, M. Necula and I. I. Vrabie, *Viability, Invariance and Applications*, North-Holland Mathematics Studies, vol. 207, Elsevier Science B. V., Amsterdam, 2007.

[5] R. Cristescu, *Order Structures in Normed Vector Spaces*, Editura Ştiin ţifică şi Enciclopedică, Bucureşti, 1983 (in Romanian).

[6] G. Da Prato and J. Zabczyk, *Stochastic Equations in Infinite Dimensions*, Cambridge University Press, Cambridge, 1992.

[7] K. Deimling, *Multi-valued Differential Equations*, De Gruyter, Berlin, New York, 1992.

[8] J. Dugundji and A. Granas, *Fixed Point Theory*, Springer-Verlag, New York, 2003.

[9] N. Dunford and J. T. Schwartz, *Linear Operators Part I: General Theory*, Interscience Publishers, Inc./Interscience Publishers Ltd., New York/London, 1958.

[10] Y. Hino, S. Murakami and T. Naito, *Functional Differential Equations with Infinite Delay*, Lecture Notes in Mathematics, vol. 1473, Springer-Verlag, Berlin, 1991.

[11] A. Lasota and Z. Opial, An application of the Kakutani–Ky Fan theorem in the theory of ordinary differential equations, *Bull. Acad. Pol. Sci., Sér. Sci. Math. Astron. Phys.* **13** (1965), 781–786.

[12] B. Øksendal, *Stochastic Differential Equations: An Introduction with Applications*, 4th edn., Springer-Verlag, Berlin, 1995.

[13] C. Prevot and M. Rockner, *A Concise Course on Stochastic Partial Differential Equations*, Lecture Notes in Mathematics, vol. 1905, Springer, Berlin, 2007.

[14] D. Revuz and M. Yor, *Continuous Martingales and Brownian Motion*, 3rd edn., Springer, 1999.

[15] I. I. Vrabie, C_0-*Semigroups and Applications*, Elsevier, 2003.

https://doi.org/10.1515/9783112251898-012

Index

C_0-semigroup 107
Q-Hilbert–Schmidt operators 165
Q-Wiener process 41
γ-Lipschitz 98
σ-algebra 12

Arzelà–Ascoli theorem 136

Bihari lemma 232
Black–Scholes model 1
Bochner integrable 79
Borel tribe 343
Brownian motion 12
Burkholder–Davis–Gundy 131

Carathéodory 95
Chebychev inequality 15
completely continuous 95
cross variation 26

diffeomorphisms 142
Donsker's theorem 23

equi-continuous 89

filtration 15
fractional Brownian motion 62
fractional integrals and derivatives 69
Fubini's theorem 186
functions of bounded variation 26

Gronwall inequality 181

Hausdorff lower semicontinuous 127
Hausdorff measure 248
Hilbert space valued Wiener processes 40
Hilbert–Schmidt norm 295
Hilbert–Schmidt operator 43
Hölder inequality 15
homeomorphism map 81
Hurst parameter 165

infinitesimal generator 116
integrals for fractional Brownian motion 70
Itô formula 53
Itô isometry 76

Jensen inequality 15

Krasnoselskii theorem 206

Lebesgue dominated convergence theorem 134, 175
lemma of Bihari 79
lemma of Mazur 97
Lipschitz matrix 84
long-range dependence 62

Malliavin calculus 12
martingale 17
measurable space 12
mild solution 168
multi-valued map 94
multi-valued map convex 94

Nagumo type growth conditions 132
Nemytskii operator 316
noncompact semigroup 242
nonsmooth analysis 126

p-variation of $B^H(t)$ 66
path differentiability 65
Perov's theorem 84
probability measure 12

quadratic variation and covariation 51

Riemann–Stieltjes integrals 31

Schaefer's fixed point theorem 147
self-similar 63
semi martingale 50
semigroup strongly 107
Shwartz inequality 15
square integrable martingales 19
standard Brownian motion 21
stochastic product rule 39
stopping time 19
sub-differential 126
submartingale 18
supermartingale 18

The Borel–Cantelli lemma 25
the Minkowski functional 86
the normal cone 126

https://doi.org/10.1515/9783112251898-013

theorem of Arzelà–Ascoli 89
theorem of Bachelier 21
theorem of Burholder–Davis–Gundy inequalities 38
theorem of Covitz and Nadler 106
theorem of Hille–Yosida 117
theorem of Kolmogorov 23
theorem of Perov fixed point 98
theorem of Schauder fixed point 99

uniformly equi-continuous 89
upper semi-continuous 94

weak solutions 184
Wiener helix 61
Wiener process 20, 249

www.ingramcontent.com/pod-product-compliance
Lightning Source LLC
Chambersburg PA
CBHW061157240926
51846CB00006B/1036
* 9 7 8 3 1 1 2 2 5 1 8 8 1 *